MEMORY AND TRANSFER
OF INFORMATION

MEMORY
AND TRANSFER
OF INFORMATION

Edited by
H. P. Zippel
*Physiologisches Institut der Universität
Göttingen, West Germany*

PLENUM PRESS • *NEW YORK–LONDON* • *1973*

Proceedings of a symposium sponsored by the
MERCK'SCHE GESELLSCHAFT für KUNST und WISSENSCHAFT
held at Göttingen, May 24-26, 1972

Library of Congress Catalog Card Number 73-80325
ISBN 0-306-30743-X

© 1973 Plenum Press, New York
A Division of Plenum Publishing Corporation
227 West 17th Street, New York, N.Y. 10011

United Kingdom edition published by Plenum Press, London
A Division of Plenum Publishing Company, Ltd.
Davis House (4th Floor), 8 Scrubs Lane, Harlesden, London, NW10 6SE, England

Printed in the United States of America

ACKNOWLEDGEMENTS

The editor of these Proceedings wishes to express his gratitude to the "MERCK'SCHE GESELLSCHAFT FÜR KUNST UND WISSENSCHAFT" (Darmstadt, W. Germany) whose generous sponsorship enabled the planning and realization of the Symposium. Thanks are also due to the "MAX-PLANCK-INSTITUT FÜR BIOPHYSIKALISCHE CHEMIE" (Göttingen, W. Germany) for placing its attractive rooms at the disposal of the members. In addition the editor would like to thank the PLENUM PUBLISHING CORPORATION for providing him with the opportunity to publish the Proceedings of the Symposium.

My special gratitude is extended to my assistant Mrs. U. LINAREZ for her kind aid before and during the Symposium, to Mr. A. BURT for his friendly assistance in correcting the manuscripts and to Miss Ch. EINECKE for her careful typing of the same.

Finally, the editor wishes to express his thanks to the participants of the Symposium for supplying their manuscripts so promptly and thus ensuring rapid publication of the Proceedings.

PREFACE

The contents of this book are the presentations of a Symposium on "Memory and Transfer of Information", held at Göttingen, May 24-26, 1972 .

One of the main reasons for organizing this Symposium was to stimulate interdisciplinary discussion between scientists working in the field as a whole. Most of the previous meetings dealing with memory and transfer of information have tended to be rather limited in scope. The present Symposium covered a wide range of topics, including neurophysiological, neuropharmacological, neurochemical, behavioral and clinical aspects of learning and chemical transfer of information, presented by specialists in these areas.

The Proceedings of the meeting present a large number of previously unpublished results, e.g., recent experiments in neurophysiology and neurochemistry, new approaches to chemical transfer of learned information, experiments using synthetic scotophobin and drugs influencing learning and behavior. The importance of interdisciplinary discussion is perhaps most clearly emphasized by the advances in neurochemical micromethods which are of particular interest to scientists working on the chemical transfer of information. Only such interdisciplinary collaboration between highly specialized scientists guarantees further progress and deeper insight into the complex, and until now little understood, mechanisms of that most intricate of organs, the brain.

Hans Peter ZIPPEL

vii

CONTRIBUTORS

BRADLEY, P.B.
>   Department of Pharmacology, Medical School,
>   Birmingham B15 2TJ, England

BYRNE, W.L.*
>   Department of Biochemistry, University of Tennessee,
>   College of Basic Medical Sciences,
>   Memphis, Tennessee 38103, U.S.A.

CREUTZFELDT, O.D.
>   Neurobiologische Abteilung, Max-Planck-Institut für
>   biophysikalische Chemie, 34 Göttingen, Am Faßberg,
>   W. Germany

CRONHOLM, B.
>   Department of Psychiatry, Karolinska Sjukhuset,
>   10401 Stockholm, Sweden

DOLCE, G.
>   Medizinische Forschung, E. Merck, 61 Darmstadt,
>   Frankfurter Straße 250, W. Germany

DOMAGK, G.F.
>   Physiologisch-Chemisches Institut der Universität,
>   34 Göttingen, Humboldtallee 7, W. Germany

ETTLINGER, G.
>   Institute of Psychiatry, De Crespigny Park, Denmark
>   Hill, London S.E.5 8 AF, England

FJERDINGSTAD, E.J.
    Department of Anatomy B, University of Aarhus,
    DK-8000 Aarhus C, Denmark

GLASSMAN, E.
    Division of Chemical Neurobiology, Department of
    Biochemistry, School of Medicine, University of
    North Carolina, Chapel Hill, North Carolina 27514,
    U.S.A.

GOLDSTEIN, Leonide
    New Jersey Neuro-Psychiatric Institute, Box 1000,
    Princeton, New Jersey 08540, U.S.A.
        Present address:
        Department of Psychiatry, Rutgers Medical School,
        New Brunswick, New Jersey, U.S.A.

GUTTMAN, H.N.
    Department of Biological Sciences, University of
    Illinois at Chicago Circle, Chicago, Illinois 60680,
    U.S.A.

HYDEN, H.
    Institute of Neurobiology, University of Göteborg,
    Göteborg 33, Medicinaregatan 5, Sweden

KANIG, K.**
    Abteilung für Neurochemie, Universitäts-Nervenklinik,
    665 Homburg/Saar, W. Germany

KORNHUBER, H.H.
    Abteilung für Neurologie, Universität Ulm (MNH),
    79 Ulm, Steinhövelstr. 9, W. Germany

MATTHIES, H.**

    Institut für Pharmakologie und Toxikologie, Medizi-
    nische Akademie Magdeburg, 301 Magdeburg, Leipziger
    Straße 44, E. Germany

McCONNELL, J.V.

    Brain Research Institute, University of Michigan,
    Ann Arbor, Michigan 48107, U.S.A.

MÜLLER-CALGAN, H.

    Medizinische Forschung, E. Merck, 61 Darmstadt,
    Frankfurter Straße 250, W. Germany

NEUHOFF, V.

    Max-Planck-Institut für experimentelle Medizin,
    34 Göttingen, Hermann-Rein-Straße 3, W. Germany

PARR, W.

    Department of Chemistry, University of Houston,
    Cullen Boulevard, Houston, Texas 77004, U.S.A.

RAHMANN, H.

    Zoologisches Institut der Universität, 44 Münster,
    Hindenburgplatz 55, W. Germany

SCHAEFER, K.-P.

    Universitäts-Nervenklinik, 34 Göttingen,
    von-Siebold-Straße 5, W. Germany

THINES, G.

    Department of Comparative Physiology, Université
    de Louvain, Pellenberg, Belgium

UNGAR, G.
    Department of Anesthesiology, Baylor College of
    Medicine, Houston, Texas 77025, U.S.A.

DE WIED, D.
    Rudolf Magnus Institute for Pharmacology, University
    of Utrecht, Vondellaan 6, Utrecht, The Netherlands

WESTERMAN, R.A.
    Department of Physiology, Medical School, University
    of Bristol, University Walk, Bristol, BS8 1TD, England
        Present address:
        Department of Physiology, Monash University,
        Clayton, Victoria, Australia 3168

ZIPPEL, H.P.
    Physiologisches Institut der Universität, Lehrstuhl II,
    34 Göttingen, Humboldtallee 7, W. Germany

  *   Paper presented at the Symposium but not included in
       the book
 **   Paper included in the book but not presented at the
       Symposium

CONTENTS

NEUROPHYSIOLOGY, NEUROPHARMACOLOGY
AND BEHAVIOR

    H.H. KORNHUBER

    Keywords: brain areas, channel capacity, ex-
    perience, memory (quantitative limitation of),
    motivation, reduction of data, selection unit,
    storage capacity (language)

    B. CRONHOLM and D. SCHALLING

    Keywords: aging (normal, pathological), con-
    solidation, digit span, ECT, memory scores,
    memory tests, Raven's coloured matrices, reg-
    istration, retention, vocabulary

    G. ETTLINGER

    Keywords: electric shock, equivalence, lan-
    guage, learning (bi-modal, conditional, gen-
    eral), matching (cross-modal, within-modal)

TRANSFER OF ACQUIRED INFORMATION

NEUROCHEMISTRY

OPENING ADDRESS

On behalf of E. Merck I should like to welcome you
most cordially to the Symposium "Memory and Transfer of
Information". I am very grateful that you were willing
to accept our invitation to participate in the discussion
on this highly interesting field of current research. My
special thanks are due to Dr. Zippel for his efforts and
the organisational work necessary to assemble this prominent group of experts here today.

I need not remind you of the purpose and necessity of
a Symposium on Memory and Transfer of Information. We are
all aware that in recent years new neurobiological knowledge concerning input, processing and storage of information in the central nervous system, has resulted in very
encouraging progress in this field. Thanks to neurophysiology and especially to neurochemistry, the problem of memory
begins to show tangible contours for the first time.

I believe that during such a phase of scientific
evolution, personal contact within a small group of successful investigators may be of decisive importance. These
days in Göttingen will not only enable you to learn more
about the latest research developments from other laboratories and clinics, but you will also have an opportunity
to discuss them among yourselves. I am very grateful indeed that our colleagues from Merck's Research Division
(Darmstadt, W. Germany), who have been interested in this
field for a number of years, are also able to participate.

I hope that this Symposium will be a success and, in addition to an inventory of our knowledge, will produce new impetus and ideas for research. Most probably, learning and learning ability will have a decisive effect on the fate of our society and of humanity. Let us hope that the prophecy of Bovet will be fulfilled:

"Je pense que nous pourrons donner à l'humanité dans le futur prochain quelque chose pareil à un élixir d'esprit ou aux pilules de mémoire".

Professor Dr. J. THESING
Vice President of Research

NEUROPHYSIOLOGY, NEUROPHARMACOLOGY

AND BEHAVIOR

NEURAL CONTROL OF INPUT INTO LONG TERM MEMORY: LIMBIC SYSTEM AND AMNESTIC SYNDROME IN MAN

H.H. Kornhuber

79 Ulm/Donau (W. Germany)
Steinhövelstraße 9
Universität Ulm

ABSTRACT

The importance of different brain regions, in particular the limbic system, in selecting and transforming external information either for storage in short term memory or transmission into long term memory is described. Various experimental findings in animals as well as observations on human patients are discussed from a neurophysiological and neurological point of view.

The necessity for a control of input into long term
memory arises from the quantitative limitations of memory.
With an unlimited memory, storage would perhaps be easier,
but recall much more difficult, the recall time longer and
the certainty of storage and recall much less.

Examining the literature, one could gain the impression that there is no need for a reduction of data prior
to storage in long term memory, since values as high as
10^{21} bit have been assumed as the storage capacity (for
a review see SCHAEFER, 1960). This value does not, however, stand up to critical examination. This may be ad
equately demonstrated by a consideration of the storage
of language. It has been calculated that in order to write
50,000 words of the German language a storage capacity of
1.5×10^6 bit is needed; and to select words for sentences,
0.4×10^6 bit (KÜPFMÜLLER, 1958, 1971). Probably less information is required to understand a spoken language if
one knows the words and their spelling; and certainly less
information for speaking, if one knows the written, and
understands the spoken language; furthermore, the rules
of grammar and syntax and the idioms require less storage
capacity than the vocabulary. Therefore, a total of about
$4 - 5 \times 10^6$ bit is probably sufficient for a perfect knowledge of a language. The capacity of human memory to store
languages in such a complete manner is of the order of 10.
Therefore, the total storage capacity for languages is of
the order of $4 - 5 \times 10^7$ bit. This information is stored
in roughly one tenth of the human cortex, and since the
number of neurons in the human cortex is about 3×10^9

(PARKENBERG, 1971), about 10 neurons are necessary to store one bit of information. There are, of course, uncertainties in this estimation, but it seems unlikely that the error is more than one order of magnitude.

The ratio of 10 neurons to 1 bit of information reflects the amount of redundancy in our nervous organization which makes a brain more certain than any technical computer so far built. If we extrapolate this ratio to the whole cortex, the total information capacity for storage in long term memory is of the order of 3×10^8 bit.

It is possible that not all of the cortical neurons are concerned with permanent storage of information and that this proportion might be higher in the language areas than, for example, in the motor fields. In this case, the total cortical capacity for long term storage of information would be somewhat less than 3×10^8 bit.

Let us now consider how much of the total flow of information through consciousness can be stored in such a long term memory. The flow of information through conscious perception in man for all kinds of sensory activity so far investigated is below 50 bit $\times$ sec^{-1}. For instance the channel capacity for quiet reading is about 40 bit $\times$ sec^{-1}, for mental arithmetic, 12 bit $\times$ sec^{-1}, and for counting, 3 bit $\times$ sec^{-1} (KÜPFMÜLLER, 1958). 20 bit $\times$ sec^{-1} is probably a good average for the channel capacity, that is to say for the maximal flow of information through short term memory per unit time. If we assume that in a busy life this flow is maintained for 16 hours per day through 70

years, the total flow of information per life is 3×10^{10}
bit. This means, with a storage capacity in long term memory of 3×10^{8} bit, that the flow of information through
short term memory has to be reduced by a factor of 100 for
storage in long term memory.

This result (i.e. that only 1 % of the information
flowing through consciousness is stored in a more or less
permanent way) corresponds well to our every day experience:
we forget many details and sometimes even important points.
Learning would be indeed much simpler and the life of
teachers easier if every perception could be stored immediately into permanent memory. Permanent storage of everything with secondary partial supression by unconscious
psychic processes belongs to Freudian mythology. As EBBING-
HAUS found in 1885, 50 repetitions are necessary in adult
man to learn a list of 35 syllables even with the best motivation for learning; that is to say, only about 2 % of
the whole task per repetition (when the learning material
is free of redundancy). The difficulty of learning is hidden in everyday life by the fact that we are usually dealing with highly redundant material.

The selection and transfer ratio of about 1:100 between short term and long term memory is, of course, only
a small part of the total data reduction taking place between the collection of raw data via the sensory receptors
and storage in long term memory: most of the data reduction
occurs between the receptors and short term memory, since
maximal information flow thorugh our, perhaps, 10^{9} recep-

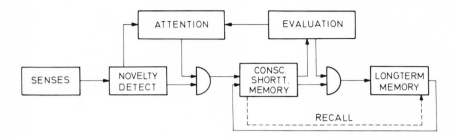

Fig. 1: <u>Information flow diagram from sensory receptors to</u>
<u>long term memory</u>
Mechanisms of novelty detection include adaptation,
motion detection and presynaptic inhibition in the
spinal cord. Between the conscious short term memory and the long term memory is the selection unit.
The pathway for recall from long term memory is
independent of the selection unit.

tors may reach 10^{10} bit x sec^{-1}. The enormous amount of
data reduction from 10^{10} to 2×10^{1} bit x sec^{-1} occurs in
the sensory channels (which are equipped with novelty detectors) and in the conscious short term memory with its
capacities for evaluation and attention (see Fig. 1).

If we were engineers having the task of constructing
an apparatus for the reduction of information before the
storage in long term memory - how would we do this job?
Would we choose a diminution by chance as has been suggested sometimes in psychological literature of the last
years? Or would we try to provide a more meaningful data
reduction, namely, a selection of the important facts at
the expense of the less important? We would, of course,
choose the latter, since storage of information is as important for the survival of the individual and the species

as food or useful behavior, like the rearing of infants.

There is a system in the brain (to which the hypothalamus, amygdala and orbital cortex belong) which receives
messages from the internal milieu of the organism and from
the external environment. It is specialized to remind us
of the important points regarding behavior and environment. It is the system concerned with motivation: the crea-
tion of drives, values and emotions. As this system is
available, it could perhaps also be used for the selection
of information prior to storage in long term memory.

The fact is that drives, values, and emotions play an
important role in perception as well as in learning, memory and the development of intelligence in man. Experimental
subjects, who were asked to count objects distributed between colored letters on colored paper, remembered very
little about the letters or the colors of either the letters or the paper (MAYERS, 1913). An investigation of the
development of intelligence in children between the ages
of 3 and 12 (SONTAG et al., 1958) showed that the intelligence found at the age of 3 has only a low correlation with
the intelligence at the age of 12 ($r = 0.46$); however,
children who improved their I.Q. were characterized by a
certain motivation, namely by initiative and competition.
According to the theory of HULL (1943), motivation is an
important factor for learning and retention. Thus the memory system evolved just as we would have designed it:
having built a system for selection of the important points
in behavior, the same system was used to evaluate the con-

tents of short term memory, regarding its importance for
long term storage.

It is not my intention to present a full theory of
motivation, but in order to avoid a common misunderstanding,
let me discuss some important points. The Freudian theory
of motivation is evidently fallacious. As ethology, the
science of animal and human behavior, shows (for a review
see EIBL-EIBESFELD, 1967) there are many more drives in
man, as well as in other higher animals, than just the
one or two acknowledged by psychoanalysis - these two being "libido" (which is an exaggerated sexual drive) and
aggression. With no more than these two drives, man would
neither survive phylogenetic selection nor the first day
of his life. As a highly differentiated animal, man has
more, rather than less, special drives than other animals.
Just to mention a few, there are: hunger, thirst, the
breathing-drive, sleepiness, micturation drive, defaeca-
tion drive, grooming, territorial drive, distaste, nausea,
shame, fear of heights, aggression, flight, jealousy, envy,
gathering drive, drive to move, drive to play, drive to
speak, exploratory drive, sexual drive, maternal drive,
compassion, drive of social rank, gregarious drive, con-
fidence/mistrust, and tenderness. All of these drives have
their inborn basis, although many of them are not present
at birth but develop and mature during childhood and youth,
being influenced or imprinted by environmental factors. In
addition to these specialized drives there are general
motivations like malaise (I avoid the term displeasure,
for I do not refer just to the contrary of pleasure -

pleasure is a mechanism of positive feedback for certain
drives (KORNHUBER, 1967) while malaise is subject to the
common principle of negative feedback). Another generalized motivation is the general drive to be active, which
has nothing to do with "libido"; it is, rather, connected
to alertness. By means of such generalized drives the organism is able to learn and store into long term memory,
even in situations which have never occurred in phylogenetic development. For instance, rats are able to feed
themselves successfully, choosing their menus from 17
different substances (including the pure components of the
vitamin B complex - RICHER, 1942), obviously using the
principle of trial and error, and changing their behavior
according to their hunger drive or state of health. In
short, the system of motivation is on the one hand specific
but diversified enough to cope with all the common needs
of the organism and the survival of the species, and on
the other hand open and unspecialized enough to handle new
situations.

After this digression, let us come back to the problem of evaluating experience in terms of storage for long
term memory. How should this evaluation be made? Obviously
the primary afferent signals from the receptors should be
preprocessed to give a picture of the whole sensory situation, and this picture of a high level of abstraction and
synthesis should be presented to the evaluation system. It
should contain integrated information from all the senses.
Sensory information (in other words the contents of short
term memory) and the biological value of this information

are compared in a selection unit where these values filter
the sensory information prior to storage in long term memory.

What would be the consequences of destroying the selection unit? The short term memory would be undisturbed,
and the long term memory with all its contents from former
life would be intact, with normal recall. By contrast,
transmission of new information from short into long term
memory would be impossible, that is to say, there would be
no retention of new information for more than a few seconds
or minutes.

Such a disturbance exists. It is called the amnestic
syndrome or Korsakow-syndrome (sometimes also called the
Wernicke-Korsakow-syndrome). In patients with an amnestic
syndrome there is normal recall from long term memory acquired prior to the disease, and conversation with such a
patient seems at first to be quite normal, for they have
good access to short term memory, which is itself obviously undisturbed, with a normal attention span. However, to
uncover the memory transmission defect, a good test is to
ask the patient to hide his wallet. When one returns after
5 minutes and enquires after the wallet, the patient is not
aware of what he has hidden or where, in fact he does not know
that he has hidden anything at all. While forgetting is partial in normal man at the transition from short into long
term memory (that is within a few seconds to minutes), it
is complete in cases of amnestic syndrome. There are, of
course, cases of amnestic syndrome with an additional de-

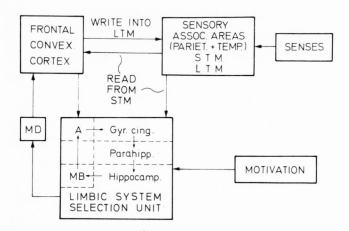

Fig. 2: <u>Scheme of anatomical structures involved in selection of information between short term memory
(STM) and long term memory (LTM)</u>

MB = mamillary body, A = anterior thalamic nucleus,
MD = mediodorsal thalamic nucleus.

terioration of the long term memory itself, but it is clear
from my investigations and from other reports, that there
are also cases with an excellent recall of old events in
long term memory despite a complete amnestic syndrome.

The lesions which produce the amnestic syndrome are
located in a specialized part of the limbic system (Fig. 2),
in a circuit including the cingulate gyrus (PRIBRAM
et al., 1962), the hippocampus (PENFIELD and MILNER, 1958;
SCOVILLE and MILNER, 1959), the fornix (HASSLER and RIE-
CHERT, 1957), the mamillary body (GAMPER, 1929), the anterior nucleus of the thalamus and the medio-dorsal nucleus of the thalamus (VICTOR et al., 1971; THOMPSON, 1964).

The loop from the cingulate gyrus through the parahippocampus to the hippocampus and then via the fornix, mamillary body, anterior thalamic nucleus back to the cingulate
gyrus had been recognized as a functional unit by PAPEZ in
1937. In order to produce the amnestic syndrome, the lesion in this hippocampal loop must be bilateral.

The short term memory and the long term memory are
not located in this loop. Rather, they are located topic-
wise in different parts of the cortex; for instance, the
area involved in understanding language is located in the
left temporal lobe. Short term memory and long term memory
for a given topic are located in the same part of the brain.
Moreover, for normal functioning of the short term memory
and for recall from long term memory, the thalamus is essential and a state of alertness is required which depends
also on brain stem structures.

The input and output connections of the hippocampal
loop, important for the transmission from short into long
term memory, have been discovered only recently. The output of the fornix to the limbic region of the midbrain has
been known for a long time and is probably not the essential one for the amnestic syndrome, since midbrain lesions
do not cause an amnestic syndrome nor are they prominent
in amnestic patients.

Sensory afferent signals from the primary sensory
projection areas undergo further information processing in
the sensory association areas; in some of them multisensory

convergence occurs (KORNHUBER, 1965; KORNHUBER and DA
FONSECA, 1964). The input to the cingulate gyrus - hippocampus system comes from the sensory association areas in
the temporal and parietal lobes, in part directly from
the parietal lobe to the cingulate gyrus and parahippocampus, in part, however, via the cortex at the convexity

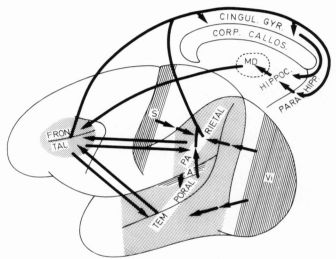

Fig. 3: Scheme of pathways in the monkey brain involved
in the flow of information from primary sensory
areas via the sensory association areas of the
temporal and parietal lobe and the cortex of the
frontal convexity to the limbic system and then
the loop back via the medio-dorsal nucleus of the
thalamus (MD) and the frontal cortex to the
temporal and parietal areas for long term storage.

Primary sensory areas: Vi = visual, A = auditory,
S = somatosensory; the vestibular area is in the
lower part of S. Based, in part, on data of PANDYA
and KUYPERS (1969) and PANDYA et al. (1971).

of the frontal lobe (see Fig. 3). From the cingulate gyrus
the afferents pass via the parahippocampus to the hippocampus, the output of which is through the fornix. The afferents from the sensory association areas have been established in two ways: electrophysiologically (NIEMER et al.,
1963) and histologically (PANDYA and KUYPERS, 1969; PANDYA
et al., 1971).

The output of the hippocampal system, important for
transmission from short into long term memory, is so far
less well established histologically. The well known connection of the hippocampus through the fornix to the mamillary body seems to feed back mainly to the cingulate
gyrus via the anterior nucleus of the thalamus, thus forming a closed loop and not an output of the system. However,
according to electrophysiological, as well as clinical and
pathoanatomical data, one important output of the system
is via the medio-dorsal nucleus of the thalamus to the cortex at the convexity of the frontal lobe. On electrical
stimulation of the hippocampus, potentials are evoked in
the medio-dorsal thalamic nucleus (GREEN and ADEY, 1956).
In clinical and pathoanatomical studies the lesion most
often related to the amnestic syndrome is bilateral in the
mediodorsal nucleus (VICTOR et al., 1971). The medial (par-
vo-cellular) part of the medio-dorsal nucleus projects to
the cortex at the frontal convexity (AKERT, 1964), from
where association fibres go back to the sensory association
areas of the temporal and parietal lobes (PANDYA et al.,
1971; PANDYA and KUYPERS, 1969). Frontal lesions, however,
result only in an incomplete amnestic syndrome (see below);

therefore, one might speculate that apart from the efferent
connection via the frontal lobe, there are other efferents,
perhaps from the cingulate gyrus, back to the temporal and
parietal association areas (or to the pulvinar which projects to these areas). Another argument for the existence
of such efferents is the fact that the cingulate gyrus receives direct fibres from the posterior parietal cortex
and additionally, many (though not all) cortico-cortical
associations are bidirectional.

In summary: short term memory as well as long term
memory are located largely in the sensory association
areas, while the "selection unit" is a specialized part
of the limbic system, the cingulate gyrus - hippocampus
loop. The motivation system feeds into this selection unit
via the septal region and the cingulate gyrus. It is located in the hypothalamus, the amygdaloid complex and the
orbital cortex.

If the input, as well as the output, of the cingulate
gyrus - hippocampus loop travels in part via the frontal
lobe convexity, there should be a partial amnestic syndrome resulting from lesions of the latter area. This is
indeed so. BIANCHI (1922) concluded from his early behavioral observations in monkeys with frontal lesions: "Memory .. becomes enormously reduced. The mutilated monkey
does not utilize past experience". JACOBSEN found in 1935
that after lesions of the convexity of the frontal lobe,
monkeys show a persistent defect in the delayed response
task and he interpreted this as a defect in recent memory.

While these early investigations were leading in the right
direction, the problem later became confused, and this early interpretation was rejected (for a review see WARREN
and AKERT, 1961) for two reasons: 1. extensive conclusions
on the basis of negative findings and 2. lack of understanding of the nature of the frontal memory defect. In
my opinion, the interpretation of GROSS and WEISKRANTZ
(1964) for the frontal memory defect was the one most compatible with the theory presented here. They assumed that
frontal lesions interfere with retrieval from short term
memory, but do not affect either storage itself or retrieval from long term memory - an interpretation certainly not tenable for the amnestic syndrome in man.

In the meantime, two findings have made the situation
clearer: 1. in the monkey, it is the cortex on both sides
of the middle third of the sulcus principalis in which bilateral lesions result in maximal impairment in the delayed alternation response (BUTTERS and PANDYA, 1969).
2. it is just this part of the frontal convexity cortex
of the monkey which projects to the cingulate gyrus and
parahippocampus (PANDYA et al., 1971). Obviously the defect in the delayed response or delayed alternation tests
is a mild or partial amnestic syndrome. Findings compatible
with this interpretation have been obtained in man following frontal lesions (BENTON, 1968; BUTTERS et al., 1970;
MILNER, 1963; 1965). Of course the interpretation presented
here for the delayed response and delayed alternation defects does not mean that input control into long term memory is the only function of the frontal convexity cortex.

In summary, it is clear from normal as well as pathological data that one of the main factors in the physiology of memory is selection of information for storage into long term memory. For this purpose, the contents of
consciousness, present in the short term memory, must be
evaluated with respect to their importance for survival
of the individual and the species.

Inborn as well as acquired drives useful for this
evaluation, meet the preprocessed sensory information in
a selection unit which is a specialized part of the limbic
system; the sensory information and the storage commands
of the selection unit are relayed in part by way of the
frontal convexity cortex. Lesions of the cingulate gyrus
- hippocampus-mediodorsal nucleus-frontal convexity system
result in a disturbance of transmission from short into
long term memory, while the retrieval from short term as
well as from long term memory remains intact.

The theory presented here leads to the conclusion
that at a given neuron there might be special synapses
responsible for the formation of long term storage. In
the sensory association areas, the synapses of afferents
coming (directly or indirectly) from the limbic system
are essential for forming long term memory, while other
synapses on the same neurons are essential for information
processing and for recall.

The mechanisms of transmission into long term memory
could involve coincidence of thalamic and cortico-cortical

afferents at a given cortical neuron or cell column. For the cellular or molecular biologist, one of the consequences of these thoughts could be to investigate the formation of polypeptides, the content of synaptic vesicles, and the growth and morphology of dendrites and synapses in the sensory association areas, before and after lesions of the limbic system or frontal lobes.

FOOTNOTE

This paper was first presented at the British-German Neurological Meeting, Göttingen, June 13, 1971.

LITERATURE

AKERT, K.: Comparative anatomy of the frontal cortex and thalamocortical connections.
In: The Frontal Granular Cortex and Behavior (J.M. WARREN and K. AKERT, eds.), pp. 372-396, McGraw-Hill Book Comp., New York, San Francisco, Toronto, London (1964).

BENTON, A.L.: Differential behavioral effects in frontal lobe disease.
Neuropsychologia 6, 53-60 (1968).

BIANCHI, L.: The mechanism of the brain and the function of the frontal lobes. (Transl. by J.H. MacDonald). Wood and Co., New York (1922).

BUTTERS, N., and D. PANDYA: Retention of delayed alterna-
    tion: effect of selective lesions of sulcus princi-
    palis.
    Science 165, 1271-1273 (1969).

BUTTERS, N., J. SAMUELS, H. GOODGLASS, and B. BRODY: Short
    term visual and auditory memory disorders after parie-
    tal and frontal lobe damage.
    Cortex 6, 440-459 (1970).

EBBINGHAUS, H.: Über das Gedächtnis. Untersuchungen zur ex-
    perimentellen Psychologie.
    Duncker und Humblot, Leipzig (1885).

EIBL-EIBESFELD, J.: Grundriß der vergleichenden Verhaltens-
    forschung. Ethologie.
    Piper, München (1967).

GAMPER, E.: Schlaf - Delirium tremens - Korsakow'sches Syn-
    drom.
    Zbl. Neurol. 51, 236-239 (1929).

GREEN, J.D., and W.R. ADEY: Electrophysiological studies of
    hippocampal connections and excitability.
    EEG clinical Neurophys. 8 245-262 (1956).

GROSS, C.E., and L. WEISKRANTZ: Some changes in behavior
    produced by lateral frontal lesions in the Macaque.
    In: The Frontal Granular Cortex and Behavior (J.M.
    WARREN and K. AKERT, eds.), McGraw-Hill Book Comp.,
    New York, San Francisco, Toronto, London (1964).

HASSLER, R., and T. RIECHERT: Über einen Fall von doppel-
    seitiger Fornicotomie bei sogenannter temporaler
    Epilipsie.

Acta Neurochir. 5, 330-340 (1957).

HULL, C.L.: Principles of behavior. An introduction to be-
    havior theory.
    Appleton Crofts, N.Y. (1943).

JACOBSEN, C.S.: Functions of the frontal association area
    in primates.
    AMA Arch. Neurol. Psychiat. 33, 558-569 (1935).

JACOBSEN, C.S.: Studies of cerebral function in
    I. The functions of the frontal association areas in
    monkeys.
    Comp. psychol. monogr. 13, 3-60 (1936).

KORNHUBER, H.H.: Zur Bedeutung multisensorischer Integra-
    tion im Nervensystem.
    Dtsch. Zschr. Nervenheilk. 187, 478-484 (1965).

KORNHUBER, H.H.: Zur Funktion der Lust. Antrittsvorlesung,
    Freiburg i.Br., 1963.
    In: Psychiatrie der Gegenwart B I/1 Teil A (H.W.
    GRUHLE, R. JUNG, W. MAYER-GROSS, M. MÜLLER, Hersg.),
    zit. nach R. Jung: Neurophysiologie und Psychiatrie,
    S. 580, Springer, Berlin, Heidelberg, New York (1967).

KORNHUBER, H.H., and J.S. DA FONSECA: Optovestibular inte-
    gration in the cat's cortex: a study of sensory con-
    vergence on cortical neurons.
    In: The Oculomotor System (M.B. BENDER, ed.), pp. 239-
    277, Hoeber, New York (1964).

KÜPFMÜLLER, K.: Informationsverarbeitung durch den Menschen.
    Nachrichtentechn. Z. 12, 68-74 (1958).

KÜPFMÜLLER, K.: Grundlagen der Informationstheorie und der
    Kybernetik.
    In: Physiologie des Menschen, Bd. 10: Allgemeine Neu-
    rophysiologie (O.H. GAUER, K. KRAMER und J. JUNG,
    Hrsg.), pp. 195-231, Urban und Schwarzenberg, München,
    Berlin, Wien (1971).

MAYERS, G.C.: A study in incidental memory.
    Arch. Psychol. N.Y. 5, Nr. 26 (1913).

MILNER, B.: Effects of different brain lesions on card
    sorting. The role of frontal lobes.
    AMA Arch. Neurol. 9, 90-100 (1963).

MILNER, B.: Visually-guided maze learning in man: effects
    of bilateral hippocampal, bilateral frontal and uni-
    lateral cerebral lesions.
    Neuropsychologia 3, 317-338 (1965).

NIEMER, W.T., E.F. GOODFELLOW, and J. SPEAKER: Neocortico-
    limbic relations in the cat.
    EEG clin. Neurophysiol. 15, 827-838 (1963).

PAKKENBERG, H.: The number of nerve cells in the cerebral
    cortex of man.
    J. comp. Neur. 128, 17-20 (1966).

PANDYA, D.N., P. DYE, and N. BUTTERS: Efferent cortico-
    cortical projections of the prefrontal cortex in the
    Rhesus monkey.
    Brain Res. 31, 35-46 (1971).

PANDYA, D.N., and H.G.J.M. KUYPERS: Cortico-cortical con-
    nections in the Rhesus monkey.
    Brain Res. 13, 13-36 (1969).

PAPEZ, J.W.: A proposed mechanism of emotion.
    AMA Arch. Neurol. Psychiat. 38, 725-743 (1937).

PENFIELD, W., and B. MILNER: Memory deficit produced by
    bilateral lesions in the hippocampal zone.
    AMA Arch. Neurol. Psychiat. 79, 475-497 (1958).

PRIBRAM, K.A., W.A. WILSON, and J. CONNORS: Effects of
    lesions of the medial forebrain on alternation be-
    havior of Rhesus monkeys.
    Exp. Neurol. 6, 36-47 (1962).

RICHTER, C.P.: Total self regulatory functions in animals
    and human beings.
    Harvey Lect. 1942-43, p. 63-103.

ROSSVOLD, H.E.,and M.K. SZWARCBART: Neural structures in-
    volved in delayed-response performance.
    In: The Frontal Granular Cortex and Behavior (J.M.
    WARREN and K. AKERT, eds.), pp. 2-15, McGraw-Hill
    Book Comp., New York, San Francisco, Toronto, London
    (1964).

SCHAEFER, E.: Das menschliche Gedächtnis als Informations-
    speicher.
    Elektronische Rundschau 14, 79-84 (1960).

SCOVILLE, W.B., and B. MILNER: Loss of recent memory after
    bilateral hippocampal lesions.
    J. Neurol., Neurosurg. Psychiat. 20, 11-21 (1957).

SONTAG, L.W., C.T. BAKER, and V.L. NELSON: Mental growth
    and personality development: A longitudinal study.
    Monogr. Soc. Res. Develop. 23, 61 (1958).

THOMPSON, R.: A note on cortical and subcortical injuries
    and avoidance learning by rats.
    In: The Frontal Granular Cortex and Behavior (J.M.
    WARREN and K. AKERT, eds.), McGraw-Hill Book Comp.,
    New York, San Francisco, Toronto, London (1964).

VICTOR, M., R.D. ADAMS, and G.H. COLLINS: The Wernicke-
    Korsakoff-Syndrome.
    Blackwell scientific publications, Oxford (1971).

WARREN, J.M., and K. AKERT (eds.): The frontal granular
    cortex and behavior.
    McGraw-Hill Book Comp., New York, San Francisco,
    Toronto, London (1964).

A STUDY OF MEMORY IN AGED PEOPLE

Börje Cronholm and Daisy Schalling

Department of Psychiatry
Karolinska Sjukhuset
104 01 Stockholm, Sweden

ABSTRACT

Memory performances were studied in a group of 22 old
(65-74 years) and 29 very old (75-91 years) women, and com-
paired with performances of a group of 31 middle-aged (35-
49 years) women, matched for socio-economic and cultural
level. The intellectual level did not differ significantly
as estimated by the Terman Vocabulary test, but in Raven's
Coloured Matrices there was a considerable decline with
age. Digit span forwards and still more backwards was reduced with age.

To study memory performances, a test comprising three
subtests, 30 figures, 30 word pairs and 30 personal data
were used. Remembering was studied both immediately after
presentation of the material (immediate memory score) and
three hours later (delayed memory score), the difference
between the two being forgetting scores. A considerable de-

23

cline of immediate and delayed memory scores with age was
found, but forgetting measures did not increase. The regression lines between immediate and delayed scores were
also quite similar. The findings indicate a pronounced decline of registration with age but do not show any decline
of retention, such a decline of course not being disproved.

It is a commonplace that old people quite often com-
plain of "bad memory" and also that this forgetfulness may
be quite evident to others. Without being conspicuously
deteriorated in other respects, they may for instance show
difficulties in recalling names and telephone numbers, have
to make notes about everything that they want to remember,
etc. However, "memory" is a very broad concept, "memory
disturbances" thus comprising a wide variety of conditions,
differing both as regards functional type of impairment,
pathogenesis and etiology.

Before reporting our own investigations, I think it may
be useful briefly to describe the model we have applied
when constructing our tests and interpreting our findings.

We assume a series of memory phases or processes.
First of all, the registration (acquisition) of material
or tasks is a necessary prerequisite, itself depending on,
inter al., perception and cognitive interpretation (apper-
ception). After that, a period of consolidation (or rehearsal) ensues, accompanied by "short term storage". These
processes may be subsumed under the heading of "learning".
Retention (or long term storage) then follows. Remembering

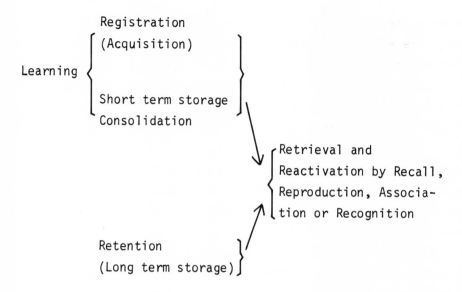

Fig. 1: Hypothetical memory processes

by recall, reproduction, association or recognition implies
retrieval and reactivation. All these processes may be described as psychological constructs - they are inferred or
postulated, but not directly observable. It may be assumed
that they depend - at least in part - on different neuro-
anatomical structures and/or on different neuro-physiologi-
cal activities. The model is schematically represented in
Fig. 1.

    By means of systematic clinical observation and psy-
chological examination or experimentation on patients with
various cerebral injuries, it is possible to elucidate specific disturbances in the psychological processes described,
and their relation to cerebral injuries with, for instance,
different localization. The amnestic syndrome (Korsakow

syndrome) is of special interest in this context. The patients are able to reproduce immediately a series of numbers read to them as well as are "normals" (i.e., their
memory span for digits is approximately normal). This may
be interpreted to mean that at least some aspects of registration and "short term storage" are intact. On the
other hand, they virtually forget from one minute to another. This highly specific disturbance has been related
to lesions in the limbic system (especially in the mamillary bodies) probably interfering with consolidation processes, with retention per se or with retrieval (see CRON-
HOLM, 1963; BARBIZET, 1963; BARBIZET and CANY, 1969);
TALLAND (1965) pointed to "premature closure of activation"
as the central disturbance.

Earlier investigations have shown that aged people
display conspicuous difficulties in registration (CANES-
TRARI, 1968). However, their memory span for digits may
be rather little affected (see TALLAND, 1967). During the
last few years, the importance of stress sensitivity and
anxiety for reduced registration with aging has repeatedly been underlined (EISDORFER , 1968). On the other hand,
little is known about possible retention, retrieval or reactivation impairment involved in the memory complaints
and disturbances of aged people. Based on studies comparing memory scores with recall and recognition, some
authors claim that "retrieval" is more disturbed than
"storage" (SCHONFIELD and ROBERTSON, 1966) but this conclusion has been criticized by others (MC NULTY and CAIRD,
1967). The results of studies concerning possible impair-

ment of retention with normal aging by WIMER and WIGDOR, 1958; WIMER, 1960, and HULICKA and WEISS, 1965, have been inconclusive.

The aim of the present study was to elucidate further the influence of "normal" aging on memory processes, especially on "registration" and "retention".

It may be instructive to discuss in some detail the difficulties involved in such a study.

First, the definition of "normal" aging is far from self-evident. We know that mental decline with age is, like age changes in general, to a large extent genetically determined. On the other hand, environmental factors are also of importance. Thus, the aging that we observe is a function both of the genetical code and of the "wear and tear of life". In this we have to include the effects of general infections and minor head injuries, but also of such diseases as mild and benign essential hypertonia and other mild cardiovascular disturbances, etc. If not, we would meet "normal" aging only as a Platonic idea. On the other hand, we have to mark a dividing-line - of course arbitrary - against mental decline, due to pathological aging. Typical examples are mental disturbances after cerebro-vascular accidents, with incompensated heart disease, with presenile dementia such as Mb Pick and Mb Alzheimer and with senile dementia.

Secondly, there is the difficulty of finding a group

of aged people suitable for such a study, and still worse,
to find a control group of younger people similar to the
aged group in all relevant respects except age.

We chose to study a group of women aged 65 or more,
living at two apartment houses for the aged, and at a community home for the aged, all within the same parishes of
the city of Stockholm. There was a conspicuous loss of
subjects; 79 of the 179 women available were not able or
willing to participate, and 34 were excluded for medical
reasons on the basis of anamnestic data. Thus, the group
finally comprised 66 women. All women were interviewed by
a psychiatrist; 15 women were judged to be pathologically
deteriorated, and were excluded from the group of "normally" aged women. The material reported thus comprises 51
women, 22 were "old" (65-74 years) and 29 were "very old"
(75-91 years).   It is difficult to estimate what effects
the great loss of subjects may have had.

The final group was quite homogeneous from a socioeconomic aspect, belonging to a rather low social stratum.
The women had themselves had unskilled jobs, and so had
their husbands if they were or had been married. They had
only passed primary school, and many of them had only a
few years of schooling.

As a control group we chose 31 women 35-49 years old,
at a similar socioeconomic and cultural level as the experimental group, who were employed for unskilled work at
the community school board or at a cleaning company. In

spite of our efforts to equalize the experimental and
the control group, there may be differences relevant for
cognitive performances, including "memory". Due to improved schooling and the increased influence of mass media,
the educational level of the middle-aged group is higher
than that of the aged groups. On the other hand, opportuni-
ties for intelligent women from lower classes to get higher education have also considerably improved, and in that
way they will not be included in a group consisting of
middle-aged women having only primary school, like our
group. Thus, the education level of the old women is <u>low-
er</u>, but their level of innate capacities may be <u>higher</u>.

Our study comprised a series of psychological tests,
only a few of which will be reported here. The testing
was made by clinical psychologists, and divided into several sessions. With the old women the testing was rather
time-consuming, and much time had to be spent on "pep-
talk", coffee-drinking, etc.

Before describing the memory tests and the results
in them, we will briefly mention a few other tests.

To get an idea of different aspects of the general
cognitive level of the women we used a Vocabulary test,
taken from the well-known Terman-Merrill Intelligence
Scale. The task is to define the meaning of a series of
nouns, and the score is the number of correct answers.

As can be seen from Tab. I, the performance is rather

Tab. I: <u>Cognitive performances in three groups of women, differing in age</u>

MA = Middle Aged, 35-49 years, n = 31; O = Old, 65-74 years, n = 22, VO = Very
Old, 75-91 years, n = 29.

	MA		O		VO		F	
	$\overline{X}$	s	$\overline{X}$	s	$\overline{X}$	s	(df 2:79)	p
Age	43		70		79			
Terman's Vocabulary test	29.0	5.8	29.0	5.6	26.5	6.2	1.61	n.s.
Raven's Coloured Matrices	30.0	3.7	24.0	5.0	20.5	5.0	31.80	<.001

similar in the three groups. Earlier investigations have
shown that the vocabulary declines very little or not at
all with normal aging. The equality of our groups in Vocabulary may thus mean that we had been fairly successful in
equalizing our groups as regards their earlier intellectual
level.

On the other hand, there was a considerable decline
in Raven's Coloured Matrices, where the task is to denote
among 6 alternatives the figure that fits into a pattern.
The score is the number of such items, correctly solved.
For problem solving in this test, good reasoning and spa-
tial abilities are necessary. These abilities are known to
decline considerably with age, and as seen from Tab. I,
our result is in line with earlier investigations.

"Digit span" forwards is often used as a test of "short
term memory". As can be seen from Tab. II, in our material
there is a decline with age, but it is not very conspicuous.

On the other hand, there is a more marked decline from
middle age to old age as regards "digit span" backwards. This
is not astonishing, as it requires a much more complicated
handling of the material in making the reversal.

More complicated memory performances were studied by
means of a test, originally used to measure memory disturbances after ECT. The main aim of the test is to get
estimates, both of "registration" and of "retention". It

Tab. II: "Memory span" for digits in three groups of women, differing in age

MA = Middle Aged, 35-49 years, n = 31; O = Old, 65-74 years, n = 22; VO = Very old, 75-91 years, n = 29.

	MA Md	O Md	VO Md	p*	
Digit span				MA - O	O - VO
forwards	5.4	4.7	4.8	< 0.001	n.s.
backwards	4.9	3.3	3.3	< 0.001	n.s.

*Mann-Whitney's U-test

consists of three subtests, each one with 30 items, and it exists in two parallel versions. It is common to the three subtests, that the subjects are requested to remember both immediately after the presentation and after an interval, as a rule three hours later. Thus, in all three tests we get an "immediate memory score" and a "delayed memory score"; the difference between the two scores being a "forgetting score". These operationally defined variables may be regarded as functions of underlying, hypothetical memory variables. Thus, the "immediate memory score" is assumed to be a function mainly of registration, but the "forgetting" score mainly of retention. We then assume that retrieval and reactivation are of the same importance and enter into the scores to the same extent both immediately after presentation of the material and three hours later.

A. Immediate memory scores   = f (Registration,
                                        Short term storage,
                                        (Retention),
                                        Retrieval and
                                        Reactivation,
                                        ---------------)

B. Delayed memory scores   = f (Registration,
                                        Consolidation and
                                        Retention,
                                        Retrieval and
                                        Reactivation,
                                        ---------------)

C. Forgetting scores (A - B)   = f (Consolidation and
                                        Retention,
                                        ---------------)

Fig. 2: Operationally defined memory variables as functions
        of hypothetical constructs

A schematical representation of these ideas is given in
Fig. 2.

    The memory material and the procedure differ in the
three subtests. In the first of them, "30 figures", a
picture with drawings of 30 common objects is shown and
the objects named as they are pointed out to the subjects.
The task is to point out these objects among 30 others on
a new picture. After 3 hrs, another picture with the same

60 drawings is shown, and the subject is requested to
point out the 30 ones first shown. The next subtest, "30
word pairs", is a typical paired associates test. It comprises three series, each with 10 word pairs. They are
read aloud to the subject, and the text is shown at the
same time. Immediately after each series the 10 stimulus
words are given in different order and the subject is
asked to say the other word of the pair. After 3 hrs the
whole series of 30 stimulus words is again read in a different order and the subject asked to give the associate.
In the "30 personal data" test, six drawings are shown,
and for each of them five data are given concerning, e.g.,
age, occupation and hobbies. After presentation of all six
drawings, they are shown in a different order and the subject is told to tell all data he remembers about each per-
son. After 3 hrs, the drawings are again shown in a different order.

As it may be assumed that all three tests at least to
some extent measure the same underlying variables (registration and retention) the sum of the scores is also reported. This measure is more reliable than scores in the
separate subtests (CRONHOLM and OTTOSSON, 1963). (For a
detailed description of the test, see also CRONHOLM and
MOLANDER, 1957; OTTOSSON, 1960, and D'ELIA, 1970).

The test has proved to be a very valuable tool for
studying memory disturbances after ECT. Forgetting as
measured by the test is considerably increased, and the
measure is sensitive enough to differentiate between dif-

ferent modifications of ECT (see OTTOSSON, 1960,CRONHOLM,
1969, and D'ELIA, 1970). On the other hand, immediate
memory scores are much less lowered even a few hours after
a treatment, and after a series of treatments the measures
may even increase if the patient has improved (CRONHOLM
and OTTOSSON, 1961). This has been interpreted to mean,
that after ECT there is an amnestic syndrome with conspicuous retention disturbances, whereas registration is
much less or even not at all impaired (CRONHOLM, 1963,
1969).

In the present study, we have used the design described above, but we have also studied delayed memory
scores without first having obtained the immediate memory
scores. We thought, that it was more correct to compare
these scores with the immediate memory scores, as the questioning immediately after presentation involves rehearsal
and reinforcement of the items then remembered. To this
purpose, both test forms were used on each subject, and
rotated at random.

As can be seen from Tab. III, there is a highly significant and quite conspicuous decline from the middle-
aged, via the old to the very old women. This indicates -
in line with earlier findings - a marked decline in "registration" ability with age.

The same decline holds true, of course, for the memory scores obtained three hours later. But the difference
between the immediate and delayed memory scores, the

Tab. III: <u>Memory performances in three groups of women,</u>
          <u>differing in age</u>
          MA = Middle Aged, 35-49 years, n = 30; O = Old,
          65-74 years, n = 22; VO = Very Old, 75-91 years,
          n = 27.

	MA		O		VO		F	
	X̄	s	X̄	s	X̄	s	(df 2:76)	p
A. Immediate memory scores								
30 figures	25.0	3.4	23.1	3.4	20.4	5.2	8.78	< .001
30 word pairs	20.4	4.5	17.6	4.6	14.6	5.3	9.89	< .001
30 personal data	18.2	5.0	14.0	4.1	12.0	5.7	10.97	< .001
Combined test	63.7	8.2	54.7	9.6	47.0	13.7	16.38	< .001
B. Delayed memory scores I								
(3 hrs after A)								
30 figures	21.5	3.7	19.8	4.1	16.7	5.2	8.64	< .001
30 word pairs	14.2	5.2	10.6	4.1	8.5	4.8	9.93	< .001
30 personal data	16.7	5.3	11.6	3.8	8.8	4.4	20.74	< .001
Combined test	52.4	9.3	42.1	8.7	34.0	11.8	23.12	< .001
C. Forgetting scores I (A-B)								
30 figures	3.5	3.2	3.3	3.2	3.7	2.8	0.12	n.s.
30 word pairs	6.2	3.5	6.9	2.9	6.1	2.6	0.01	n.s.
30 personal data	1.5	3.4	2.4	2.0	3.2	3.1	2.33	n.s.
Combined test	11.2	6.3	12.6	5.2	13.0	6.0	0.74	n.s.
D. Delayed memory scores II								
(3 hrs after presentation,								
no immediate testing)								
30 figures	18.5	4.5	18.6	3.5	16.6	4.6	-	n.s.
30 word pairs	7.0	4.1	5.8	3.7	3.3	2.5	-	< .001
30 personal data	12.6	4.2	6.5	3.0	4.2	3.0	-	< .001
Combined test	38.2	8.4	30.9	7.3	23.8	8.0	-	< .001
E. Forgetting scores II (A-D)								
30 figures	6.5	4.3	4.6	4.2	3.9	5.4	-	-
30 word pairs	13.4	4.9	11.8	3.7	11.3	5.0	-	-
30 personal data	5.6	6.3	7.5	3.8	7.9	4.9	-	-
Combined test	25.5	7.7	23.9	8.0	22.9	11.5	-	-

forgetting score, an estimate of retention,does not show
but a minor, non-significant increase. We have also an-
alyzed the regression of delayed on immediate memory
scores. As may be seen from Fig. 3 the slopes and levels
of the regression lines of the combined test in the different groups are quite close to each other. An analysis
of co-variance was made in order to adjust for differences
in immediate memory scores. This analysis (possible only
with the 30 figure and the 30 word pair tests) showed
that the level, another estimate of retention, did not
differ between the groups.

As already mentioned, we also studied delayed memory scores three hours after presentation of the material
but without any immediate questioning. As expected, performances are lower in all groups, and there is a conspicuous and significant decline with age. Forgetting expressed as the difference between immediate memory scores
and these delayed scores in the three groups shows, to
our astonishment, a slight decrease with age! But as the
two measures were obtained with different forms of the
tests and on different days, error variance will be considerable.

However, our findings support the assumption of no or
at least only minor disturbances of retention performances
in old people: less was registered, but what was learnt, was
retained nearly as well as by middle-aged people. All the
same we have to be careful when trying to draw conclusions
as regards changes in retention ability with aging. Where

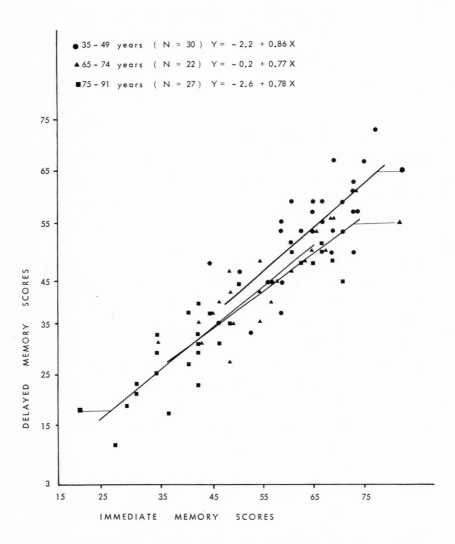

Fig. 3: <u>Delayed memory scores over immediate memory scores</u>
<u>(combined scores) in three groups of women, dif-</u>
<u>fering in age</u>

only little is registered, lower retention ability may be
needed to retain than when more is registered - and thus
forgetting scores at different levels of registration may
have a non-linear relation to an underlying retention ability. However, as the test has proved to be highly valid to
retention disturbances after ECT, we think that we may all
the same conclude, that with "normal" aging, decline of
retention is less important than registration disturbances.
- It may be mentioned that in an earlier study, a weak but
significant negative correlation was found between immediate memory scores and age in a group of mostly middle-aged
psychiatric patients, but not between forgetting and age
(CRONHOLM et al., 1970).

To sum up, registration was shown to become considerably impaired with aging. Probably, this impairment is related to impaired perception, cognitive analysis and ap-
perception, and to a decreased stress tolerance. On the
other hand, no increase of forgetting and thus no impairment of retention could be shown. The memory disturbances
of aged people thus differ from those found in amnestic
syndromes such as the Korsakow syndrome and after ECT.

ACKNOWLEDGEMENT

This study was supported by Foundations' Fund for
Research in Psychiatry, grant 61-225.

LITERATURE

BARBIZET, J.: Defect of memorizing of hippocampo-mamillary
    origin: a review.
    J. Neurol. neurosurg. psychiat. 26, 127-135 (1963).

BARBIZET, J., and E. CANY: A psychometric study of various
    memory deficits associated with cerebral lesions.
    In: The Pathology of Memory (G. TALLAND and N. WAUGH,
    eds.), pp. 49-64, Academic Press Inc., New York
    (1969).

CANESTRARI, R.: Age changes in acquisition.
    In: Human Aging and Behavior (G. TALLAND, ed.), pp.
    169-188, Academic Press Inc., New York (1968).

CRONHOLM, B.: L'application des méthodes psychologiques
    dans l'analyse des troubles de la mémoire.
    Rev. psychol. appl. 13, 171-188 (1963).

CRONHOLM, B.: Post-ECT amnesias.
    In: The Pathology of Memory (G. TALLAND and N. WAUGH,
    eds.), pp. 81-89, Academic Press Inc., New York
    (1969).

CRONHOLM, B., and L. MOLANDER: Memory disturbances after
    electroconvulsive therapy. 1. Conditions 6 hours after
    electroshock treatment.
    Acta psychiat. scand. 32, 280-306 (1957).

CRONHOLM, B., and J.-O. OTTOSSON: Memory functions in
    endogenous depression before and after electrocon-
    vulsive therapy.
    Arch. gen. psychiat. 5, 193-199 (1961).

CRONHOLM, B., and J.-O. OTTOSSON: Reliability and validity
    of a memory test battery.
    Acta psychiat. scand. 39, 218-234 (1963).

CRONHOLM, B., J.-O. OTTOSSON, and D. SCHALLING: The mem-
    ory variables learning and retention in relation to
    intelligence and age in adults.
    Acta psychiat. scand. Suppl. 219: Studies Dedicated
    to Erik Essen-Möller, pp. 50-58, Vol. 46 (1970).

D'ELIA, G. (ed.): Unilateral electroconvulsive therapy.
    Acta psychiat. scand. Suppl. 215 (1970).

EISDORFER, C.: Arousal and performance: Experiments in
    verbal learning and a tentative theory.
    In: Human Aging and Behavior (G. TALLAND, ed.), pp.
    189-216, Academic Press Inc., New York (1968).

HULICKA, I., and R. WEISS: Age differences in retention as
    a function of learning.
    J. Consult. Psychol. 29, 125-129 (1965).

MCNULTY, J.A., and W.K. CAIRD: Memory loss with age: An
    unsolved problem.
    Psychol. Rep. 20, 283-288 (1967).

OTTOSSON, J.O. (ed.): Experimental studies of the mode of
    action of electroconvulsive therapy.
    Acta psychiat. scand. Suppl. 145, 103-131, Vol. 35
    (1960).

SCHONFIELD, D., and B.A. ROBERTSON: Memory and storage and
    aging.
    Canad. J. Psychol. 20, 228-236 (1966).

TALLAND, G.: Deranged memory.
    Academic Press Inc., New York (1965).

TALLAND, G.: Age and the immediate memory span.
    The Gerontologist 7, 4-9 (1967).

WIMER, R.E.: A supplementary report on age differences in
    retention over a twenty-four hour period.
    J. Geront. 15, 417-418 (1960).

WIMER, R.E., and B.-T. WIGDOR: Age differences in retention
    of learning.
    J. Geront. 13, 291-295 (1958).

THE TRANSFER OF INFORMATION BETWEEN SENSE-MODALITIES: A NEUROPSYCHOLOGICAL REVIEW

George   Ettlinger

Institute of Psychiatry

London SE5   8AF, England

ABSTRACT

This paper deals with a special case of the transmission of information within the normal brain: the exchange of information between sense-modalities. Two main kinds of relevant behaviour can be distinguished: (1) cross-modal matching or recognition; and (2) cross-modal transfer. The findings indicate that man can match across sense-modalities even in the absence of language as a mediator or bridge; that apes probably possess this ability; but monkeys probably do not (although insufficient experiments have been undertaken with both apes and monkeys). Specific learning apparently fails to transfer between sense-modalities in man and the monkey (when verbal mediation is precluded for man); and no relevant findings exist for apes. Curiously, some non-primates succeed where primates apparently fail, but the so-called specific "cross-modal transfer" of non-primate mammals should perhaps be regarded instead as wide

stimulus generalisation. In contrast, general learning has
so far been shown to transfer across sense-modalities only
in man, but this claim must be qualified: apes have not
yet been assessed; and non-primates may transfer a primitive form of general learning.

These experiments on intact animals (a) advance the
classification of cognitive processes in behaviour; (b)
taken with other lines of evidence, should help to establish
meaningful groupings of different species of mammal; and
(c) permit predictions to be made (independently of any
other behavioural data) of the existence of different kinds
of neural system in different species. For example, the behavioural evidence suggests that man and apes may have a
higher-order cortical system for cross-modal matching and
general transfer, but monkeys and non-primates may not have
this system; on the other hand, non-primates may possess a
lower-order sub-cortical system (which, if present, is functionally inhibited in man and most primates) for cross-modal
stimulus generalisation.

MATCHING AND RECOGNITION

A spoken message has the same meaning for most of us
as a written message. Indeed, we hardly notice that we hear
the former and see the latter. Information derived through
different sense-organs is treated as "equivalent" by the
brain (and presumably reaches a single neural system from
different points of origin). Likewise, we recognise objects

interchangeably through different sense-modalities. BURTON
and ETTLINGER (1960) proposed language as a cross-modal
"bridge" for this human ability. Additionally this ability
could result from learning, probably in childhood, that a
certain sight usually corresponds with a certain feel and
sound or smell, that is from "bi-modal" or "multi-modal"
conditional learning (ETTLINGER, 1961).

However, BRYANT (1972) has recently indicated that in
man information can perhaps be exchanged between sense-modalities in yet a further way. By using as subjects infants
below the age of 12 months he was able to exclude any intermediary role of language; and his objects (e.g. ellipses)
as shown in Fig. 1 were probably sufficiently unfamiliar to

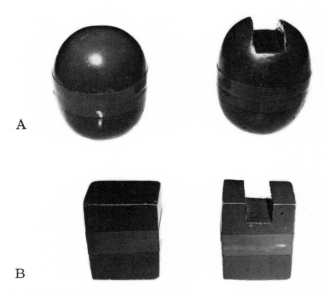

Fig. 1: Two pairs of shapes used in an unpublished cross-
modal matching experiment by P.E. BRYANT

have precluded prior experience of similar objects concurrently in two modalities. The infants received in their
hands an object which could not be seen but emitted sound.
This object (now not emitting sound) together with a new
object was then placed on a table where both were visible
but could not be touched. (In earlier work it was shown as
indicated in the within-modal column of Tab. I that infants
prefer to reach for the one of a pair of visible objects
that had previously emitted sound.) In two of three different object problems the infants reached significantly
more frequently for that object on the table that had been
held in the hand (and had then emitted sound). These results are given in the cross-modal column of Tab. I. The

Tab. I: The number of subjects reaching for the one of two
visible objects that had previously emitted sound
or been silent

| Objects | Single Experimenter | | | | Two Experimenters Blind | |
| | Within-modal V-V | | Cross-modal T-V | | Cross-modal T-V | |
	Noisy	Silent	Noisy	Silent	Noisy	Silent
Pair A	21	9	23	7	24	7
Pair B	22	8	19	11	18	12

Unpublished observations from an incomplete experiment by
P.E. BRYANT

visual choice could only have been based on a comparison
of the visible object qualities with the remembered tactile qualities of the sound emitting object. To the extent
that there had been no prior concurrent visual and tactile
experience of the test objects and no acquisition of general
rules governing the correlation between visual and tactile
qualities, BRYANT's findings seem to me to imply a direct
perceptual "equivalence" between some visual and tactile
object quality or qualities, e.g., shape. (My interpretation
will be contraindicated by evidence of bi-modal conditional
learning below 12 months of age).

BRYANT is continuing these experiments. Meanwhile they
extend other claims that man can achieve cross-modal recognition without language. These have been reviewed by DREWE
et al. (1970, pp. 150-151). For instance GAZZANIGA et al.
(1965) found cross-modal recognition through the minor hemisphere of an adult split-brain patient who has never given
any evidence of language capacity in that hemisphere. This
suggested that the human brain can exchange information between the senses without the intermediary of language. However, this instance of cross-modal recognition need not have
been directly perceptual. The patient would probably have
previously learnt that particular objects usually have certain visual qualities V (e.g., the yellow colour of a banana)
associated with often unrelated tactile qualities T (e.g. a
slippery skin). When tested with a visible sample having
quality V, he might then have chosen whatever object from
a tactile selection possessed quality T only because he had
learnt that qualities V and T are common to the same object,

not because V and T are directly analogous.

Few experiments of this kind have been done in animals.
Using monkeys ETTLINGER and BLAKEMORE (1967) obtained negative findings, that is no better performance on "true"
than on "false" matching tasks. (Identical visual and tactile objects are made available in the former, but only
non-identical objects in the latter.) These experiments
were greatly extended by MILNER (1972) but with the same
results. Prior training to match within vision was ineffective in eliciting cross-modal matching (see Tab. II),
as was use of single sample cross-modal problems (see Fig.
2). However, DAVENPORT and ROGERS (1970, 1971) have reported significant although not perfect performance at
cross-modal matching in a proportion of their apes. Their
method (prolonged training on very many cross-modal problems followed by testing on 40 new problems) has not been
tried with monkeys. Therefore it remains uncertain whether
apes are superior on this task to monkeys. Moreover MILNER
(1971, p. 101) has described a conditional strategy which
could have gained the apes their about 75 % correct performance: " if the sample is wider than the mid-point in
the range of widths, select the wider choice-object; and
conversely". Nevertheless, given that genuine cross-modal
matching occurs in apes as claimed by DAVENPORT and ROGERS,
language would seem not to be necessary. Moreover, since
the objects in the 40 test problems were new, perceptual
matching of analogous visual and tactile qualities may have
occurred (unless a general system of correlational rules
for transforming visual and tactile qualities had been
learnt).

Tab. II: Learning scores (or, in brackets, errors made on trials 1-30) on five cross-modal matching tasks by monkeys with previous training of within modal (i.e., visual) matching

Previous visual matching:-

| | TRUE | | | | | | FALSE | |

Present cross-modal matching:-

| | TRUE | | | FALSE | | | FALSE | |

Animals:-

TASKS	SCT1	SCT2	CST1	CST3	SCT3	SCT4	CST2	SCF1	SCF3
I	1000+ (14)	420 (18)	1000+ (12)	1000+ (19)	900 (15)	180 (16)	280 (19)	490 (14)	1000+ (16)
II	750+ (11)	100 (4)	-	-	320 (15)	80 (15)	90 (16)	140 (16)	750+ (18)
III	500+ (17)	60 (8)	-	-	220 (14)	100 (20)	70 (16)	40 (9)	500+ (14)
IV	500+ (15)	70 (6)	-	-	130 (15)	180 (13)	200 (16)	120 (15)	500+ (17)
V	-	110 (9)	-	-	200 (15)	90 (17)	110 (14)	110 (15)	-

Unpublished observations by A.D. MILNER

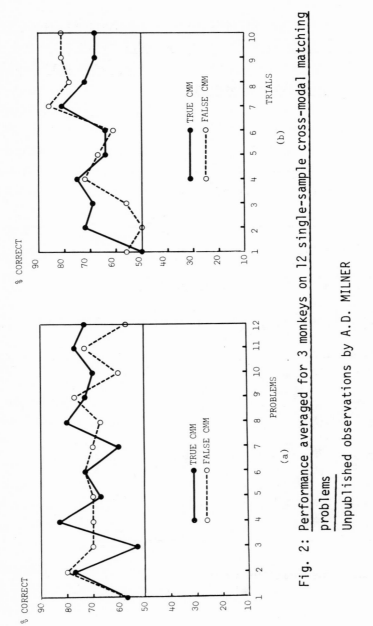

Fig. 2: Performance averaged for 3 monkeys on 12 single-sample cross-modal matching problems

Unpublished observations by A.D. MILNER

TRANSFER

Historically, the transmission of information between
sense-modalities has been chiefly investigated by a behavioural method termed "transfer" testing. The subject learns
to respond differentially to stimuli presented only in one
sense-modality. The same (or closely analogous) stimuli are
subsequently made available only through another sense. Will
the subject now again make the same differential response he
learnt to make in the first modality? I have previously
(1967, pp. 58) defined the formal differences between the
matching and the transfer procedures; and I have also argued
that the ability to match across sense-modalities is a necessary pre-condition for the ability to transfer. Some six
years later, the distinctions made in 1967 still seem to be
valid.

However, it can now be seen that transfer testing,
although particularly convenient and time-saving in animal
experiments, suffers from being too special a case of the
cross-modal exchange of information. Its basic defect lies
in the ambiguity of the requirements when the stimuli are
introduced in the second modality: no indication is given
to the subject that the previous experience with these (or
analogous) stimuli is to be taken into account. Cross-modal
matching, on the other hand, is generally assessed (DAVEN-
PORT and ROGERS, 1970; MILNER, 1972) only when the subject
has given some prior indication of having learnt that it is
expected to match. (It is of course easy to be wise after
the event. My strictures apply only because transfer ex-

periments have given negative results. Such negative findings were totally unexpected).

Specific Learning
     When verbalisation can readily take place man gener-
ally transfers learning across sense-modalities. However,
there exists as yet no clear evidence that specific learning transfers without verbalisation in man. The only strong
positive findings are those of BLANK et al. (1968) who obtained good transfer of a shape discrimination from vision
to touch, but not conversely, in 3-4 year old children.
Although these children were unable to name the objects,
some elementary verbal labelling might have been used.(Alternatively the preliminary training given at cross-modal
matching might have induced the expectation that equiva-
lences would occur in the two senses.) Other experiments
reviewed by DREWE et al. (1970, pp. 149) indicate weak or
absent transfer when the stimulus material precluded verbalisation.

     Apes have so far not been assessed for cross-modal
transfer. With monkeys the findings have been generally
negative, as described by ETTLINGER (1967). To overcome
the ambiguities of the transfer method ETTLINGER and BLAKE-
MORE (1969) tried to teach their monkeys that they were
required to transfer between vision and touch. We used 11
pairs of objects, and gave 8 or 2 alternations of modality
with each pair: training being continued in each modality
until learning had taken place. Then at the end of each
problem 60 trials in the light alternated with 60 in the

dark. In all, 3500 - 5150 trials of such repeated alternation between sense modalities were given to each of the 5
animals. Nonetheless there was little if any evidence for
cross-modal transfer.

Curiously, when different training methods are used
cross-modal transfer may take place in prosimians (bush
babies, Galago senegalensis) and in non-primate mammals (cat,
rabbit, rat and mouse). (These training methods are known
to psychologists as classical conditioning, avoidance learning or discriminative operant conditioning.) In many of these
experiments electric shock was given for errors, whereas monkeys are generally rewarded with food for responding correctly. Further details can be found in MILNER (1971). At present it remains uncertain whether the different outcome with
monkeys and non-simian species reflects merely differences
in test procedure or genuine differences between species. On
balance the latter seems more likely. For example, OVER and
MACKINTOSH (1969) trained rats to depress a lever to changes
in the intensity of a background light or sound. Correct responses were rewarded with food. Significant (but weak)
cross-modal transfer was found. In contrast, a similar experiment by ZIELER (1968) was negative for squirrel monkeys.
We must therefore entertain the possibility that certain
prosimian and non-primate species can achieve transmission
of information between the senses despite the apparent failure of man and monkey on such (non-verbal) tasks.

General Learning

As patients in a hospital, we learn to identify our own

doctor, our own nurse and ward-cleaner; but we may also
learn what features can be used to distinguish one doctor
from another, or from others who wear white coats but are
not doctors; we learn how to recognise one nurse from another, from a nursing sister or orderly; and so forth. It
was shown many years ago that animals not only learn to
discriminate between stimuli and objects but also may learn
the principle that objects can be readily distinguished (or
that other tasks can be solved by attending to certain features of the problem).

General learning of this kind seems to transfer between
sense-modalities in man, even in the absence of verbalisation. Two convincing experiments are those of PICK et al.
(1966) and of BLANK and KLIG (1970). In the latter, children
aged under 5 were compared for their ability to transfer
"dimensional" learning across modalities (touch to vision)
and within one modality (vision to vision). Thus the first
task was tactile for some children, visual for others. Those
children who were first trained on a task (either tactile of
visual) in which differences between the stimuli in texture
were relevant (but differences in shape were irrelevant)
performed better than did other children trained first
(tactually or visually) with shape relevant (and texture
irrelevant) when new visual stimuli were given with texture
(but not shape) relevant. When the children were afterwards
questioned, no verbal labels were elicited for the object
- and indeed only verbalisation of the relevant dimension
could have been effective.

Significant but weak cross-modal transfer of general learning has been claimed for the monkey in a few experiments. However, when MILNER and ETTLINGER (1970) introduced an additional control group it emerged that animals previously trained on an unrelated task showed as much "transfer" as the experimental group. In monkeys the non-specific improvement which occurs during the course of the rather lengthy training required to test for general learning in two senses, seems to be of the same magnitude as the average "transfer" effects that have been claimed.

J.G. WEGENER (personal communication) has introduced important modifications to the procedures traditionally used with monkeys. He is training monkeys to avoid electric shock by moving from one end to the other of a box; the stimuli to be discriminated are temporal patterns, either visual or auditory; and the principle to be learnt is complex: "if red and not green has signified 'go' in a simple discrimination, then red-green-red and not green-red-green will signify 'go'; analogously, if a low-pitched tone has signified 'go', then low-high-low (instead of high-low-high) will signify 'go'". So far transfer effects (if any) are small and unidirectional (from vision to audition) as shown in Tab. III. However, WEGENER has pioneered in varying the basic training procedures with monkeys.

Responding to the onset of a single stimulus may be a primitive form of general learning, but probably differs from the kind of general learning already discussed. The only studies of cross-modal transfer of general learning

Tab. III: <u>Learning scores on cross-modal transfer tests. The symbol + indicates failure to learn</u>

Animals:-

TASKS		Experimental Order I			Control		Experimental Order II		
		84	85	91	Avge	Range	93	92	90
Flashing Green Light	= Go	50	60	110					
Flashing Red Light	= No-go								
Green-Red-Green	= Go	330	70	360					
Red-Green-Red	= No-go								
Tone Bursts of 1200 Hz	= Go	30	50	40	48	25 to 115	40	50	115
Tone Bursts of 800 Hz	= No-go				n=16				
1200-800-1200 Hz	= Go	45	0	155	158	25 to 330	330	200	136
800-1200-800 Hz	= No-go				n=11				
Flashing Green Light	= Go				73	50 to 110	505	1001+	440
Flashing Red Light	= No-go				n= 3				
Green-Red-Green	= Go						+	+	+
Red-Green-Red	= No-go								

Unpublished observations by J.G. WEGENER. Pulse duration = 850 msec; pulse interval = 150 msec; interval between groups of three pulses = 1500 msc.

in non-primates known to me have used single-stimulus dis-
criminations, generally with positive results.

CONCLUSIONS

Behaviour

Despite possible and desirable variations of procedure,
the matching and transfer experiments remain distinct:
stimuli are directly compared in the former but only through
the intermediary of arbitrary learnt responses in the latter. The need is now to extend each experiment. Cross-modal
matching could be assessed in monkeys with the procedures
used successfully for apes; conversely, apes could be tested with the methods that have failed with monkeys. Cross-
modal transfer could be looked for in the monkey with the
methods that have proved successful with non-simians. Ul-
timately the same animals might be trained under four or
more conditions, e.g., discrimination learning and operant
conditioning, each with or without electric shock.

Species Differences

The possibility must be entertained that apes, like
man, can match across sense-modalities whereas only with-
in-modality matching is accomplished by monkeys and non-
simian species. If verified by future research, this outcome would not be too surprising in that it corresponds
with other evolutionary differences.

Quite unexpected, however, would be the finding that

non-primates (and a prosimian) are able to transmit specific
learning between sense-modalities whereas neither the monkey
nor man (when verbalisation is precluded) can accomplish
this. Only further research can establish such a remarkable
failure by most primates on the one hand, and success by
the other mammals (including one prosimian) on the other.

A priori, the brain might be expected to accomplish
specific and general learning separately. Another difference
between species - the cross-modal transfer of general learning (even without verbalisation) in man, but its absence in
the monkey - would then at least accord with other evidence of
evolutionary progression. Nonetheless, there are grounds
for extreme caution: so far the monkey has failed on all
cross-modal tasks whereas the higher primates succeed at
matching (and man at general transfer), and the non-primates at specific transfer. What evolutionary forces could
conceivably be held to account for the monkey's falling between all (cross-modal) stools?

Neural Mechanisms

It is important to establish whether man can match
across modalities by the direct perceptual comparison of
qualities in different modalities; or only by the acquisition, through bi-modal experience, of a set of rules which
govern the correlation of qualities in separate modalities.
Also, whether the apparent concordance across species for
cross-modal matching and transfer of general learning is a
coincidence. If not, it can be predicted that apes will yet
show transfer of general learning (since they can apparently

match). We may then suppose that man and the apes possess
a higher-order cortical system, relatively undeveloped or
absent in monkeys and non-primate mammals, which mediates
cross-modal matching and transfer of learnt principles.
Such a cortical system would be capable of transmitting
between the sensory system of the brain information dealing
either with stimulus qualities (in matching); or with stimulus dimensions, attentional factors or other strategies
of problem solution (in transfer). On the other hand, the
prosimians and non-primate mammals may possess a lower-
order sub-cortical system (possibly still present in man
and most primates but functionally inhibited by cortical
systems) which mediates the cross-modal transfer of specific learning. If this formulation were to be correct,
then this "transfer" would be more closely allied to wide
stimulus generalisation than to discriminative transfer
proper: information about modality would be lost in such
a system.

ACKNOWLEDGMENTS

Drs. P.E. BRYANT, A.D. MILNER and J.G. WEGENER kindly
commented on the first draft of this paper; and have per-
mitted me to refer to their unpublished observations.

ADDENDUM TO THE PAPER

Important modifications to the conclusions relating
to the absence of cross-modal transfer in the monkey are
needed as a result of observations made after the meeting
held in Göttingen. As already mentioned at Göttingen under
"Transfer of specific learning", cross-modal transfer
takes place in prosimians. This was reported by J.P. WARD,
A.L. YEHLE and R.S. DOERFLEIN (1970): Cross-modal transfer
of a specific discrimination in the bushbaby (Galago
senegalensis), J. Comp. Physiol. Psychol. 73, 74-77. These
authors trained 8 animals to discriminate between lights
and sounds interrupted at 3 and 18 cps, using shock avoidance for errors with a self-correction (escape) procedure.
Significant transfer effects were found. On balance this
positive outcome, taken with other evidence previously
discussed, then appeared more likely to reflect a species
difference between the monkey and non-simians than a difference in test procedures.

It is now probable that this interpretation was incorrect, since it has recently been claimed that when
appropriate training procedures are used cross-modal transfer can also be found in the monkey. In unpublished experiments E.H. YETERIAN and W.A. WILSON, Jr. have obtained
a positive outcome in the rhesus monkey when using the
methods of training found to be successful in the bushbaby
by WARD et al. (1970), see Tab. IV. In another unpublished
and still incomplete experiment, G. FRAMPTON, A.D. MILNER
and G. ETTLINGER have now obtained evidence for some cross-

Tab. IV: Total errors to criterion in acquisition and errors in first five sessions (100 trials total) of transfer by monkeys receiving either the direct or reversal transfer condition

	DIRECT TRANSFER				REVERSAL TRANSFER			
	Auditory		Visual		Auditory		Visual	
	3 cps	18 cps	3 cps	18 cps	18 cps	3 cps	18 cps	3 cps
Total Errors in Acquisition	83	99	89	37	89	62	52	68
	Visual		Auditory		Visual		Auditory	
	3 cps	18 cps	3 cps	18 cps	3 cps	18 cps	3 cps	18 cps
Total Errors in First Five Sessions of Transfer Testing	50	24	10	18	55	58	53	55

Unpublished observations by E.H. YETERIAN and W.A. WILSON, Jr. Auditory stimulus consisting of intermittent clicks. Visual stimulus consisting of intermittent flashes of light. In both cases rate refers to go-stimulus. Cps indicates cycles per second.

modal transfer between vision and touch in at least a
proportion of their monkeys without the use of electric
shock. (It is, however, likely that significant transfer
will be found on only one measure.) These new findings
again emphasise the need for caution when invoking differences between species - at least until such time as
test procedures are truly comparable. They also reinforce
other lines of evidence which suggest that discrimination
learning in the monkey is not always unitary: learnt information may be transferred across sense-modalities when
learning is of one kind, but not with conventional training procedures.

LITERATURE

BLANK, M., L.D. ALTMAN, and W.H. BRIDGER: Cross-modal
    transfer of form discrimination in preschool chil-
    dren.
    Psychon. Sci. 10, 51-52 (1968).

BLANK, M., and S. KLIG: Dimensional learning across sen-
    sory modalities in nursery school children.
    J. Exp. Child Psychol. 9, 166-173 (1970).

BRYANT, P.E.: Paper read at E.P.S., London (1972).

BURTON, D., and G. ETTLINGER: Cross-modal transfer of
    training in monkeys.
    Nature 186, 1071-1072 (1960).

DAVENPORT, R.K., and C. M. ROGERS: Inter-modal equivalence
    of stimuli in apes.
    Science 168, 279-280 (1970).

DAVENPORT, R.K., and C.M. ROGERS: Perception of photo-
    graphs by apes.
    Behaviour 39, 318-320 (1971).

DREWE, E.A., G. ETTLINGER, A.D. MILNER, and R.E. PASSING-
    HAM: A comparative review of the results of neuro-
    psychological research on man and monkey.
    Cortex 6, 129-163 (1970).

ETTLINGER, G.: Learning in two sense-modalities.
    Nature 191, 308 (1961).

ETTLINGER, G.: Analysis of cross-modal effects and their
    relationship to language.
    In: Brain Mechanisms Underlying Speech and Language

(C.G. MILLIKAN and F.L. DARLEY, eds.), Grune and
Stratton, New York (1967).

ETTLINGER, G., and C.B. BLAKEMORE: Cross-modal matching
in the monkey.
Neuropsychologia 5, 147-154 (1967).

ETTLINGER, G., and C.B. BLAKEMORE: Cross-modal transfer
set in the monkey.
Neuropsychologia 7, 41-47 (1969).

GAZZANIGA, M.S., J.E. BOGEN, and R.W. SPERRY: Observations on visual perception after disconnexion of the
cerebral hemispheres in man.
Brain 88, 221-236 (1965).

MILNER, A.D.: Cross-modal transfer and matching in primates.
Ph.D. Dissertation, London University (1971).

MILNER, A.D.: Matching within and between sense modalities
in the monkey. J. Comp. Physiol. Psychol., in press (1972).

MILNER, A.D., and G. ETTLINGER: Cross-modal transfer of
serial reversal learning in the monkey.
Neuropsychologia 8, 251-258 (1970).

OVER, R., and N.J. MACKINTOSH: Cross-modal transfer of
intensity discrimination by rats.
Nature 224, 918-919 (1969).

PICK, A., H.L. PICK, and M.L. THOMAS: Cross-modal transfer
and improvement of form discrimination.
J. Exp. Child Psychol. 3, 279-288 (1966).

ZIELER, R.: Quoted by W.A. WILSON, Abstract in Proc. XIX
Int. Congr. Psychol. p. 168, Br. Psychol. Soc.,
London (1971).

THE SIGNIFICANCE OF EXOGENOUS AND ENDOGENOUS FACTORS IN THE HEREDITARY DIFFERENCES IN LEARNING ABILITY OF RATS

H. Müller-Calgan and E. Schorscher

Department of Pharmacology
Medical Research Division
E. Merck, Darmstadt, W. Germany

ABSTRACT

Rats from the strain Wistar/WU/Ivanovas were trained to jump over a hurdle to avoid an electric shock. Learning experiments performed on rats selected for good (GL) and bad (BL) learning, (plus unselected control groups), yielded the following results:

1. F_7 - F_{14} BL no longer react to the selection bias and the corresponding GL react only very feebly.

2. In comparison with the F_1 and F_2 generations of GL and BL the variability decreases. The animals become less sensitive to exogenous factors.

3. Details are given on exogenous (and endogenous) factors with or without influence on the learning results.

INTRODUCTION

According to W. CORRELL (1970) learning in human beings is never determined by the subject alone, but always
by the totality of the existential circumstances of the
subject and the conditions of its environment. It is generally acknowledged that this formulation is applicable to
all more differentiated animal species and consequently also to the majority of our experimental animal material.

Numerous investigations of recent years have demonstrated to what extent variables may influence the experimental results of studies on animal behavior and learning
procedures. In rodents, e.g., differences in emotionality
between or within strains (LA BARBA and WHITE, 1971; JAY,
1963; LINDZEY and WINSTON, 1962; DE NELSKY and DENENBERG,
1967; RAPAPORT and BOURLIERE, 1966; REYNIERSE, 1970; survey
by BROADHURST, 1960) as well as age-dependent differences
(DOTY and DOTY, 1964; DYE, 1969; FEIGLEY and SPEAR, 1970;
GOODRICK, 1968; RAPAPORT and BOURLIERE, 1966; VERZAR and
MC DOUGALL, 1957) have frequently been described. A series
of investigations also concerned the influence of litter
size variations, of population and living space density
(LA BARBA and WHITE, 1971; BELL et al., 1971; LATANE et
al., 1970; SCHREIBER, 1971). However, many of the experimental factors are grouped under the heading environmental
influences (CAMPBELL and BLOOM, 1965; CHOCHOLOVA, 1966;
LATANE et al., 1970; ROSENZWEIG et al., 1968; survey by
BARRY and BUCKLEY, 1966; GLICK, 1969), and consequently
also often include variables such as stimulus parameters

pertaining to a certain experimental technique.

The results of our own investigations on this topic have previously been communicated and discussed in smaller internal circles (MÜLLER-CALGAN, 1971). Since this material has not been published as yet, part of it will be presented here, supplemented by the data meanwhile obtained. Details of the genetical model will be extensively reported elsewhere (MÜLLER-CALGAN, 1973) and not be dealt with here. Moreover, a contribution to the possible mechanism of the phenomenon of good and bad learning in our active two-way avoidance model by one of us is in preparation (MÜLLER-CALGAN, in prep.). If the present results should be confirmed, the social connections of the animals to the foreign species, the human experimenter,   should also gain importance, more than hitherto has been admitted (MÜLLER-CALGAN, in prep.).

METHOD

Experimental Animals

Our experimental animals originate from a consignment of rats of the strain Wistar/WU/Ivanovas (Breeder: S. Ivanovas GmbH., Med. Versuchstierzucht, Kisslegg im Allgäu) acquired early in 1966. From these rats, by means of brother-sister matings (inbreeding) we selected certain lines. Our selection was aimed at producing strains homozygous for the genes determining the characteristics of positive and negative avoidance training. At the time of communication (24.5.72) we are well on the way to homozygosity with two strains each of GL and BL in the 14th and 16th F generation (P_{II} and P_{III}), and in the 15th and 12th generation (P_{IV} and P_V) respectively.

All external conditions were kept constant as far as possible. We only varied the age at the onset of training, the time of day and the cage population density. Larger cage populations were obtained by putting together animals of the same sex from two (or more) litters. After weaning and segregation of sexes, the animals were kept in Makrolon (type III) cages of 900 sq.cm. bottom surface area for adaptation and subsequently, until the end of training, in wire cages of 34 x 30 x 24 cm (unselected animals were only kept in wire cages). During training normal food and water were provided ad lib.

Avoidance Training and Selection

Selection for good and bad learning involved two handicaps. First of all the animals, when jumping over the hurdle, continuously had to change compartments in order to avoid the unpleasant electrical stimulus (active two-way avoidance). As second handicap we considered the very long intervals between experimental days (Fig. 1). We usually tested only three times a week, i.e., on Tuesday, Wednesday and Friday. The animals resolved their entire programme within three weeks with alternating intervals of once,

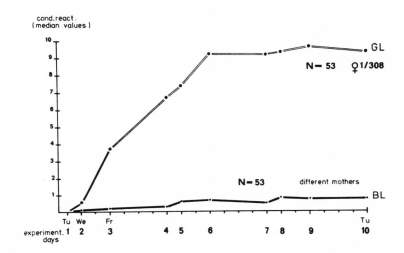

Fig. 1: <u>Performance curves of GL (F_6) and BL (F_7)</u>

twice and four times 24 hours (normal training). In this
way we ensured the procedure also involved elements of a
long-term memory for the combination conditioned/uncon-
ditioned stimulus. Only longer intervals of more than 24
hours without frequent repetitions provide the possibility
of extinguishing an initiated or developing association between stimulus and reaction. Repetitions were only presented in the form of a ten-fold application of the stimulus
combination (with intervals of 20 seconds between the moment
the conditioned or unconditioned jumping is performed, and
the next CST) within a period of about 3.5 to 5 min at each
of the 10 trials.

In this way a total of 100 individual trials had to be resolved. The first 10 of these trials (on the 1st experimental day) merely involved application of the unconditioned stimulus (UST), in this case a foot shock elicited by the bottom grid in order to familiarize the animals with escape across the hurdle. The UST was discontinued immediately after jumping and did not last for more than 10 sec. On the 2nd to the 9th experimental day the conditioned stimulus (CST) was given, viz, a buzzer sound of median loudness to which in 5 sec time the UST was added (overlapping procedure). CST or the combination CST-UST immediately stopped once the animal climbed or jumped over the hurdle into the safe compartment in which, for the moment, it was not threatened by the electrical stimulus. Reactions to the buzzer tone only are designated as conditioned (CR), reactions to the stimulus combination as unconditioned (UR). As a measure of performance we took the total number of CR scored on the 2nd to the 9th experimental day.

For the selection of good learners it was also required that the partners mated for inbreeding possibly had at least once reached the criterion of the so-called correct response, i.e., they should have obtained a score of 9 or 10 out of 10 possible reactions per trial. For the selection of bad learners, however, partners were mated for inbreeding which, at a possibly low number of CR, did not score 9 out of 10 CR (for particulars see MÜLLER-CALGAN, 1973).

Evaluation of Data

All conceivable variables were registered and, together with the learning results, collected on IBM punched cards.
The evaluation was performed partly by hand, partly by machine. Histograms, series of median values and comparisons
of the latter by means of the MANN-WHITNEY U-test were made.
A subsequent evaluation of the collective data in the form
of a factorial analysis is envisaged. Presented here are
data from a total of 5249 individual learning experiments
on unselected (N = 427) selected (F_7 - F_{14}) GL (N = 1087)
and BL (N = 691) rats.

RESULTS

Results Obtained with Unselected Rats

The first histogram (Fig. 2) suggesting a two-peaked
distribution demonstrates that some of the rats, under the
conditions described above, had extreme difficulty in making correct responses. About 55 % of the groups scoring
0 - 30 reactions could not unfalteringly combine the CST
and UST and scarcely 19 % could be classified as good per-
formers (groups 51 - 90). As demonstrated elsewhere (MÜL-
LER-CALGAN, 1971), two different Wistar strains did not
only show strain and sex differences (Tab. I, Fig. 3) but
even differences between individual families, i.e., in the
offspring of two series of homologous matings. The upper
part of Tab. I exhibits the strain differences between WU/

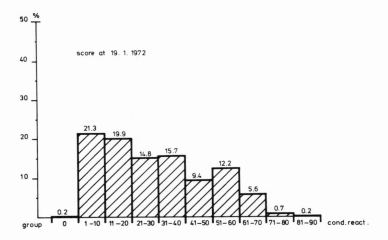

Fig. 2: <u>Relative frequency distribution of unselected rats</u>

Tab. I: <u>Number of conditioned reactions (median values) in</u>
<u>unselected rats</u>

strain differences	Wistar WU/ Ivanovas	Wistar AF/ Han/EMD	significance
males	26.3 (N=102)	15.3 (N= 75)	*****
females	49.0 (N= 52)	20.3 (N= 65)	*****
total	32.6 (N=154)	19.2 (N=140)	

sex differences	males	females	significance
Wistar WU/ Ivanovas	26.3 (N=102)	49.0 (N= 52)	*****
Wistar AF/ Han/EMD	15.3 (N= 75)	20.3 (N= 65)	(*)

(*) 2α= 0.1

*****   < 0.002

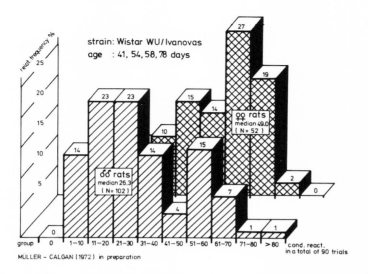

relat frequency %

strain: Wistar WU/Ivanovas
age : 41, 54, 58, 78 days

25

20

15

10

5

♂♂ rats
median 26.3
(N = 102)

14

23 23

14

10

0

15

14

15

4

♀♀ rats
median 49.0
(N = 52)

27

19

7

2

0

1   1

group   0   1—10   11—20   21—30   31—40   41—50   51—60   61—70   71—80   >80   cond. react.
in a total of 90 trials

MÜLLER - CALGAN (1972) in preparation

Fig. 3: <u>Sex difference in the frequency distribution for unselected rats</u>

IVANOVAS and another strain: AF/Han/EMD. The former strain shows better learning in both sexes. On the whole 1.7 times more CR were performed. Besides, sex differences may be more or less distinct as shown in the lower part of Tab. I. In our original strain (Fig. 3) the sex differences are particularly striking. Moreover, the males exhibit a pro- nounced two-peaked distribution of the CR (good and bad). From this collective we selected males and females with best and worst learning results for mating (parent generation) and started inbreeding in two directions.

Results Obtained with Selected Rats

Fig. 4 gives a survey of the selection results so far
obtained up to the 14th F generation. It shows already in
the $F_{1, 2}$ generations of GL a good reaction to the selection bias, whereas in the other direction a considerable
resistance preponderates with an inclination to heterosis
in the corresponding F generations. As we have demonstrated elsewhere (MÜLLER-CALGAN, 1971; MÜLLER-CALGAN, 1973), a
back cross of siblings of a GL and BL line in the F_5 gen-

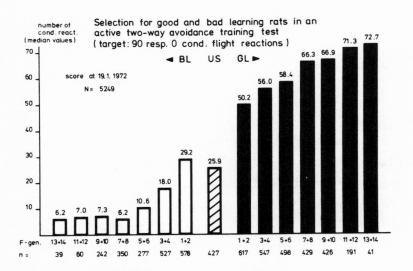

Fig. 4: Selection for good and bad learning rats in an
active two-way avoidance training test

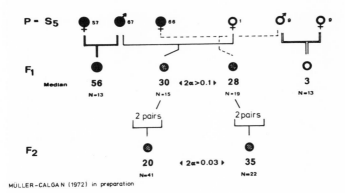

MÜLLER-CALGAN (1972) in preparation

Fig. 5: <u>Cross (breed) with GL (P II) and BL (P IV)</u>

eration also rendered proof of heredity (Fig. 5). The cumu-
lative values of CR for the P generation are indicated to
the right of the sex symbols and below for the F generations.
They show median values as expected in the unselected animals: 20 - 35, whereas the offspring of the control matings
(Fig. 5 top left and right) inherited great and small learning ability. Remarkable is the fact that in the F_2 generation of the cross breed a higher score appears only then
when in the P generation matings occurred between female
GL and male BL, and not vice versa (2 α = 0.03). It could
be tentatively suggested, that this phenomenon is linked
up with the X-chromosome. More detailed investigations on
the sex chromosomes have not been conducted as yet. However,

our colleague Dr. Bauer (Institute of Toxicology, E. Merck,
Darmstadt) made a preliminary chromosome analysis of the
four strains, indicating that all individuals have normal
chromosome counts (N = 42) and haploid or triploid forms
did not occur.

The survey in Fig. 4 also shows that in the GL from
F_7 - F_{14} only a small increase in the number of CR (viz.
less than 1 CR/generation) could be obtained. The BL in the
corresponding range of F generations did no longer react
with certainty to the selection bias. The same impression
is obtained from the next 2 figures (Figs. 6 and 7). Fig. 6
shows that only 2 % of the GL was still learning badly while
91 % could be designated as good learners. In the BL (Fig.
7) this proportion is practically inverted: 91 % to 3 %.

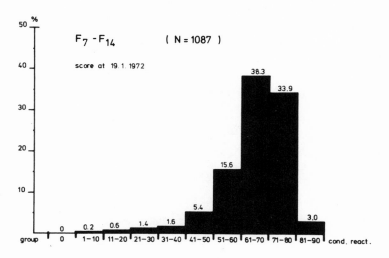

Fig. 6: Relative frequency distribution of GL

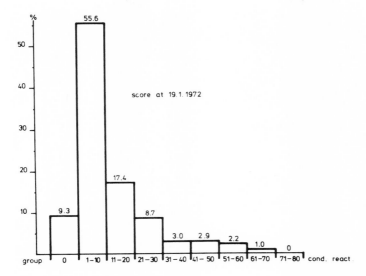

Fig. 7: Relative frequency distribution of BL

Other Exogenous and Endogenous Factors

Summarizing the results of the 1st analysis of the collective data obtained from the F_7 - F_{14} generations, the following can be established:

(1) In comparison with the conditions in the F_1 and F_2 generations (MÜLLER-CALGAN, 1971) of GL and BL the variability decreases. This implies that the animal material subjected to progressive selection becomes more insensitive to exogenous factors.

(2) In the GL and BL remained <u>uninfluenced</u>: the number of

the litter as well as the "rank" within the litter (according to weight at weaning).

(3) Continuing   influence on the learning results were: sex, age, serial number[+], cage population density, time of day as well as the seasonal month during which the 1st to 5th tests were performed.

(4) The sex differences in GL and BL are dominant (Fig. 8). The females exhibit more CR than the males. As shown in Fig. 8 decoded for sex and age, the sex differences are age-dependent. In GL 35 - 40 days old sex differences do not occur whereas from 58 days onward they are rather marked. No sex differences were noticed during the winter months

[+] i.e. animal's order of removal from cage

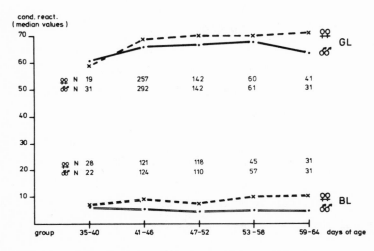

Fig. 8: Influence of sex and age in GL and BL

(November - February), for the lower cage densities (2 - 4
animals), in the higher serial numbers (7 - 10 x) as well
as during the hours of the forenoon (7 - 10 o'clock). Sexual differences are largest in GL between 1 and 3 o'clock
p.m., during spring and summer months, with median cage
populations (5 - 8 animals) and in the serial numbers 5 -
6.

(5) GL males are not influenced by serial numbers, neither
are the females by the cage population densities.

(6) Apparently diurnal influences exhibit a time shift in
both sexes of the GL, the males having their first performance minimum at noon (11 a.m. to 12.30 p.m.) and thus
1 h earlier than the females.

                    Possible Applications

        The survey in Tab. II indicates a series of possible
applications. For our own purpose the most interesting of
these are the methodical possibilities for a more directed
future research on psychopharmaca.

        Particularly fascinating are three aspects which are
here presented as a working hypothesis:

(1) Apart from heredity, the results of direct observations
in our rats during the conditioning procedure as well as
sporadic references in the literature suggest that the abil-

Tab. II: <u>Survey of possible applications of avoidance-</u>
<u>test selected rat strains with genetically</u>
<u>determined GL and BL factors</u>

Study object	concerning	applicable in
Genetic relationships	both	General genetics
Biochemistry, Electrophysiology and Behavior of Learning	both	Mechanism of learning
Environmental factors Social interaction	both	Laboratory animal science
Suggested assign- ment to other learning procedures	both	Discrimination learning Classical conditioning Instrumental learning
Development of spe- cial models with other psychic para- meters:		Psychopharmacology
1. Special excitation	Intertrial activity — GL	Major and new Minor tranquilizers
2. Anxiety for fear	GL (BL?)	Anti-anxiety drugs, Antiphobics
3. "Depression", Inhibition	BL	Antidepressants, Anticompulsants
Learning and memory	$\dfrac{BL}{GL}$	Memory $\dfrac{\text{promoting drugs}}{\text{extinguishing drugs}}$

ity of good or bad learning in avoidance conditioning would exert its influence in a positive or negative way in other learning processes as well. On watching the behavior of the animals under the influence of CST and in the 20 sec pauses between trials, various types of rats may be distinguished. Besides excited and a few chaotically reacting animals there are also quiet, attentive rats among the GL. Occasionally animals with distinctly depressed or inhibited behaviour will be noticed among the BL. These inhibited rats start responding to the buzzer sound with an orientating reaction, set off to jump and then shy away showing no obvious signs of fear.

(2) The phenomenon of inhibition just described is of particular interest to us as for the development of anti-depressants (thymoleptics) and anticompulsants, we had no specific experimental model at our disposal as yet. If it could be proved that our model is transferable to other learning processes, e.g.,discrimination or instrumental learning, better and well-founded arguments could be put forward against the frequently repeated criticism on avoidance procedures, remarks on negative motivation, methodical waste (BOVET and GATTI, 1965) or failing discrimination (BOOTH, 1970).

(3) One could imagine an experimental model using old GL rats with diminished ability of acquisition and retention owing to age-dependent neuronal and biochemical catabolic processes in spite of a hereditary good disposition, permitting the testing of memory-promoting pharmaca.

LITERATURE

LA BARBA, R.C., and J.L. WHITE: Litter size variations
and emotional reactivity in Balb/c mice.
J. comp. phys. Psychol. 75, 254-257 (1971).

BARRY, H. and J.P. BUCKLEY: Drug effects on animal performance and the stress syndrome.
J. Pharmac. Sci. 55, 1159-1183 (1966).

BELL, R.W., C.E. MILLER, and J.M. ORDY: Effects of population density and living space upon neuroanatomy,
neurochemistry, and behavior in the C57Bl/10 mouse.
J. comp. phys. Psychol. 75, 258-263 (1971).

BOOTH, D.A.: Neurochemical changes correlated with learn-
ing and memory retention.
In: Molecular Mechanisms in Memory and Learning (G.
UNGAR, ed.), pp. 54-55, Plenum Press, New York (1970).

BOVET, D., and G.L. GATTI: Pharmacology of instrumental
avoidance conditioning.
In: Pharmacology of Conditioning Learning and Retention, M.Ya. MIKHELSON and V.G. LONGO, eds.), pp. 75-
89, Pergamon Press Ltd., Czechosl. Med. Press, Praha
(1965).

BROADHURST, P.L.: Experiments in psychogenetics.
In: Experiments in Personality (H.J. EYSENCK, ed.),
Vol. I: Psychogenetics and Psychopharmacology, pp.
3-102, Routledge and Kegan Paul, London (1960).

CORRELL, W.: Lernpsychologie. Grundfragen und pädagogische
Konsequenzen, 8. Auflage,
Verlag Ludwig Auer, Donauwörth (1970).

CAMPBELL, B.A., and J.M. BLOOM: Relative aversiveness of
    noise and shock.
    J. comp. physiol. Psychol. 60, 440-442 (1965).

CHOCHOLOVA, L.: The reactibility of rats with defined
    susceptibility to an acoustic epileptogenic stimulus,
    to some acoustic and optic stimuli.
    Physiol. bohemoslov. 15, 337-343 (1966).

DOTY, B.A., and L.A. DOTY: Effect of age and chlorpromazine
    on memory consolidation.
    J. comp. physiol. Psychol. 57, 331-334 (1964).

DYE, C.J.: Effects of interruptions of initial learning upon
    retention in young, mature and old rats.
    J. Geront. 24, 12-17 (1969).

FEIGLEY, D.A., and N.E. SPEAR: Effect of age and punish-
    ment condition on long-term retention by the rat of
    active- and passive-avoidance learning.
    J. comp. physiol. Psychol. 73, 515-526 (1970).

GLICK, St.: Pharmacological, surgical and environmental in-
    fluences upon arousal and learning.
    Yeshiva University, Univ. Microfilsm, Inc. Ann Arbor,
    Mich. Nr. 70-1457, Ph.D. (1969).

GOODRICK, Ch.L.: Learning, retention, and extinction of a
    complex maze habit for mature-young and senescent
    Wistar albino rats.
    J. of Gerontol. 23, 298-304 (1968).

JAY, G.E.: Genetic strains and stocks.
    In: Methodology in Mammalian Genetics (W.J. BURDETTE,
    ed.), Holden-Day, San Francisco (1963).

LATANE, B.,

Social deprivation, housing density and gregariousness in rats.

J. comp. phys. Psychol. 70, 221-227 (1970).

LINDZEY, G., and H. WINSTON: Maze learning and effects of pretraining in inbred strains of mice.

J. comp. phys. Psychol. 55, 748-752 (1962).

MÜLLER-CALGAN, H.: Einige verhaltensphysiologische Grundlagen des Lernens und Gedächtnisses.

Interne Symposia, E. Merck, Mai 1970 und Januar 1971.

MÜLLER-CALGAN, H.: A model for inherited learning in the rat.

Z. Versuchstierk. (in prep.) 1973.

MÜLLER-CALGAN, H.: Does active 2-way avoidance learning (jumping a hurdle) mean pure avoidance learning or conditioned fear and/or conditioned social response to the experimentator? In prep.

DE NELSKY, G.Y., and V.H. DENENBERG, Infantile stimulation and adult exploratory behaviour in the rat.

Anim. Behav. 15, 568-573 (1967).

RAPAPORT, A., et F. BOURLIERE: La facilitation sociale de l'apprentissage d'une tâche opérationelle chez le rat âgé.

Gerontologia 12, 74-78 (1966).

REYNIERSE, J.H.: Differences in emotionality and avoidance in two stocks of rats.

J. comp. physiol. Psychol. 72, 233-237 (1970).

ROSENZWEIG, M.R., W. LOVE, and E.L. BENNETT: Enriched ex-
    perience for two hours a day alters brain chemistry
    and anatomy in rats.
    Fed. Proc. <u>27</u>, 2 (1968).

SCHREIBER, R.A.: Effect of housing density on the incidence
    of audiogenic seizures in DBA/2 J mice.
    J. comp. physiol. Psychol. <u>76</u>, 300-304 (1971).

VERZAR, E.J., and B. MCDOUGALL: Studies in learning and
    memory in ageing rats.
    Gerontologia <u>1</u>, 65-85 (1957).

THE REACTIVITY OF WISTAR RATS HIGHLY SELECTED FOR GOOD AND BAD LEARNING, OBSERVED IN VARIOUS PHYSIOLOGICAL AND PHARMACOLOGICAL TEST MODELS. 1st. COMMUNICATION

H. Müller-Calgan, K.H. Becker, H.J. Enenkel,
H.J. Schliep and A.J.N. Wild

Department of Pharmacology, Medical Research
Division, E. Merck
61 Darmstadt, Germany

ABSTRACT

Preliminary investigations on selected rats, bred from Wistar WU/Ivanovas since 1966, presented the following results:

1. No differences could be detected between GL and BL as far as the pharmacological action on peripheral organs and receptors are concerned, viz. inflammation by means of Freund's adjuvant, sensitivity to nor-adrenalin of the vascular system and sensitivity to carbachol and acetyl-choline of cholinergic receptors in the gastrointestinal tract.

2. Without detectable differences in GL and BL were the effects of centrally acting substances such as myotonolytic tranquilizers, cataleptics, convulsants and narcotics.

3. Equally undifferentiated were pain sensitivity and exploration behaviour in both strains. These factors, therefore, do not come into consideration as causal agents of the differences in learning capacity of GL and BL.

4. Contrary to this, GL generally appeared to be slightly more sensitive to central stimulants and the sympathomimetic amphetamine,BL,however, distinctly more sensitive to the cholinergic tremorine(induced tremor). This is in accordance with the concepts of KHAVARI (1971) and LEITH and BARRETT (1971). KHAVARI assumes "implication of a dichotomous CNS adrenergic-cholinergic neurotransmitter mechanism in the control of learned behaviour". LEITH and BARRETT conclude that "differences between strains in the avoidance performance are, at least partially, related to variations in the relative activity of the cholinergic and adrenergic systems". Our own observations, however, demonstrated here a marked overlapping with the sex differences.

5. Distinct sex differences often exist.
5.1. The female animals are more sensitive to the chemical pain stimulant phenylbenzoquinone and explore slightly more frequently than the males. This could explain, at least in part, a difference in learning ability between the two sexes of the same strain.

5.2. In contradistinction to this, the females are more insensitive to two centrally acting muscle relaxants (at a higher level of the basic tonus), two narcotics and two cataleptics. A prominent sex difference was brought

about by the cataleptic tetrabenazine. This sex difference could also be demonstrated in the control animals of
two unselected strains. No sex differences in the narcotic
action could be observed after hexobarbital administration.
It is envisaged to extend our studies with GL and BL.

The last contribution contains a literature survey as
well as an extensive discussion on the results of our own
investigations.

INTRODUCTION

In the present paper dealing with a series of investigations that have been planned for detecting the presence
or absence of reactive differences in our GL and BL laboratory animals, the following objects were pursued:

1. A search for principles influencing the learning
results. This question is closely connected with the fun-
damentals of good and bad avoidance learning as such:
what in particular is inherited in our test animals at
all?

2. Early recognition and countering of the well-known
dangers pertaining to intensive inbreeding is necessary.
Hereditary predispositions could arise from other psychic
or metabolic traits changing simultaneously with the selection for good and bad performance.

We also select particularly carefully for high fecundity, persistent thriving and general insusceptibility to stress caused by the training procedure. This second item again is inseparable from the third:

3. (which is a pharmacological and a more pragmatical one). For our routine investigations of psychopharmaca we used rats of two Wistar strains trained in the avoidance test, as well as their ancestry (WU/Ivanovas; Breeder: S. Ivanovas GmbH., Med. Versuchstierzuchten, Kisslegg/Allgäu) and AF/Han-EMD SPF). From a series of investigations (BENESOVA and BENES, 1968; COLLINS and LOTT, 1968; CRAIG and KUPFERBERG, 1971; FULLER, 1970; ISOM et al., 1969; SCHLESINGER et al., 1968) it is known that in various strains of rodents sex differences as well as strain differences in the sensitivity to pharmaca may occur. The present paper deals primarily with the more general pharmacological problems and attempts to solve the urgent question of emotionality and pain sensitivity.

METHOD

Test Animals

A total of 454 selected trained and untrained rats of both sexes derived from a population of Wistar WU/Ivanovas, and, for direct comparison, another 115 control animals were used in a total of 1320 individual tests. In various experiments the animals were repeatedly used.

For testing pain sensitivity, central muscle relaxa-

tion and excitation animals of the lower F-generations
(F_3 - F_6) were at first employed. A total of 403 untrained
animals of the higher F-generations was obtained by en-
larging the breeding stock of these generations, and segregation into GL, BL and other strains according to the
following scheme:

	GL		BL	
	P_{II}	P_{III}	P_{IV}	P_V
F_7	20		9	
F_8	8		4	
F_9	69	13	44	
F_{10}	34	5	51	
F_{11}	29	1	38	10
F_{12}	20		11	
F_{13}	9		28	
total N	189	19	185	10

Procedure and Statistics

In our laboratories we proceeded by means of conventional pharmacological tests according to elaborate trial
plans. Determination of an effect was based on the evaluation of scores (MÜLLER-CALGAN, unpubl.), obtained in blind

trials or from a sequential (staircase) process, computing ED_{50} values and fiducial limits according to DIXON and MOOD (1948) and MÜLLER-CALGAN et al (1968). The subsequent statistical analysis was based on non-parametric criteria (U-Test of MANN-WHITNEY, 1956; WILCOXON's signed-ranks test, 1956).

EXPERIMENTS

Exploration Behaviour Versus Avoidance Training

In previous tests we observed that avoidance-trained animals in the unrewarded Y-labyrinth according to RUSHTON and STEINBERG (STEINBERG et al., 1961) showed an abnormal behaviour with a kind of "freezing" and an above average emotional defecation during the period of training, viz. 3.5 min. For this reason we adopted another technique in order to solve the question whether GL and BL exhibit a different exploration behaviour that might explain varying learning abilities.

Method. 36 rats of both sexes (GL: F_7 of P_{II}; BL: F_6 and F_7 of P_{IV}), 50 - 52 days old, weights ranging from 94-142 g (males) and 88 - 109 g (females), were prepared as for the normal 3 weeks avoidance training. However, every second individual remained untrained. For a period of 3.5 min, during which neither buzzer nor electrical stimulation was applied, the untrained animals were examined on their exploration behaviour in the training cage. In each test the number of times the animals changed sides by climbing

over the hurdle as well as the frequency of rearing was
recorded for 3 weeks. On each day of the experiment the
median value for the explorations as well as that for the
number of conditioned avoidance reactions in the GL and BL
was compared with the corresponding ones of the trained
siblings.

Results. Fig. 1 shows that the different learning
ability of GL and BL is not influenced by the faculty of
more or less exploring. In GL and BL the inclination to
explore, which is more evident in the females than in the
males, differs at no time during training. In perambulation
and rearing there is a slight increase in both groups until the 6th day of training, dropping afterwards. As could

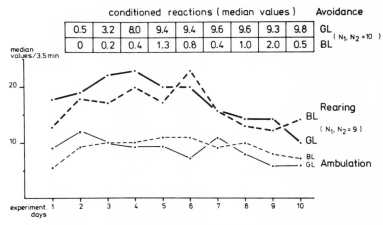

Fig. 1: Exploration versus training

be expected, the results of avoidance training in highly selected siblings differ very distinctly (median values are 69.3 for GL and 9.0 for BL; $2 \alpha \ll 0.002$).

Excitation and Hyperthermia Elicited by d-Amphetamine and Aggregation

Method. Subsequent to the method of aggregation testing (MÜLLER-CALGAN, unpubl.), the aggregation syndrome was now elicited by grouping 5 rats in a glass jar and, in order to avoid injury to the animals, injecting only half the dosage s.c., viz. 2 mg/kg of d-amphetamine in a 0.02 % solution of physiological saline.

The experimental animals were rats experienced in training, viz. 20 males and 20 females of all four strains (GL: F_4 and F_5 of P_{II}; F_3 and F_4 of P_{III}; BL: F_3 of P_{IV} and F_2 and F_3 of P_V). At the onset of the experiment the animals were 2 1/2 to 7 months old; the males weighing 200-354 g, the females 140 - 191 g. The animals had been trained in the avoidance test between 14 days and 4 months previously. Several weeks later they were used again in other experiments (see below).

The animals were kept in metal cages at room temperature of $25 \pm 1^{\circ}$ C, and were given food and water ad lib., until the onset of the experiments. Lots of 5 animals were put into transparent yellowish Makrolon jars, 24 cm in diameter and 24.5 cm high, covered by a wire netting lid.

Half an hour later, as described below, the rectal temperature was taken (initial value). The application of d-
amphetamine followed immediately afterwards. The animals
were blindly tested everyhour for a period of 6 h. Rectal
temperatures were not taken prior to determining the degree
of excitation of all animals. Individual inspection of the
5 animals in each collective took place within 10 sec after
lifting the lid of the jar and classification occurred by
means of a scale from 0 - 4, allowing possible subdivisions
(MÜLLER-CALGAN, unpubl.). Evaluations from 0 - 1.5 corresponded to a normal unexcited behaviour, from 2 and over,
to some state of excitement. Indicated by 4 was the full
amphetamine effect, characterized by intensive perambulation
and particularly by rearing, stereotyped   licking or gnaw-
ing jaw movements (at the rim of the jar) and periodic
fighting behaviour.

Results. The most important results are summarized in
Tab. I. Combined with the separate evaluations according
to strain and sex, these data permit the following conclusions:

1. With adult animals in general, the reduced doses
of 2 mg/kg s.c. - preventing over-excitation injuries or
exitus from critical hyperthermia - bring about the fully
developed amphetamine aggregation syndrome.

2. The differences in eliciting excitation are largely
dependent on sex rather than on strain. Initially the excitation generally develops to a larger extent in the females
and persists for a considerably longer time in the males,

Tab. I: <u>Excitation in the d-amphetamine aggregation test</u>
<u>(median scores after 2 mg/kg s.c.)</u>

hours	total of GL 1	total of BL 2	male 3	female 4	significance strain 1:2	sex 3:4
1	3.7	3.6	3.5	3.8	n.s.	*
2	3.6	3.5	3.2	3.7	n.s.	**
3	3.1	2.5	2.3	3.3	n.s.	*****
4	2.3	2.0	1.7	3.0	n.s.	*****
5	2.3	1.5	1.2	2.8	n.s.	*****
6	1.7	1.3	0.8	2.3	n.s.	*****

n.s. 2 $\alpha > 0.1$ Mann-Whitney   U-test
** ≤ 0.02
*****   0.002
*   0.05

viz. for more than 6 h in the former and 4 h in the latter.
Strain differences seem to be of essentially less importance
(Tab. I), not becoming obvious before 3 - 4 h, when the ex-
citation in the BL recedes more rapidly than in the GL.

3. Generally speaking, excitation is more pronounced
than hyperthermia, the latter being of shorter duration:
1 - 3 h, rarely and only briefly reaching values exceeding
39.5^{o} C, except in the male BL. This seems to be contrary
to excitation: the BL react with more hyperthermia than the
GL.

Tab. II: <u>Central muscle relaxation after 1 mg/kg diazepam</u>
         <u>s.c.</u>

		total of		total of		significance	
		GL	BL	male	female	strain	sex
		1	2	3	4	1:2	3:4
basic value		-0.8	-0.1	-0.7	-0.4	*	n.s.
duration in h		5	4	>5	4		
score differences with respect to basic value	after h						
	1	-1.8	-2.0	-2.4	-1.5	n.s.	n.s.
	2	-0.6	-0.8	-0.9	-0.6	n.s.	n.s.
	3	-1.2	-1.3	-1.6	-0.8	n.s.	(*)
	4	-0.6	-1.4	-1.3	-0.8	(*)	n.s.
	5	-1.3	-0.3	-1.3	-0.6	n.s.	(*)

n.s. $2 \alpha > 0.1$
(*) $= 0.1$
 * ≤ 0.05

Sensitivity to Central Muscle Relaxants

We selected two minor tranquilizers of different chemical structure and pharmacological action profile, viz.
diazepam and EMD 16 165 (MÜLLER-CALGAN, unpubl.), that elicit central muscle relaxation of various degree, though as
principal action. Apart from a central (e.g., cortical) ac-

tion these substances affect spinal areas as well.
The experimental animals used were the same 40 rats experienced in training and drug effects that 6 weeks previously had been employed in the above experiments.

An experienced investigator is able to notice the body
tonus and to estimate the muscle relaxing components of a
substance by picking up the animal and, carefully groping
in the fur with the full hand particularly in the abdominal
area, to classify them blindly according to the following
scale (MÜLLER-CALGAN, unpubl.): 0 = normal tonus, no relaxation; -1 **to** - 5 = faint to quite strong relaxation and +1
to +2 = very slight to medium increase of tonus. The experiments were performed as crossed tests at weekly intervals; each animal - its basic tonus being determined previously - was immediately thereafter subcutaneously injected
with 1 mg/kg diazepam, or EMD 16 165. The muscle tonus
was tested hourly for 5 hours after application of the
drugs.

Results. The results shown in Tab. II and Tab. III
permit the following conclusions:

1. Surveyed in general, the whole collective reacted
with more or less equal sensitivity to both muscle relaxants as far as intensity and duration is concerned.

2. The basic tonus, estimated in advance, is slightly
at variance between the 4 groups, tending to below average
in the GL, and above average in the BL (difference 2 α =
0.05). Moreover, the females exhibited rather a raised, the

Tab. III: <u>Central muscle relaxation after 1 mg/kg EMD</u>
          <u>16 165 s.c.</u>

	total of		total of		significance	
	GL	BL	male	female	strain	sex
	1	2	3	4	1:2	3:4
basic value	-0.8	-0.1	-0.7	-0.4	**	n.s.
duration in h	5	4	4	5		
after h						
1	-1.5	-1.8	-2.1	-1.2	n.s.	*
2	-1.3	-1.0	-1.2	-1.1	n.s.	n.s.
3	-1.4	-0.8	-1.2	-0.9	n.s.	n.s.
4	-0.7	-0.6	-0.5	-0.8	n.s.	n.s.
5	-1.2	-0.3	-0.6	-0.8	n.s.	n.s.

score differences with respect to basic values

n.s. 2α >0.1
  *   <0.05
 ** $\overline{}$0.02

males a lowered muscle tonus.

3. Obviously then, sex differences dominate over
strain differences, also under the influence of the pharmaca. As shown by the individual evaluations as well as
the sex and strain groupings, the BL females reacted least
sensitively to both pharmaca, the BL males most sensitively to diazepam, whereas the GL males were particularly sen-

sitive to EMD 16 165. The differences in sensitivity were
more   pronounced by the duration than the extent of the
effect. Besides, alternating fluctuations in the strength
of the action could be observed (Tab. II and Tab. III),
possibly due to the consecutive way of testing with repeated handling.

Pain Sensitivity in the Writhing Test

We investigated the question of pain sensitivity at
an early stage as it was theoretically possible that by selecting good and bad learning rats pain sensitive or pain
insensitive animals were simultaneously segregated. For
this purpose we used trained animals in order to be able
to correlate the evaluation of the expressions of pain
directly with that of the avoidance learning. In our experience the writhing test is also applicable to rats, provided that the techniques in sound and blind tests are performed, although this species reacts to chemical pain stimuli somewhat less sensitively than the mouse (MÜLLER-CALGAN,
unpubl.).

Method. In a well-planned experimental schema that also takes diurnal and treatment influences into consideration,
the same 20 male and 20 female animals of all 4 strains
(GL: P_{II} and P_{III}; BL: P_{IV} and P_{V}) experienced in training
and drug effects were used as previously in the experiments
above. They were tested for expressions of pain at weekly
intervals for a total of 5 tests. At the onset of the experiment the animals were 5 1/2 to 10 months old; the male

animals weighed 160 - 357 g, the females 163 - 207 g. Avoidance training for the majority of these rats had been performed between 3 1/2 and 7 months previously.

Immediately prior to the beginning of the experiment
the typical pain reactions (writhings) are elicited by
means of intraperitoneal injection of 2 ml/kg body weight
of a saturated 0.02 - 0.023 % watery solution of 2-phenyl-
1,4-benzoquinone and continuously registered for a period
of 30 min. Single animals are observed in glass jars and
are unable to see each other.

Results. Evaluation of the results (Tab. IV and Tab.
V) permits the following conclusions:

1. A difference in pain sensitivity, even with regard
to the unconditioned stimulus, cannot be of much import for
the inception of such differentiated learning results as
shown by BL and GL. Tab. IV shows that (with one exception)
neither the weekly results nor their summation differ from
each other (2 α > 0.1). Likewise, no correlation (r = 0.002,
2 α > 0.5; x = number of conditioned reactions; y = pain
reactions during 5 observations of 30 min each) is found by
comparing the training results with the pain valuations
within the whole group (N = 40).

2. Contrary to this it is shown that sex differences
exist in the pain sensitivity to phenyl benzoquinone (Tab.
V). In general, the female animals show pain reactions more
frequently (2 α < 0.02 to < 0.002). This dissociation between sexes is, however, still more marked in the BL than

Tab. IV: <u>Pain reactions in the writhing test (median values)</u>

	total of GL	BL	total of male	female	significance strain	sex
	1	2	3	4	1:2	3:4
week I	42	39	34	66	n.s.	*****
II	47	62	43	66	*	**
III	39	40	25	52	n.s.	**
IV	57	31	20	56	n.s.	*****
V	38	32	15	44	n.s.	**
I - V	209	227	148	259	n.s.	*****

n.s. 2α >0.1
*   <0.05
**   ⁻0.02
*****   0.002

in the GL. The female BL reacted about twice as often as
the male BL (median values, 275 and 131 respectively, for
5 tests) and were, particularly in the first 3 tests, also
markedly more sensitive than the female GL (222 reactions
in 5 tests). These results also speak against any essential
importance of a differential pain sensitivity for the inception of the learning results in our avoidance test procedure.

3. From the third and also fourth test onwards BL
of both sexes showed a marked decrease in the number of

Tab. V: <u>Pain sensitivity in the writhing test (median values)</u>

	GL male BL		GL female BL		significance			
					strain		sex	
	1	2	3	4	1:2	3:4	1:3	2:4
week I	33	34	44	57	n.s.	n.s.	n.s.	(*)
II	43	46	50	74	n.s.	**	n.s.	*****
III	34	15	39	65	(*)	n.s.	(*)	**
IV	37	19	59	48	(*)	n.s.	(*)	***
V	31	11	44	44	(*)	n.s.	n.s.	***
I - V	173	131	222	275	*	(*)	***	*****

n.s. 2 α > 0.1
(*)   = 0.1
*   < 0.05
**   0.02
***   0.01
*****   0.002

pain reactions, which on the average amount up to 1/3 of the
initial frequency in the males, up to 2/3 in the females.
Irrespective of obvious diurnal fluctuations, none of the GL
tested showed any change in sensitivity. For this phenomenon
we have no plausible explanation as yet. It might be possible
that a correlation exists with conditioning, in such a way
that in the BL a particular excitation is also responsible
for an initial rise in pain reactions. Alternatively one could
imagine modifications in the assimilation of pain stimuli in

the course of time. It is envisaged to supplement these
results with investigations on untrained animals.

Reactivity in the Adjuvant Arthritis Test

Method. In rats (GL: F_9 of P_{II}, F_{10} of P_{III}; BL: F_9
+ F_{10} of P_{IV}) 0.05 ml of a suspension of 50 mg heat-killed
Mycobacterium tuberculosis in 10 ml of liquid paraffin
(heavy type: E. Merck, Darmstadt, Nr. 7160) was s.c. injected into the sole of the foot on the right hind paw, according to the method of NEWBOULD (1963). Within three
days the control animals (selected from the strain Wistar
AF/Han-EMD) developed a swelling on the injected paw which
up to the 10th day post injectionem remains practically unchanged, though it may slightly decrease in size. After
ten days the volume of the treated leg is distinctly increased while the left (untreated leg) also starts swelling.
Maximum swelling is reached between the 17th and 21st day
followed by a slow decrease, whereas the chronic pathological changes develop into a deforming arthrosis. The leg
volume was measured according to the method of HILLEBRECHT
(1954).

Results. 1. In GL and BL rats derived from the strain
Wistar WU/Ivanovas no differences were observed in the
average increase of the leg volume after administration of
Freund's adjuvant.

2. Fourteen days p.i. the average increase in volume

of the right and left paws was distinctly larger in the
control animals as compared with the corresponding measurements of the GL and BL.

Reactivity of Cholinergic Receptors

Central and peripheral cholinergic effects in the
tremorine test. Method. A collective of 20 adult rats from
each of 2 strains (10 males and 10 females each of GL: F_8
and F_9 from P_{II}; F_9 and F_{10} from P_{III}; and BL: F_9 from
P_{IV}), aged 4 - 5 1/2 months and weighing 231 - 371 g (males)
and 165 - 239 g (females) were used. The experiments were
carried out in 2 sections during 2 consecutive months (May
and June) according to a well-balanced experimental scheme
permitting technically conditioned (e.g., sequence of selection) and diurnal influences to even up. Single animals were
put into transparent yellowish Makrolon jars of 17 cm diameter with wire netting lids, the bottom thinly covered with
chaff. For adaptation to handling, administration of injections, and measurement of body temperature, all animals
were subjected to a preliminary test one day prior to the
actual experimental day. Physiological saline was twice
s.c. injected (1.0 ml/100 g and 30 min later 0.1 ml/100 g)
and, similar to the actual test, the body temperature was
measured after 1/2, 1, 2 and 3 h, by inserting an electrical
universal thermometer, type TE 3, of ELLAB, Copenhagen, 7
cm deep into the rectum. The preliminary temperature measurement after placebo also served as an initial value for
determining lowering of the body temperature elicited by
tremorine. On the experimental day the animals obtained, in-

stead of the usual test substance in the tremorine test
(MÜLLER-CALGAN, unpubl.), a control injection and 1/2 an
hour later the cholinergic, tremorine, in the form of a
4 % solution in physiological saline at the dose of 40 mg/
kg s.c. At equal time intervals as on the previous day the
tremorine actions were recorded in the following sequence:

1. (Without touching the animals and without lifting
the lid of the jar). The intensity of spontaneous tremor
and lachrymation.

2. Induced tremor. For this purpose, the animals were
carefully taken out of the jars, put on a table, and without further disturbance, observed for 15 sec.

3. Intensity of salivation.

4. Consistency of the excreta.

5. Measurement of rectal temperature. For its evaluation a scale from 0 - 5 was used (0 = no action, 1 - 5 =
very weak to very strong action) and for the fecal consistency a scale from 0 - 4 (0 = dry as normal, 1 = normally formed and moist, 2 - 4 = soft to liquid).

Results. Only the more important central actions of
tremorine will here be dealt with in some detail (see Tab.
VI and Fig. 2):

1. In the quiet or spontaneous tremor we noticed no
strain differences (Tab. VI above), but almost exclusively

Tab. VI: __Tremorine tremor (median values)__

differences of

| | strain | | | | sex | | |

spontaneous tremor

hours	total of GL	BL	significance	hours	total of male	female	signific.
0.5	3.0	2.9	n.s.	0.5	2.5	3.3	*
1	3.0	2.8	n.s.	1	1.9	4.1	*****
2	1.8	2.3	n.s.	2	2.0	2.2	n.s.
3	0.7	1.1	n.s.	3	0.3	1.8	*****

induced tremor

hours	total of GL	BL	significance	hours	total of male	female	signific.
0.5	3.0	4.2	*****	0.5	3.3	3.8	*
1	3.5	4.3	**	1	3.5	4.4	***
2	2.9	3.6	**	2	2.9	3.8	**
3	1.8	3.1	***	3	2.0	3.4	***

sex differences. The female animals are the most sensitive
in this respect as in all other tremorine actions, excluding
lowering of the body temperature. On the average, females
exhibit the strongest spontaneous tremor which reaches its

maximum after one hour and persists for more than 3 h. In
the male animals this tremor usually reaches its maximum
after half an hour and lasts for 2 - 3 h.

2. In the induced tremor, owing to handling and transferring the animals from the jars, distinct strain differences seem to occur (Tab. VI below). These differences,
however, arise as a result of an extreme increase in the
intensity of the tremor in the BL. In the male BL this increase is so strong that practically no sex differences can
be detected. In both sexes of the BL, however, a difference remains.

3. Neither pronounced sex nor strain differences exist
in the lowering of the body temperature (Fig. 2), which in
comparison to tremor is generally distinctly retarded, not
reaching its maximum before 2 h and lasting for more than
3 h; practically independent from sex, this lowering of the
body temperature arises a little earlier in the BL, compared
to the GL, and is also terminated more rapidly.

4. Within the scope of the experiment, quantitative
differences could readily be detected in lachrymation but
not in salivation. Nevertheless, although with larger individual variance than in the case of tremor, all 40 animals
presented increased lachrymation and 39/40 = 98 % also increased salivation. BL tended to more intensive lachrymation
and salivation than GL, but here, too, the sex differences
were of greater importance. Female animals reacted with
stronger cholinergic secretion.

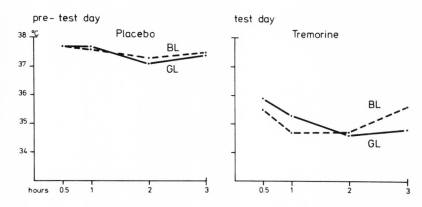

Fig. 2: <u>Rectal body temperature</u>

<u>Sensitivity of peripheral cholinergic receptors.</u>
<u>Method.</u> According to the technique of MAGNUS (1904) seg-
ments of the proximal duodenum and terminal ileum, 4 cm
long, were taken from a total of 7 male controls and 5 ani-
mals each of 2 strains: GL (F_8 - F_{10} of P_{II}; F_{11} of P_{III})
and BL (F_9 and F_{10} of P_{IV}), 3 1/2 - 5 months old, weighing
215 - 363 g and starved for 24 h. In an aerated Tyrode
solution of 37° C the pieces of intestine were adapted to
a pre-tension of 1 g for 30 min. As parasympathomimetics
watery solutions of either carbachol (Doryl[R]) or acetylcholine chloride were added in the maximum volume of 0.46
ml/10 ml of Tyrode for 1 min at intervals of 5 min. Starting
from 1 x 10^{-9} g/ml we increased the concentration of the

pharmacon to the segment's maximum capacity of contraction
which was equated to 100 %. From each intestinal segment we
recorded two concentration-effect series, 15 - 25 min apart.
During this interval the segments were washed 3 - 4 times.
ED_{50} values were graphically determined.

Results. There are no indications of marked sensitivity
changes of peripheral cholinergic receptors in the duodenum
and ileum of GL and BL brought about by either carbachol or
acetylcholine. The ED_{50} values for eliciting contractions
for both intestinal sections are in the same range of action, viz. 1.7 to 5.5 x 10^{-8} for carbachol and 1.1 to 3.1
x 10^{-8} for acetylcholine.

Norepinephrine in Demedullated Rats

Method. In other experiments performed on 18 male pithed
rats according to SHIPLEY and TILDEN (1967), dose-response
curves of the norepinephrine-induced rise in blood pressure
were plotted. The subjects were 6 controls and 12 selected
rats (GL: F_7 and F_8 of P_{II}; F_9 of P_{III} and BL: F_8 - F_9 of
P_{IV}), 4 to 7 months old, weighing 280 - 420 g.

Results. From the results obtained in this preparation
which consists of a cardiovascular system deprived of its
connections to the central nervous system, the conclusion
can be drawn:
GL and BL show no difference in the response of the
peripheral cardiovascular system to test injections of
catecholamine.

Cardio-Toxic Action (Isoprenaline-Induced Cardiac
Infarct)

Method. Five adult rats of both sexes from each of two
strains (GL: F_{12} + F_{13} of P_{II}; BL: F_{11} of P_V) 75 - 90 days
old, and weighing from 143 - 202 (males) and 101 - 166 g
(females), were put into Makrolon cages (type III) for
adaptation, 3 days prior to the beginning of the experiments.
For the duration of the experiment food and water was provided ad lib. In contradistinction to the methods of SELYE
(1961) and FLECKENSTEIN (1969), each of the rats received
a s.c. injection of 10 mg/kg of a 1 % solution of isoprenaline sulphate (Aludrin[R]). The animals were killed with CO_2
48 h after the last injection and dissected. The hearts
were macroscopically inspected and fixed in formalin. Sections for histological investigation were made by means of
a freezing microtome. Of each heart 4 sections of 10 µ thickness were made from 2 levels 400 µ apart and strained with
hematoxylinosin. For the microscopical evaluation of the
severity of the extent of necrosis we applied a scale of 7
degrees (0 - 3 with intermediate divisions of 0.5; 0 = no
necrosis, 1 = slight, 2 = medium, 3 = heavy necrosis).

Results. All animals developed necrosis and survived
the 3 days experimental time. The severity of the necrosis
corresponded to that of the controls. Sex differences could
not be detected in this small sample, although the females
tended to a slightly stronger necrotization. Irrespective
of this possible sex difference, the GL generally reacted
more sensitively to the ß-sympathomimetic, isoprenaline,
than the BL (M = 1.3 and 0.8 respectively; 2 α <0.02). These

investigations require further confirmation.

Reaction to Convulsants

Method. In the test model of a 3 x 3 latin square, 63 adult rats of both sexes from each of 2 strains of good (GL) and bad learning (BL) rats (F_9 - F_{13} of P_{II} and F_{10} - F_{13} of P_{IV}), 5 - 12 months old and weighing from 210 - 390 g (males) and 153 - 250 g (females), were allocated at random to the 3 convulsants, metrazole, nicotine hydrogentartrate and cocaine hydrochloride. At weekly intervals each animal was injected with 0.1 ml/100 g of one of the convulsants dissolved in distilled water and i.v. applied within 1 sec. Successive doses increasing at the rate of 1 : $\sqrt{2}$ were administered in sequential order (staircase). During an observation period of 2 min presence or absence of fully developed convulsions as well as the duration of the fits were recorded. The convulsive ED_{50} was evaluated according to DIXON and MOOD (1948).

Results. A survey of the individual convulsive ED_{50} values in Tab. VII reveals the following:

1. GL and BL of both sexes show differences in the type of reaction to the 3 convulsants.

2. However, uniform differences in sensitivity to all convulsants exist in both sexes. The females being less sensitive throughout, require doses 1.3 (1.0 - 1.7) to 1.6 (1.1 - 2.3) times higher ($p < 0.05$). Contrary to this the

Tab. VII: Convulsive action

ED_{50} values in mg/kg i.v.

| | GL male | BL male | GL female | BL female | significance | | | |
| | | | | | strain | | sex | |
	1	2	3	4	1:2	3:4	1:3	2:4
Metrazole	12.0 (10.4 - 13.9)	13.7 (11.0 -17.1)	17.8 (14.7 -21.5)	20.8 (18.2 -23.8)	n.s.	n.s.	s.	s.
Nicotine	0.58 (0.41- 0.81)	0.55 (0.44- 0.69)	0.91 (0.71- 1.16)	0.70 (0.57- 0.87)	n.s.	n.s.	s.	s.
Cocaine	5.4 (4.6 - 6.3)	5.3 (4.5 - 6.2)	7.7 (6.2 - 9.7)	7.7 (6.4 - 9.3)	n.s.	n.s.	s.	s.

s. significant at the level of $p = 2 \alpha \leq 0.05$
n.s. not significant

mortality shows no sex difference but was rather dependent
on the substance. In one collective receiving metrazole,
8.2 %, and, in another receiving nicotine, 2.7 % did not
survive a period of one week. No acute deaths occurred
during an observation time of 2 hours.

Pharmacogenic Catalepsy

It is known that neuroleptics with antipsychotic properties, when applied in high doses, bring about catatonic
states and a fully developed Parkinsonoid syndrome (e.g.,
KUHLENKAMPF and TARNOW, 1956; MÜLLER-CALGAN and SOMMER,
1968) in human beings and in apes. The symptoms are ascribed
to an influence on the extrapyramidal system (COURVOISIER
et al., 1957; KUHLENKAMPF AND TARNOW, 1956) and the dopamine metabolism of dopaminergic neurons in the globus
pallidus and substantia nigra (survey in HORNYKIEWICZ, 1964).
In the rat the same pharmaca elicit cataleptic states (MÜL-
LER-CALGAN et al., 1968; TAESCHLER, 1962, survey in STILLE,
1971) presented by anomalies in posture and behaviour. On
account of their extra-cortical primary point of attack we
tested the neuroleptics, perphenazine and EMD 16 139, as
well as the standard cataleptic, tetrabenazine, on our animal material.

Method. In the test model of a 3 x 3 latin square,
155 adult rats of both sexes from each of two strains of
GL and BL (F_9 - F_{11} of P_{II} and F_9 - F_{11} of P_{IV}), 4 to 10
months old, males weighing 167 - 352 g and females 131 -
250 g, were subjected at random to the action of the 3

cataleptics in the 3 treatment groups for each sex and
strain. Moreover, for comparing the sensitivity to tetrabenazine we tested 102 slightly younger unselected male and
female rats of 2 strains, Wistar WU/Ivanovas and AF/Han-
EMD. These rats were about 50 - 60 days old, animals of the
former strain weighing 121 - 150 g (females) and 120 - 160 g
(males); those of the latter strain 131 - 150 g (females)
and 140 - 185 g (males). Doses increasing at the rate of
1 : $\sqrt{2}$ or 1 : 2 were i.v. administered within 5 sec. The
substances dissolved in saline were always applied in the
uniform volume of 0.1 ml/100 g body weight. The animals were
kept in glass jars until the beginning of the tests after
30 min. For determining the catalepsy the animals were put
on a table. The right hind paw of each rat was placed on
the slightly roughened surface of a round wooden log, 3 cm
high and 4 cm across. As a criterion for catalepsy the animals had to stay for at least 15 sec on the log, observations not exceeding 5 min. The catalepsy effect was evaluated according to a sequential procedure (staircase), ED_{50}
values being computed according to DIXON and MOOD(1948)
and MÜLLER-CALGAN, unpubl. In addition the duration of postural (cataleptic) states were recorded.

Results. From a summary of the cataleptic ED_{50} values
as seen in Tab. VIII and Tab. IX the following can be concluded:

1. Female GL and BL show no differences in their reaction to the cataleptics tested with reference to the cataleptic ED_{50} as well as to the mean postural duration of the

Tab. VIII: <u>Cataleptic action</u>

ED_{50} values in mg/kg i.v.

| | GL male | BL male | GL female | BL female | significance | | | |
| | | | | | strain | | sex | |
	1	2	3	4	1:2	3:4	1:3	2:4
Perphenazine	0.22 (0.13-0.36)	0.12 (0.08-0.17)	0.30 (0.23-0.39)	0.31 (0.22-0.43)	n.s.	n.s.	n.s.	s.
EMD 16 139	6.4 (4.9-8.5)	3.3 (2.5-4.4)	6.2 (4.9-7.9)	4.6 (3.7-5.9)	s.	n.s.	n.s.	n.s.
Tetrabenazine	0.7 (0.5-0.9)	1.1 (0.5-2.5)	9.4 (5.7-15.5)	9.9 (7.5-13.0)	n.s.	n.s.	s.	s.

s. significant at the level of p = 2 α $\leq$0.05
n.s. not significant

Tab. IX: Sex differences in the tetrabenazine induced
         catalepsy

cataleptic ED_{50} in mg/kg i.v.[*]

strain	male 1	female 2	efficacy ratio 1:2	procedure
un-selected	Wistar AF/Han/EMD	3.0	11.2	3.7 (2.5- 5.5)	conventionally tested
	Wistar WU/Ivanovas	1.5	6.8	4.5 (2.7- 7.5)	
selected	GL (P_{II})	0.7	9.4	14.5 (8.1-26)	latin square route
	BL (P_{IV})	1.1	9.9	9.3 (3.9-22)	

[*] Confidence limits not indicated

reacting individuals (Tab. VIII). In the males the condition varied according to the cataleptic applied. The dose-
activity curves varied in slope after administration of
tetrabenazine; the ED_{50} values did not differ significantly.
The male BL showed slightly more sensitive reactions to the
2 other cataleptics as compared to the corresponding GL.
With EMD 16 139 doses 1.9 (1.3 - 2.8) times lower sufficed
for eliciting catalepsy (p < 0.05).

2. Sex differences occurred in GL and BL in approximately the same way after 2 of the 3 cataleptics tested.
With applications of perphenazine the cataleptic doses were
slightly less in GL and in BL strikingly lower, viz. 2.6
(1.6 - 4.2) times.

3. An exception is presented by the surprisingly high
sex difference after tetrabenazine administration (Tab.
VIII). As far as the authors are aware, this particularly
marked dissociation has not been reported as yet. In male
rats doses 9 (4 - 22) and also 14 (8 - 26) times lower
suffice for eliciting catalepsy. Tab. IX shows that similar
results were also obtained in unselected rats of the two
strains Wistar AF/Han - EMD and Wistar WU/Ivanovas. Here,
too, a significant difference in reaction was observed:
3.7 and   4.5 times lower doses proved to be sufficient
for the male animals. As our highly selected male GL and
BL with respect to drug sensitivity, are more like their
unselected parentage, it may be assumed that the sensitivity
of the female animals of both strains has decreased even
slightly more.

4. By comparing the sensitivity to cataleptics in
general and to tetrabenazine in particular of GL and BL
with that of two unselected rat strains, it appears that
hardly any relation exists to the varying learning ability
of our test animals.

Narcotic Action of Hexobarbital, Methitural and Pentobarbital

Method. The same collective of young adult rats that
3 weeks earlier had been treated with cataleptics (see
above), was used again. Within the scope of a 3 x 3 latin
square, the 3 narcotics were applied at random. At weekly
intervals the sodium salts of the narcotics, dissolved in

distilled water, were i.v. injected in doses of 0.1 ml/
100 g within 5 sec. As a criterion for narcosis, 2 min post
injectionem, the animals had to remain in the supine position for at least 5 min. If unsuccessful, the supination
trial was 4 times repeated at intervals of 30 sec. Moreover, onset and duration of narcosis as well as mortality
up to 7 days after narcosis were recorded. The narcotic
ED_{50} was determined.

Results. The ED_{50} values for narcosis are compiled in
Tab. X. From these data and the recorded but not indicated
times of onset and duration of narcosis as well as from the
corresponding values for the unselected rats (MÜLLER-CAL-
GAN, unpubl.) the following conclusions can be drawn:

1. The reactivity of highly selected male and female,
adult GL and BL rats to short-acting narcotics, viz., the
barbiturate, hexobarbital, the thiobarbiturate, methitural,
and the narcotic, pentobarbital, corresponds to that of the
unselected rats. The narcotic fatalities (acute and delayed
exitus under laboratory conditions, i.e., without atropiniza-
tion and aftertreatment), were highest with 6/117 = 5.1 %
after pentobarbital, and lowest with 0.9 % after methitural.

2. There is no difference in the action of the narcotics tested in GL and BL of the same sex.

3. However, sex differences do exist: female animals
react more sensitively to methitural and pentobarbital,
requiring doses for the narcotic ED_{50} that are in the

Tab. X: <u>Narcotic action</u>

ED_{50} values in mg/kg i.v.

| | GL male | BL male | GL female | BL female | significance strain | | significance sex | |
					1:2	3:4	1:3	2:4
	1	2	3	4				
Hexobarbital	28 (25-32)	31 (-)	30 (-)	25 (19-33)	n.s.	n.s.	n.s.	n.s.
Methitural	26 (19-36)	34 (29-39)	35 (26-46)	43 (-)	n.s.	n.s.	n.s.	s.
Pentobarbital	18 (-)	20 (16-25)	25 (21-31)	27 (22-33)	n.s.	n.s.	s.	s.

(-) no confidence limits determined because of insufficient scattering

s.   significant at the level of $p = 2$ $\alpha \leq 0.05$

n.s. not significant

region of 1.3 - 1.4 times higher than those for the males.
After hexobarbital no sex differences were observed.

4. It appeared to be possible to use rats at weekly
intervals for the evaluation of the narcotic ED_{50}. The
individual experimental parts of the latin square show
neither increase nor decrement in sensitivity.

ACKNOWLEDGEMENTS

The authors wish to acknowledge the assistance of all
laboratory and technical staff. Without their versatile
pharmacological and technical skill, their painstaking
work in the rearing of highly selected experimental animals, the difficult and time-consuming training of rats,
as well as the evaluation and graphical presentation of
numerous experiments, we would not have been able to make
these contributions. Our thanks are also due to our colleague Dr. J.W.H. HOVY and Mrs. R. HOVY for their in-
spiring accomplishment of the English translation as well
as to Mr. K.H. DECKER for his valuable help in the first
analysis of the collective data.

LITERATURE

BENESOVA, O., and V. BENES: The relation between the type
    of higher nervous activity, some biochemical para-
    meters and the reactivity to drugs.
    Act. nerv. sup. 10, 223-231 (1968).

COLLINS, T.B., jr., and D.F. LOTT: Stock and sex specifity
    in the response of rats to pentobarbital sodium.
    Labor. Anim. Care 18, 192-194 (1968).

COURVOISIER, S., R. DUCROT, and L. JULOU: Psychotropic
    Drugs.
    In: S. GARATTINI and V. GHETTI), p. 373, Elsevier,
    Amsterdam 1957, cited by Taeschler et al.

CRAIG, A.L., and H.J. KUPFERBERG: Heat loss and heat pro-
    duction in d-amphetamine hyperthermia in two strains
    of aggregated mice.
    Pharmacologist 13, 2:306 (1971).

DIXON, W.J., and A.M. MOOD: A method for obtaining and
    analyzing sensitivity data.
    J. Amer. Statist. Assoc. 43, 109-126 (1948),
    and in FINNEY, D.J.: Staircase Estimation Probit Ana-
    lysis, 2nd Ed., pp. 226-235 (1952).

FLECKENSTEIN, A.: Symposium International on Drugs and
    Metabolism of Myocardium and Striated Muscle.
    Ed. by M. LAMARCHE and R. ROYE, Nancy 1969.

FULLER, J.: Strain differences in the effects of chlor-
    promazine and chlordiazepoxide upon active and passive
    avoidance in mice.
    Psychopharmacologia 16, 261-271 (1970).

HILLEBRECHT, J.: Zur routinemäßigen Prüfung antiphlogisti-
    scher Substanzen im Rattenpfotentest.
    Arzneimittelforsch. 4, 607-614 (1954).

HORNYKIEWICZ, O.: Dopamine (3-hydroxytyramine) and brain
    functions.
    Pharmacol. Reviews 18, 925-964 (1964).

ISOM, G.E., R.B. NELSON, and A.I. EDLIN: A comparison of
    the lethal and respiratory effects of morphine in
    Long-Evans and Sprague-Dawley rats.
    Arch. int. Pharmacodyn. 182, 130-138 (1969).

KHAVARI, K.A.: Adrenergic-cholinergic involvement in
    modulation of learned behavior.
    J. comp. physiol. Psych. 74, 281-291 (1971).

KUHLENKAMPF, C., and G. TARNOW: Ein eigentümliches Syndrom
    im oralen Bereich bei Megaphenapplikation.
    Nervenarzt 27, 178 (1956).

LEITH, N.J., and R.J. BARRETT: Relationship of adrenergic
    and cholinergic systems to strain differences in
    avoidance performance.
    Pharmacologist 13, 232 (1971).

MAGNUS, R.: Versuche am überlebenden Dünndarm von Säuge-
    tieren.
    Arch. ges. Physiol. 102, 123-151 (1904).

MANN, WHITNEY: The Mann-Whitney U-Test.
    In: Nonparametric Statistics (S.S. SIEGEL, ed.), pp.
    116-117, Mc. Graw Hill, Book Comp., New York (1956).

MÜLLER-CALGAN, H: Teste des psychotropen Screening, Ratte
    (unveröffentlichte Methodensammlung).

MÜLLER-CALGAN, H., S. SOMMER, and H.-J. JESDINSKY: Bewer-
    tung psychotroperPharmaka an Ratten durch ED_{50}-Schätzung
    mit der Auf- und Ab-Methode nach DIXON-MOOD.
    Naunyn-Schmiedebergs Arch. Pharmak. exp. Path. 260,
    178 (1968).

MÜLLER-CALGAN, H., and S. SOMMER: Das Reserpin-Parkinsonoid
    beim Schimpansen und seine Behandlung mit Fencamfamin.
    Arch. Pharmak. exp. Path. 260, 177 (1968).

NEWBOULD, B.B.: Chemotherapy of arthritis induced in rats
    by mycobacterial adjuvant.
    Brit. J. Pharmacol. Chemother. 21, 127-136 (1963).

STEINBERG, H., R. Rushton, and C. TINSON: Modification of
    the effects of an amphetamine-barbiturate mixture by
    the past experience of rats.
    Nature 192, 533-535 (1961).

STILLE, G.: Zur Pharmakologie katatonogener Stoffe.
    Editio Cantor, Aulendorf i. Württb. 1971,
    Arzneimittelforsch. 21, (1. - 6. Mitteilung), 225;
    997 (1971).

SELYE, H.: The pluricausal cardiopathies.
    Ch. C. Thoma Publ. Springfield/Ill., p. 114 (1961).

SHIPLEY, R.E., and J.H. TILDEN: Pithed rat preparation
    suitable for essaying pressor substances.
    Proc. Soc. exp. Biol. Med. 64, 453-455 (1967).

SCHLESINGER, K., W.O. BOGGAN, and B.J. GRIEK: Pharmaco-
    genetic correlates of pentylentetrazol and electro-
    convulsive seizure thresholds in mice.
    Psychopharmacologia 13, 181-188 (1968).

TAESCHLER, M., H. WEIDMANN, and A. CERLETTI: Zur Pharmako-
    logie von Ponalid, einem neuen zentralen Anticholi-
    nergicum.
    Schweiz. Med. Wschr. 92, 1542-1545 (1962).

WILCOXON, F. in SIEGEL, S.S.:
    The Wilcoxon-Matched-pairs signed-ranks test.
    Mc. Graw Hill Book Comp., New York, pp. 75-83 (1956).

STATISTICAL EEG ANALYSIS IN STRAINS OF RATS WITH GENETICAL-LY DETERMINED DIFFERENT LEARNING PERFORMANCE

G. Dolce, K. Offenloch*, W.G. Sannita**, H. Müller-Calgan, H. Decker

Neurophysiological Laboratory, Medical Research Department
E. Merck, Darmstadt, Germany

ABSTRACT

The EEG of the sensorimotor cortex and the hippocampus in two strains of rats with genetically determined different conditioned behaviour was recorded. The statistical evaluation of the spectral components of the EEG displayed differences in the distribution of the frequency bands and their relative intensities in "good" and "bad" learners before and after training. So it was possible to analyse quantitatively different EEG patterns possibly resulting from genetic and conditioned influences.

----

*Present address: Neurophysiologisches Laboratorium, Zentrum der Physiologie der Johann-Wolfgang-Goethe-Universität, Frankfurt/Main, W. Germany.
**Present address: Centro di Neurofisiologica cerebral (C.N.R.) -Osp., S. Martino, Genova, Italy.

During the last years, the concepts of memory and learning have been enlarged to include phenomena with a broader biological meaning, such as immunological, genetic memory, and so on.

Such a broadening of memory is perhaps prejudicial to the meaning of the definition itself, but, in any case, at present it is no more possible to use, as ECCLES suggested (1966), definitions based on psychological data, and as a consequence, with anthropomorphic connotations. Therefore, for our present research we used the method of conditioned reflexes which, notwithstanding the above reasonings is still considered the most important technique for experimental researches on learning (JUNG, 1965).

The most extensively studied electrophysiologial counterpart of learning has been until now the EEG activity correlates, even if with poor results (as reviewed by HODOS in 1963 and HUGHES in 1970). Interesting results, on the contrary, have been obtained by studying electrical activity in subcortical structures, especially in the hippocampus, a structure which on the basis of experimental and clinical reports is considered as essential in the processing of learning and consolidation of the mnemonic trace (KLÜVER and BUCY, 1939; PENFIELD, 1954, 1958; THOMAS and OTIS, 1958a,b; ADEY et al., 1962; GRASTYAN and KARMOS, 1962; MILNER, 1962; BURES et al., 1962; DRACHMAN and OMMAYA, 1964; UNGHER and PSATTA, 1965; BRAZIER, 1964).

The EEG study of this structure has been focused on the importance of the EEG-frequencies (already dealt with

by JUNG and KORNMÜLLER in 1939) including theta waves and
the interconnection occurring between the appearance of
such waves and variations and the different behaviour conditions, particularly: motivation (OLDS and OLDS, 1963;
GRASTYAN et al., 1965), arousal (GREEN and ARDUINI, 1954),
orienting (GRASTYAN et al., 1966; YOSHJI et al., 1966;
RADULOVICKI and ADEY, 1965; PICKENHAIN and KLINGBERG,
1967) and discrimination (ADEY et al., 1961; ELAZAR and
ADEY, 1967), which occurs during the complex process of
conditioning, in a special way at high levels of learning.

During the last years the spectral analysis of the EEG
with different techniques of computation, allowed a more
detailed examination of the variations in electrobiological
potentials which can be recorded from the hippocampus, so
that it is possible to quantify the differences in amplitude and distribution of the prevailing frequencies and to
define exactly their rhythmicity, analysing relatively
short samples of EEG (ADEY et al., 1960, 1962; ADEY and
WALTER, 1963; YOSHJI et al., 1966; ELAZAR and ADEY, 1967).

The results of these studies, especially those obtained by ADEY and co-workers, focused on the importance
of the hippocampus and especially of the theta wave frequencies, supporting some of the results obtained through
visual analysis of the EEG and displaying relevant increases in amplitude and rhythmicity of the theta range and
the appearance of characteristic frequencies in behavioural
situations occurring during conditioning (aroural, orienting, motivation, and so on).

We can not exclude, however, that these activities
not only depend on learning, but also on other factors,
such as experimental conditions, due to the complexity of
response given by the animal under training.

The experiments of YOSHJI et al. (1966) and LOPES DA
SILVA and KAMP (1969) which revealed similar variations
in amplitude and shape of power spectra in situations different from conditioning seem to support this thesis.

Furthermore, it is to be kept in mind that often the
animals used had not been selected (i.e., cats and monkeys),
neglecting the importance of genetic factors in defining
EEG patterns (RAMEY 1938, 1939; DUSTMAN and BECK, 1965;
DUMERMUTH, 1969; JUEL-NIELSEN and HANVALD, 1958; VOGEL,
1958). In order to avoid these two possible methodological
sources of error, our present research was made using rats
genetically selected in accordance with their performances
in learning.

Furthermore, the EEG was recorded during basal conditions in the same animals - naive and after conditioning - not recording during the training period. The spectral analysis was not confined to the hippocampus only,
but also to the cortex and was not restricted to slow frequencies only (ELAZAR and ADEY, 1967) but covered the
whole spectrum.

MATERIAL AND METHODS

Fourteen adult rats, genetically selected according to their performance in avoidance learning were used (cf., MÜLLER-CALGAN et al., this Symposium). One group of 6 animals were "good" learners, the other group of 8 were "bad" learners. Monopolar platinum electrodes were implanted stereotactically in the right and left sensorimotor cortex. Bipolar steel electrodes were implanted in the dorsal hippocampus (A 3.0, L 4.0, H 1.5) and connected with a miniconnector. Three, five and ten days after training the basal EEG was again recorded.

The animals were trained in an active two-way avoidance test, as described by MÜLLER-CALGAN et al., in this Symposium. They had to learn to jump across a hurdle in order to avoid an electric stimulus (US), which followed after 5 sec a buzzer which was the conditioned stimulus (CS) (Hürdensprung-Training). The training consisted of 10 trials with 10 single coupled CS/US each, at an interval of 20 sec twice a day during five consecutive days.

The EEG from the sensorimotor cortex and from the dorsal hippocampus was recorded simultaneously on paper and on magnetic tape (Philips ANALOG 7). The EEG was analysed off-line after passing through a band-pass filter, and processed on a PDP-12 computer. The purposes of the filtering were: (i) to remove faster frequencies than those which our program could handle and (ii) to obtain a higher resolution of the less pronounced fast frequencies by attenuation of the dominant low frequencies (less

than 3 Hz). An analysis epoch of 16 sec was used with a
resolution of 0.5 Hz. The periodograms obtained by the
fast Fourier transformation were weighed with 1/4, 1/2,
1/4 and smoothed. From 20 spectra one average spectrum
was computed. Afterwards the filter-effects were calculated and compensated and again two further smoothings were
performed. All spectra were normalized. The averages of
the 20 spectra of 16 sec duration (that is 320 sec) were
punched on paper-tape. The subsequent statistical analysis
was performed on an IBM 370/145 computer. The spectra were
divided in segments 2 Hz in length and for the values so
obtained, the intercorrelation matrix was computed. From
that intercorrelation matrix, certain frequency bands became visible since adjacent spectral segments, belonging
to the same frequency bands, are significantly correlated
with each other. A standard test for significance of the
correlation was used (the significance for some of the frequency bands was better than alpha < 0.01).

The complete correlation matrix was submitted to a
factor analysis. In order to display statistical differences of the intensity spectra under changed conditions
(before and after training) the most important variable
spectral segments were thus obtained. Finally, these
variables were used as data for discrimination analysis.

RESULTS

Avoidance Training

The results of two-way avoidance training can be summarized as follows:
The scores for the performance of good learners (median value 43.5, n = 6) (GL) are higher than those of bad learners (median value 14.5, n = 8) (BL).

This difference is significant (2 alpha < 0.01 MANN-WHITNEY rank test for unpaired samples; see Fig. 1).

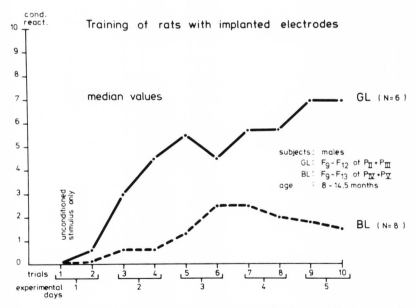

Fig. 1: Median values of the scores of the performance in the two-way avoidance test
—————— good learners (GL) ----- bad learners (BL)

EEG-Frequency Bands

<u>Before training</u>. In free moving rats, awake and in
basal conditions, the following frequency bands were
found by using intercorrelation analysis:
Left cortex   Hz 0.5-3 3-7 9-15 19-
Dorsal hippocampus Hz 0.5-3 3-5 7-9   11-15 19-23

There were no differences in these frequency bands
between the two strains of animals (Fig. 2, 3, 4, 5).

In the right cortex, however, differences between the
strains were found (Fig. 2, 3):
Right cortex
a) Good learners Hz 0.5-3 3-5 7-15 19
b) Bad learners Hz 0.5-3 3-5 13-15 19

<u>After training</u>. 3 days after training the distribution
of the frequency bands became different from that existing
before training in both good and bad learners, and there
was also a difference between the two strains of rats
(Fig. 2, 3, 4, 5).
Left cortex "GL" Hz 0.5-3 3-5   9-11   17-
Left cortex "BL" Hz 0.5-3 3-5   13-17   21-
Right cortex "GL" Hz 0.5-3 3-5   7-11   13-15   17-
Right cortex "BL" Hz 0.5-3 3-5   11-19   21-
Hippocampus "GL" Hz 0.5-3 3-5   11-15   17-
Hippocampus "BL" Hz 0.5-3 3-11   3   17-

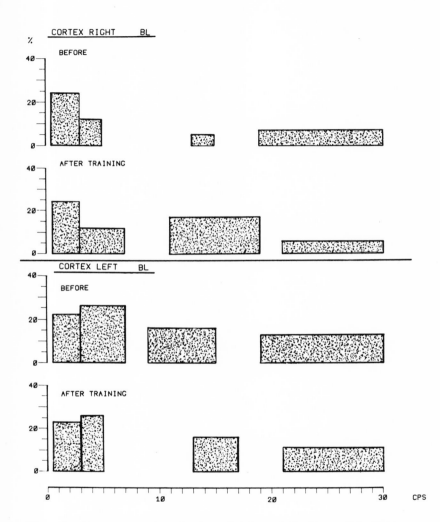

Fig. 2: Frequency bands in right and left sensorimotor
cortex of bad learners before and after training

Abscissa: frequencey (Hz). Ordinate: relative
intensity of the frequency bands expressed in percentage of the total intensity of the spectrum

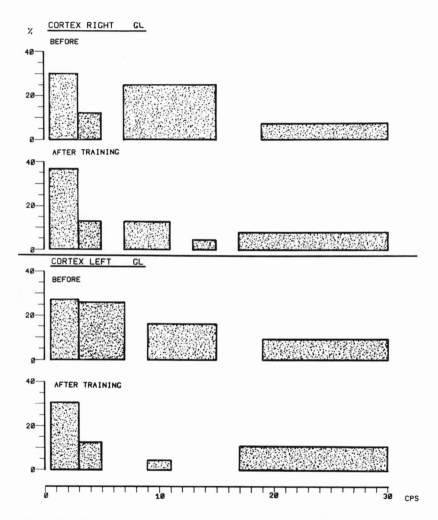

Fig. 3: Frequency bands in right and left sensorimotor
        cortex of good learners before and after training

        Abscissa and ordinate as in Fig. 2

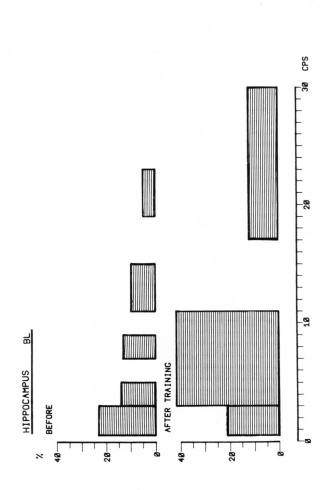

Fig. 4: Frequency bands in hippocampus of bad learners before and after training
Abscissa and ordinate as in Fig. 2

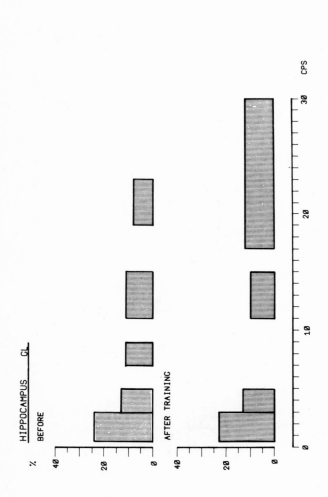

Fig. 5: Frequency bands in hippocampus of good learners before and after training
Abscissa and ordinate as in Fig. 2

Intensities. The surface areas of the frequency bands expressed in percentages after training, are shown in Tab.I.

Six and twelve days after training the newly apparent frequency bands tended to disappear and the spectrum went back to its previous state.

Statistical Computations

The frequency bands which were found before and after training are derived by intercorrelation analysis. With factor analysis the important segments of the spectra were evaluated. The differences were statistically studied by discrimination analysis.

The spectra of bad learners were significantly different in the hippocampus before and after training (alpha < 0.05).

DISCUSSION

This research was made in order to display:
i) possible modifications of EEG patterns caused by training,
ii) differences in EEG patterns related to the genetic substrate and
iii) eventual possibilities of mathematical analysis of the EEG for studying cognitive activities such as learning and/or others.

Tab. I:

Cortex right — good learners

	0.5-3		3-5		7-11		7-15		11-19		13-15		17-30		19-30		21-30 Hz	
	M	V	M	V	M	V	M	V	M	V	M	V	M	V	M	V	M	V
before	30.05	10.01	12.7	1.0	16.5	1.5	25.9	1.5	17.2	1.4	4.4	0.20	10.8	1.6	7.36	2.45	4.9	0.7
after	37.56	15.21	13.2	2.8	13.0	1.8	22.3	4.3	14.2	3.8	4.9	1.6	8.0	1.5	5.9	3.0	3.7	0.7
Diff.	7.51	21.98	0.5	2.5	- 2.0	4.5	- 5.2	5.8	-13.0	4.6	0.5	1.7	- 2.8	2.4	-1.46	3.84	-1.2	0.8

Cortex right — bad learners

	0.5-3		3-5		7-11		7-15		11-19		13-15		17-30		19-30		21-30 Hz	
	M	V	M	V	M	V	M	V	M	V	M	V	M	V	M	V	M	V
before	23.46	8.58	12.3	1.2	21.2	2.2	30.3	2.9	16.4	1.4	4.5	0.6	10.4	1.2	7.36	2.79	5.2	0.7
after	23.93	10.21	11.5	1.8	21.4	2.9	29.7	3.0	17.1	1.8	4.0	0.3	12.7	1.8	8.53	2.67	5.9	0.6
Diff.	0.47	11.46	- 0.8	1.0	0.2	2.1	0.6	3.6	0.7	1.9	-0.4	0.6	2.3	2.2	1.17	3.95	0.7	1.0

Cortex left — good learners

	0.5-3		3-5		3-7		9-11		9-15		13-17		17-30		19-30		21-30 Hz	
	M	V	M	V	M	V	M	V	M	V	M	V	M	V	M	V	M	V
before	27.13	15.1	12.1	2.0	26.61	4.51	6.1	0.5	16.15	3.98	10.6	2.0	12.6	2.5	8.95	4.48	6.2	1.3
after	30.65	11.61	12.5	2.1	27.56	2.77	4.4	0.8	14.0	5.6	8.7	1.5	11.0	1.3	8.43	2.55	5.9	0.7
Diff.	3.51	20.27	0.2	1.6	0.88	4.0	-1.6	0.6	- 1.12	5.68	-1.9	1.9	-1.5	2.1	-0.51	3.37	-0.3	0.9

Cortex left — bad learners

	0.5-3		3-5		3-7		9-11		9-15		13-17		17-30		19-30		21-30 Hz	
	M	V	M	V	M	V	M	V	M	V	M	V	M	V	M	V	M	V
before	21.43	5.95	11.4	1.3	25.52	4.88	5.8	0.4	15.7	2.64	10.4	1.5	17.1	1.6	12.71	3.17	9.4	0.9
after	16.22	5.11	9.9	1.5	23.67	5.27	7.3	1.0	17.47	3.55	10.7	1.6	20.0	2.2	14.87	4.06	10.7	0.9
Diff.	- 5.22	4.09	-1.5	0.6	-1.85	2.06	1.5	1.0	1.77	3.74	1.5	1.4	2.5	2.0	2.16	4.89	1.3	1.3

Hippocampus right — good learners

	0.5-3		3-5		3-11		7-9		11-15		17-30		19-23 Hz	
	M	V	M	V	M	V	M	V	M	V	M	V	M	V
before	23.51	7.24	13.11	3.04	47.2	2.0	11.31	2.02	11.42	1.65	11.9	1.4	7.8	3.4
after	23.01	11.75	12.83	2.98	46.9	2.4	12.11	4.3	10.4	3.48	13.2	1.7	5.0	0.6
Diff.	-6.03	8.66	-0.28	1.25	-0.3	1.6	0.9	3.34	-0.68	3.40	1.3	1.7	-2.8	3.3

Hippocampus right — bad learners

	0.5-3		3-5		3-11		7-9		11-15		17-30		19-23 Hz	
	M	V	M	V	M	V	M	V	M	V	M	V	M	V
before	23.35	6.27	13.7	2.64	49.5	1.7	13.11	2.95	9.62	1.92	11.8	1.0	4.6	0.4
after	21.35	7.31	13.57	2.47	46.6	1.7	12.08	2.61	10.76	1.82	14.2	1.1	5.5	0.4
Diff.	1.11	9.77	0.35	3.3	-1.7	1.5	-2.05	2.51	1.35	2.69	2.3	1.5	0.9	0.6

The effect of the training (performed as described) was a modification of the EEG, i.e., a variation in the distribution of frequency bands, statistically defined. Visual inspection of the EEG did not allow us to see any changes after training (Fig. 6). The study of the variation in absolute intensity of the whole spectrum or of restricted segments of the spectrum itself (as performed by GREY-WALTER, 1963; YOSHJI et al., 1966; ELAZAR and ADEY, 1967) did not show significant changes, because of a too large individual variability, as seen in the data on rats handled with different techniques of analysis

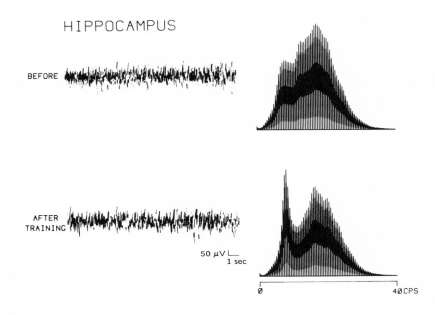

Fig. 6: EEG records and corresponding spectra (16 sec samples) of the hippocampus before and after training
See only the external contour of the spectra.

(ETEVENON and BOISSIER, 1971). Measuring the percentage
values (normalized spectra) in the different frequency
bands, we obtained no significant changes either (Fig.
7, 8). Only by using the statistical analysis of segments of averaged spectra was it possible to differentiate significantly the two populations of animals and
the training effect in the EEG of these two groups.

In order to avoid reducing the spectrum and because
it would not be convenient to use a large number of factors, we have computed only 15 factors, each one consisting
of a spectral segment 2 Hz wide (since the original resolution was of 0.5 Hz we therefore made each segment consisting of 4 points).

Like the results of the discrimination analysis, the
significance of the variation in the frequency bands after
the training was especially marked in the hippocampus
(alpha < 0.05). In the sensorimotor cortex differences
were present in the frequency bands of the right and left
side and in their changes after training. Differences in
frequency bands were also present in the right cortex,
between the two groups of animals (good and bad learners)
before training.

These changes in frequency bands in hippocampus and
cortex persisted for 3 - 4 days after training and later
on disappeared gradually. We think that it is interesting
to discuss the nature of this phenomenon.

Obviously, we cannot discuss to which kind of per-

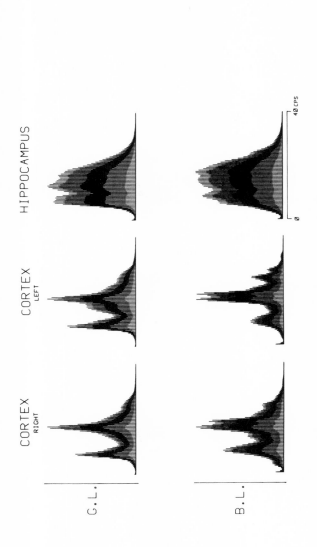

Fig. 7: Power spectra (16 sec samples) of the EEG of the right and left cortex and the hippocampus in good and bad learners before training

See only the external contour of the spectra.

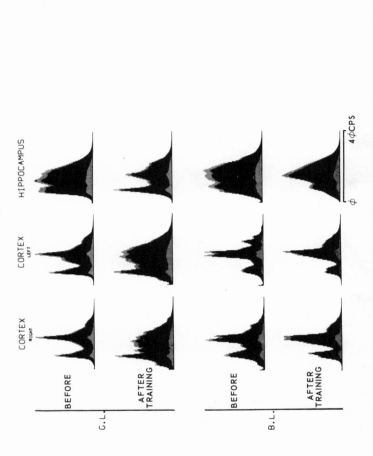

Fig. 8: Power spectra (16 sec samples) of the EEG of the right and left cortex and the hippocampus in good and bad learners before and after training

See only the external contour of the spectra.

formance (orienting, discrimination, arousal, motivation
and so on) these changes are related to, or exclude that
they were caused by something different from conditioning,
stress for example (ANOKHIN, 1964; YOSHJI et al., 1966).

We must remember that the theta-activity in the hippocampus of rats (according to PICKENHAIN and KLINGBERG
(1967), 6-12 Hz) appears in different conditions, elicited
by stimuli, only after a certain number of stimulations
(PICKENHAIN and KLINGBERG, 1967).

Furthermore, the EEG variations during cognitive activities were studied by recording during tests (GRÜNEWALD
et al., 1968, in man; ELAZAR and ADEY, 1967, in cats; and
others).

On the contrary, we recorded the spontaneous EEG without any test-situation and without external stimuli in
order to avoid any interference with the normal spontaneous
EEG-activity.

After training the frequency band from 7 to 9 Hz disappeared in the hippocampus in the good learning rats, and
in the bad learning rats there was a grouping of different
frequency bands; in other words, the hippocampal EEG was
less differentiated in its dominant frequencies after training. The disappearance in good learners of one dominant
frequency in theta range can be considered as an effect of
over-training (ELAZAR and ADEY, 1967; VANDERWOLF, 1969).

These modifications of the EEG do not consist of an

increase or decrease in percentage of some particular frequency bands which characterize the EEG of the rats before training but in a different distribution of these
bands in the spectrum. These qualitative changes are the
expression of a new organization of the EEG, caused by a
different functional and/or structural background of nervous tissue. The EEG recorded by using macroelectrodes is
the result of the total activity of the poly-compartmental
arrangement of the nervous system, involving neurons, glial
cells and extracellular compartments (ADEY, 1967).

Furthermore, the modifications we saw are related to
the activity of large cellular populations.

There is not yet any experimental evidence for a defined biological meaning of the variations we saw in the
EEG of rats after training, but - because of their relative
short duration (3-4 days) - these modifications seem to
depend on a different functional arrangement of the EEG.
The close relationship between EEG and its generators -
slow membrane potential changes (ANDERSEN and ECCLES,
1962; CREUTZFELD et al., 1964; KLEE et al., 1965; ELUL,
1964, 1968) - allows us to suppose a different organization in the connections of these generators in the intercorrelation of the neuronal populations involved in these
activities and their relationship with other structures.
This could mean a different synaptic pattern and different
degrees in their activation.

We must remember that the variations we call "functional" could be related to biochemical or anatomical var-

iations and that the EEG patterns are strongly genetically
defined (DUMERMUTH, 1969). Our research allows us to distinguish two different patterns in establishment of generators of the EEG, both conditioned by genetical background.

In fact, we must distinguish the patterns of basal
EEG and those appearing after training. The second ones
are temporary and tend to disappear some days after the
end of the training. Furthermore, while disappearing the
frequency bands return to the distribution they had before training. The functional brain organizations common
to both strains of rats are more stable than those which
divide the two groups of animals according to their different behavioural performances. This is probably due to
the close genetic definition of the patterns of the EEG
(RAMEY, 1938, 1939; JUEL-NIELSEN and HANVALD, 1958; VOGEL,
1958; DUMERMUTH, 1969), so strong that in spite of the
fact that if the two groups of animals are genetically
selected as good and bad learners during 12 generations
and each group tends to homozygosis, the structures and
the compartments involved in the genesis of EEG show some
differences only under particular conditions, requiring
particular activities and these differences are reversible.
It is possible, of course, that different training could
cause longer lasting modifications in EEG or that animals
selected in a different way could have other EEG patterns.
The reversibility of these EEG modifications focuses the
importance of this method of analysis for testing extremely differentiated physiological conditions.

In fact these frequency bands were not defined in an

arbitrary way or according to their changes during different physiological and/or pathological conditions (as
made in men, for example), but because of their mathematical evidence, their intercorrelations and their presence in
all the animals.

By measuring segments of the spectra defined in an
arbitrary way - as it is usually done - it is not possible
to discriminate dynamic changes in the organization of the
spectra and the shifts of their various dominants under
different conditions.

On the contrary, the algorhythms we used have clear
advantages for a statistical computation of the spectral
values and allow us to study subtle changes in the organization of the EEG.

                              LITERATURE

ADEY, W.R.: Intrinsic organization of cerebral tissue in
       alerting, orienting and discriminative responses.
       In: The Neurosciences (G.C. QUARTON, T. MELNECHUK
       and F.O. SCHMITT, eds.), pp. 615-633, Rockefeller
       University Press, New York (1967).

ADEY, W.R., C.W. DUNLOP, and C.E. HENDRIX: Hippocampal
       slow waves; distribution and phase relations in the
       course of approach learning.
       Arch. Neurol. (Chic.) 3, 74-90 (1960).

ADEY, W.R., and D.O. WALTER: Application of phase detec-
    tion and averaging techniques in computer analysis
    of EEG records in the cat.
    Exp. Neurol. 7, 186-209 (1963).

ADEY, W.R., D.O. WALTER, and C.E. HENDRIX: Computer tech-
    niques in correlation and spectral analysis of cere-
    bral slow waves during discriminative behavior.
    Exp. Neurol. 3, 501-524 (1961).

ADEY, W.R., D.O. WALTER, and D.F. LINDSLEY: Effects of
    subthalamic lesions on learned behavior and correla-
    ted hippocampal and subcortical slow wave activity.
    Arch. Neurol. (Chic.) 6, 194-207 (1962).

ANDERSEN, P., and J.C. ECCLES: Inhibitory phasing of
    neuronal discharge.
    Nature (Lond.) 196, 645-647 (1962).

ANOKHIN, P.K.: The electroencephalogram as a resultant of
    ascending influences on the cells of the cortex.
    Electroenceph. clin. Neurophysiol. 16, 27-43 (1964).

BRAZIER, M.A.B.: Stimulation of the hippocampus in man
    using implanted electrodes.
    In: Brain Function, II, RNA and Brain Function Memory
    and Learning, University of California Press, Berkely,
    Los Angeles (1964).

BURES, J., O. BURESOVA, T. WEISS, E. FIFKOVA, and Z.
    BOHDANECKY: Experimental study of the role of hippo-
    campus in conditioning and memory function.
    In: Physiologie de L'Hippocampe, C.N.R.S., pp. 241-
    256, Paris (1962).

CREUTZFELDT, O.D., J.M. FUSTER, H.D. LUX, and A. NACIMIENTO:
    Experimenteller Nachweis von Beziehungen zwischen EEG-
    Wellen und der Aktivität corticaler Nervenzellen.
    Naturwissenschaften 51, 166-167 (1964).

DRACHMAN, D., and A. OMMAYA: Memory and the hippocampal
    cortex.
    Arch. Neurol. (Chic.) 10, 411-425 (1964).

DUMERMUTH, G.: Die Anwendung von Varianzspectra für einen
    quantitativen Vergleich von EEG bei Zwillingen.
    Helv. paediat. acta 24, 45-54 (1969).

DUSTMAN, R.E., and E.C. BECK: The visually evoked potential
    in twins.
    EEG clin. neurophysiol. 19, 570-575 (1965).

ECCLES, J.C.: Conscious experience and memory.
    In: Brain and Conscious Experience, J.C. ECCLES, ed.),
    pp. 314-344, Springer-Verlag (1966).

ELAZAR, Z., and W.R. ADEY: Spectral analysis of low fre-
    quency components in the electrical activity of the
    hippocampus during learning.
    EEG clin. Neurophysiol. 23, 225-240 (1967).

ELUL, R.: Specific site of generation of brain waves.
    The Physiologist 7, 125 (1964).

ELUL, R.: Brain waves: intracellular recording and statis-
    tical analysis help clarify their physiological sig-
    nificance.
    Data Acquisition and Processing Biol. Med. 5, 93-115
    (1968).

ETEVENON, P., and J.R. BOISSIER: Statistical amplitude
    analysis of the integrated electrocorticogram of un-
    restrained rats before and after prochlorpemazine.
    Neuropharmacology 10, 161-173 (1971).

GRASTYAN, E., and G. KARMOS: The influence of hippocampal
    lesions on simple and delayed instrumental conditioned
    reflexes.
    In: Physiologie de L'Hippocampe, C.N.R.S., pp. 225-
    239, Paris (1962).

GRASTYAN, E., J. CZOPF, L. ANGYAN, and I. SZABO: The sig-
    nificance of subcortical motivational mechanismus in
    the organization of conditional connections.
    Acta physiol. Acad. Sci. hung., 26, 9-46 (1965).

GRASTYAN, E., G. KARMOS, L. VERECZKEY, and L. KELLENYI:
    The hippocampal electrical correlates of the homeo-
    static regulation of motivation.
    Electroenceph. clin. Neurophysiol. 21, 34-53 (1966).

GREEN, J.D., and A. ARDUINI: Hippocampal electrical activ-
    ity in arousal.
    J. Neurophysiol. 17, 533-557 (1954).

GREY-WALTER, W.: Frequency analysis.
    In: Electroencephalography (D. HILL and G. PARR, eds.),
    pp. 87-90, MacDonald, London (1963).

GRUNEWALD, G., O. SIMONOVA, and O.D. CREUTZFELD: Differen-
    tielle EEG-Veränderungen bei visuomotorischen und
    kognitiven Tätigkeiten.
    Arch. Psych. Nervenk., 212, 46-69 (1968).

HODOS, W.: Facts and artefacts in the EEG and learning.
    EEG clin. Neurophysiol. 15, 540 (1963).

HUGHES, J.R.: Electroencephalography and learning.
    In: Progress in Learning Disabilities (R. MYKLEBUST,
    ed.), p. 113, Grune and Stratton, New York (1969).

JUEL-NIELSEN, N., and B. HANVALD: The electroencephalogram
    in uniovular twins brought un apart.
    Acta ganet. 8, 57-64 (1958).

JUNG, R.: Neurophysiologie und Psychiatrie.
    In: Psychiatrie der Gegenwart, Bd. 1/1A, pp. 325-928,
    Springer (1967).

JUNG, R., and A.E. KORNMÜLLER: Eine Methodik der Ablei-
    tung lokalisierter Potentialschwankungen aus sub-
    cortikalen Hirngebieten.
    Arch. Psych. Nervenk. 109, 1-30 (1939).

KLEE, M.R., K. OFFENLOCH, and J. TIGGES: Cross-correlation
    analysis of electroencephalographic potentials and
    slow membrane transients.
    Science 147, 519-521 (1965).

KLÜVER, H., and P.C. BUCY: Preliminary analysis of the
    functions of the temporal lobes in monkeys.
    Arch. Neurol. Psych. (Chic.) 42, 979-1000 (1939).

LOPES DA SILVA, F.H., and A. KAMP: Hippocampal theta fre-
    quency shifts and operant behaviour.
    Electroenceph. clin. Neurophysiol. 26, 133-143 (1969).

MILNER, B.: Les troubles de la memoire accompagnant des
    lesions hippocampiques bilaterales.
    In: Physiologie de L'Hippocampe, C.N.R.S., pp. 257-

272, Paris (1962).

OLDS, M.W., and J. OLDS: Approach-avoidance analysis of
    rat diencephalon.
    J. comp. Neurol. 120, 259-295 (1963).

PENFIELD, W.: Studies of the cerebral cortex of man. A
    review and an interpretation.
    In: Brain Mechanismus and Consciousness (J.E.
    DELAFRESNAY, ed.), pp. 284-309, Blackwell, Oxford
    (1954).

PENFIELD, W.: Functional localization in temporal and
    deep sylvian areas.
    Res. Publ. Ass. nerv. ment. Dis. 36, 210-226 (1958).

PICKENHAIN, L., and F. KLINGBERG: Hippocampal slow wave
    activity as a correlate of basic behavioral mechanisms
    in the rat.
    Prog. Brain Res. 27, 218-227 (1967).

RADULOVACKI, M., and W.R. ADEY: The hippocampus and the
    orienting reflex.
    Exp. Neurol. 12, 68-83 (1965).

RAMEY, E.T.: Reversed lateral dominance in identical twins.
    J. exp. psychol. 23, 304-312 (1938).

RAMEY, E.T.: Brain potentials and lateral dominance in
    identical twins.
    J. exp. psychol. 24, 21-39 (1939).

THOMAS, G.J., and L.S. OTIS: Effects of rhinencephalic le-
    sions on maze learning in rats.
    J. comp. physiol. Psychol. 51, 161-166 (1958a).

THOMAS, G.J., and L.S. OTIS: Effects of rhinencephalic lesions on conditioning of avoidance respones in the rat.
J. comp. physiol. Psychol. 51, 130-134 (1958b).

UNGHER, J., and D. PSATTA: Deficits de la mobilite des processus nerveux superieurs chez le chat avec des lesions de l'hippocampe.
Rev. roum. Neurol. 2, No. 3, 137-238 (1965).

VANDERWOLF, C.H.: Hippocampal electrical activity and voluntary movement in the rat.
Electroenceph. clin. Neurophysiol. 26, 407-418 (1969).

VOGEL, F.: Über die Erblichkeit des normalen Elektroencephalogramms.
Thieme, Stuttgart, 92 p. (1958).

YOSHJI, N., M. SHIMOKOCHI, K. MIYAMOTO, and M. ITO: Studies on the neural basis of bahvior by continuos frequency analysis of EEG.
In: Progress in Brain Research, Vol. 21a (T. TOKIZANE and J. P. SCHADE, eds.), pp. 217-250, Elsevier, Amsterdam (1966).

SOME VIEWS ON THE NEUROPHYSIOLOGICAL AND NEUROPHARMACOLO-
GICAL MECHANISMS OF STORAGE AND RETRIEVAL OF INFORMATION

Leonide Goldstein and Judith M. Nelsen

Bureau of Research in Neurology and Psychiatry
Neuropsychiatric Institute

Princeton, N.J. U.S.A. 08540

One of the chief characteristics of nervous tissue is that of "memory": that is, speaking generally, a susceptibility to permanent alteration by a single process. This offers a striking contrast to the behavior of a material that allows a wave-movement to pass through it and then returns to its former condition. Any psychological theory deserving consideration must provide an explanation of memory. Now any such explanation comes up against the difficulty that ... after an excitation neurones are permanently different from what they were before, while, on the other hand, it cannot be denied that, in general, fresh excitations meet with the same conditions of reception as did the earlier ones. Thus the neurones would appear to be both influenced and also unaltered ... We cannot off-hand imagine an apparatus capable of such complicated functioning.

Sigmund Freud

ABSTRACT

Electrophysiological and behavioral studies were carried out in the rabbit, the rat, and man in an attempt to
clarify the qualitative nature, the underlying mechanisms,
and the functional consequences of drug-induced arousal.
Our data support the existence of two subcortical arousal
systems, probably mutually inhibitory, which produce the
same end product, namely cortical activation, but also
which produce functionally different forms of arousal. One
of the systems appears to be related to the reticular formation and the other, to the limbic system (particularly,
the hippocampus). Direct electroencephalographic (EEG)
measures in the rabbit showed that amphetamine can affect
concurrently the inter-relationships between the reticular
formation and cortex, and the hippocampus and cortex. However, our EEG data from human subjects indicate that the
relative effect of acute doses on the two systems is controlled, at least partially, by the pre-existing states of
these systems. The highly variable performance of rats
tested on a continuous attention task after acute doses of
amphetamine support indirectly the premise that the baseline
state of the systems dictates their response to pharmacological manipulation.

EEG measures in rabbits supplied evidence that nicotine also affects the inter-relationships both between the
reticular formation and cortex and the hippocampus and the
cortex, but to different relative degrees depending on whether the administration is acute or chronic. Acute doses act

primarily on the reticular-cortex relationships, while
chronic treatment causes a gradual shift from predominant
reticular formation control of arousal to hippocampal control. Behavioral testing of rats operating on the continuous attention task, revealed that initial doses of nicotine
cause impairment of performance, but that chronic treatment causes rats to perform more efficiently on the task
than they do under saline control conditions. Acquisition
studies with rats on this same task indicate that unlike
its effect on performance per se, nicotine disrupts the
learning of the task.

The arousal system related to the reticular formation
does appear to control the general reactivity of the organism and provide organization for response, while the
second system (related to the limbic structures, particularly the hippocampus) provides control of responses
through incentive-related stimuli . The functional consequences of the manipulation of the two arousal systems
is discussed in relation to information storage (learning)
and retrieval (performance).

INTRODUCTION

It is usually agreed that both storage and retrieval
of information are most succesfully obtained when the behavioral state corresponds to "arousal". This appears to
be the case whether arousal is spontaneous, i.e., part of
the normal circadian cycle of activity/inactivity, or wheth-

er it is induced artificially by so-called stimulant drugs,
especially at low dose levels (MC GAUGH and PETRINOVICH,
1965).

Arousal has been considered for many years to be a
discrete neurophysiological entity resulting from an "activation" of the cortex by the ascending mesencephalic reticular formation (RF). This concept was derived from the
discovery that cortical arousal (as defined by electroencephalographic patterns of activity and by behavioral
excitation) could be induced in sedated animals by electrical stimulation of the RF (MORUZZI and MAGOUN, 1949),
and by the finding that in animals in which a mid-collicular transection was performed ("cerveau isolé" preparation),
a constant state of sleep ensued (as judged on short duration EEG observations, BREMER, 1935). No other structure
in the brain had been found to have similar properties in
so far as arousal and sleep are concerned.

However, in recent years, a number of experimental
findings have cast serious doubts on the concept of the
uniqueness of the RF as the site for the induction and
maintenance of arousal. In the first place, it was shown
that in animals with mid-collicular sections, the EEG state
of sleep was not permanent. After 10 to 15 days, a clear-
cut spontaneous arousal was found to occur, followed by
alternating states of sleep and wakefulness (BATSEL, 1960;
VILLABLANCA, 1962, 1965) Such arousal periods could not
have been the result of a neuronal action from the RF on
the cortex since all the tracts had been severed. Also,

the so-called "paradoxical sleep" or REM sleep, during
which cortical recordings appear to correspond to a state
of arousal, was found by JOUVET (1961) to be inducible by
electrical stimulation of the caudal pontine nucleus and
not of the mesencephalic RF. Finally, it was pointed out
in a number of studies that electrical stimulation of parts
of the limbic system, especially the amygdaloid nucleus
and the hippocampus, produced a state of cortical arousal
even in "cerveau isolé" preparations, which suggested that
such an effect did not need to be mediated by the RF (CAR-
LI et al., 1965).

In behavioral studies, it was observed that retrieval
of stored information ("Continuous Attention Task" in rats)
was significantly impaired upon electrical stimulation of
the RF, in spite of a typical behaviorally aroused state
(KORNETSKY and ELIASSON, 1969). In studies to be described
later, concerned with retrieval of similar information, we
found also a deficit in performance following acute or subacute administrations of nicotine and D-amphetamine, two
drugs known to activate cortical electrical activity via
effects on the RF. An important aspect of such impairment
effects was that a large scale individual variability was
in evidence, with some subjects being improved, and some
others being non-affected or worsened in their behavior.
On a group basis, however, rather than to improve performance, drug-induced stimulation of the RF impaired it.

Similar conclusions were reached after a review of
the results of a number of varied experimental procedures,

leading ROUTTENBERG (1968) to set forth a schema which suggests dual, mutually inhibitory subcortical arousal systems
which control not only the presence of a state of activation of the cortex, but also functional or qualitative
features of that activation. ROUTTENBERG has proposed that
the arousal related to the RF maintains the reactivity of
the organism and provides organization for response, whereas, the second arousal system is related to the limbic
structures and provides control of responses through incentive-related stimuli.

If such a dual arousal system exists, it would lead
one to raise a number of questions regarding the relationships between the above mentioned characteristics of arousal
and the capacity for storage and retrieval of information.
In particular, it would raise serious doubts concerning the
interpretation of the results of many studies which have
focused on the storage (learning or acquisition) and retrieval (performance) of information. The determination of
the type of arousal prevailing (be it spontaneous or experimentally induced) during studies would assume critical
importance both for controlling the conditions of an experiment and for a valid interpretation of the results.

It is the purpose of this paper to review recently obtained neurophysiological and neuropharmacological experimental data supporting the proposed existence of two different arousal mechanisms and to show to what extent this
dualism may affect storage and retrieval of information.

MATERIAL AND METHODS

Measurement of the Magnitude of Association between Electrical Activity in Cerebral Structures in Rabbits

The animals used were male New Zealand rabbits weighing 2.5 to 3.0 kgs. They were prepared surgically under pentobarbital anesthesia with implanted bipolar electrodes according to the stereotaxic map of SAWYER et al. (1954). The subcortical electrodes were located in the dorsal hippocampus and the mesencephalic reticular formation. In some animals electrodes were also placed in the amygdaloid nucleus. Bipolar cortical recordings were obtained from gold-plated screws located in the cranium, above the sensory-motor cortex. Two to three weeks after surgery the animals were trained to sit quietly, without restraint, in a sound-proof, ventilated isolation enclosure. Observation of their behavior was made possible by direct viewing through a one-way mirror.

The electrical signals from the various electrode locations were recorded on an 8-channel Grass polygraph. At the same time, these signals were processed through solid-state, analog to digital amplitude integrators developed from the original design of DROHOCKI (1948). The integrators perform continuous, on-line, measurements of the amplitudes of full wave rectified signals, yielding numbers which are directly proportional to the cumulated amplitudes. These numbers can be totalized for pre-set periods of time, constituting basic epochs of measurement.

In the studies to be described the epochs were 2 seconds.
Thus, a 10 minute recording session provided 300 numerical
values for analytical and statistical computations.

In order to measure the degree of relative association
between cerebral structures, the method devised by BYFORD
(1965) was used. It consists of grouping, separately for
each source of electrical signals, the numbers representing
the cumulated amplitudes for periods of typical EEG and behavioral states. For each group a mean and a variance are
computed. Next, the variances are cumulated and least square
regression lines computed. The slopes of such lines constitute the "variance index". As shown by BYFORD, the ratio of
the indices for two different sources of electrical signals
provides a statistically valid measurement of the degree of
inter-dependency between these signals.

Measurement of the Acquisition of the Performance
in a Continuous Attention Task by Rats

Male Holtzman rats were maintained at approximately
85 % of their normal body weights. They were trained to
press a lever for a food pellet reinforcement following
the presentation of a conditional stimulus (C.S.) which
was a white cue light in a standard operant conditioning
box. The duration of the cue light was only 0.2 seconds,
but the light was followed by an available response time
of 5.0 seconds during which the first lever-press was reinforced. Failure to respond was scored as an <u>omission error</u>
<u>(o.e.)</u> and had no consequence for the animal other than the

loss of a reinforcement pellet. The inter-trial interval
(I.T.I.) was variable with a mean of 10.0 seconds. A lever
press during I.T.I. was scored as a <u>commission error (c.
e.)</u> and was punished by the imposition of a 30 second
"dark period" or time-out. Repeated indiscriminate responses (c.e.'s) could prevent the presentation of the C.S.
throughout the whole session, which was limited to 100
reinforcements, or in case of disrupted behavior, one
hour in the operant box.

This type of task in which the animal is asked not
only to make appropriate responses to a very short stimulus but to inhibit inappropriate responses is a difficult
one for rats to learn. Several months were required to
train the animals to perform efficiently. The criteria
set for efficient performance were for:

1) o.e.'s, $\dfrac{\text{o.e.'s}}{\text{reinforcements}}$ x 100 = 30 %

2) c.e.'s, $\dfrac{\text{c.e.'s}}{\text{reinforcements}}$ x 100 = 30 %

Measurement of EEG Amplitude and Variability in Human
Subjects

In human subjects (normal volunteers) the EEG was recorded monopolarly from the left occipital area with ear
lobes for common references. The sessions of 10 to 15
min duration, took place in a sound attenuated, dimly

lighted room, with the subjects supine, eyes closed.

The analysis of the electrical activity was performed with the same type of EEG integrator described in connection with experimentation on rabbits, except that instead of 2 seconds, the basic measurement epoch was 20 seconds. Thus, a 10 minute recording session yielded 30 successive measurements. From these, a mean integrated amplitude (MIA) and its variance were computed. Variability was expressed in the form of the coefficient of variability (CV) which is obtained by dividing the standard deviation by the MIA times 100.

Between subjects averaging was performed by pooling all the means and computing an overall CV from the covariances. Statistical significance for the changes was tested with the 2-tailed t-test for the means and the F-ratios for the variances.

RESULTS

Effects of D-Amphetamine on Electrophysiological Parameters and on Performance

Effects of amphetamine on cortical/subcoritcal relationships in rabbits. The experimentation in rabbits was initiated as a result of our interest in the phenomenon known as "awake drunkeness". When rabbits receive orally relatively large doses of ethanol (1 to 2 Gm/kg), they

display clear signs of behavioral drunkeness: inappropriate gait, ataxia and so on. At the same time, cortical EEG
recordings reveal deep drowsiness or sleep. When the administration of ethanol is preceded by an intravenous injection of amphetamine (2 mg/kg), there is as much behavioral disturbance; however, the cortical EEG corresponds
to a state of arousal ("desynchronization").

Measurements of the relationships between quantitated
electrical activity in the cortex and the 2 subcortical
structures (reticular formation and hippocampus), using
BYFORD's (1965) procedure, produced the data represented
in Tab. I. (Most of these data were previously published:
GREENBERG and GOLDSTEIN, 1969).

Ethanol produced marked changes in the relationships
between the electrical activity in the cortex and the RF,
as manifested by a 4-fold increase in the ratio of the
slopes of the cumulated variances. No change occurred in
so far as the relationships between cortex and hippocampus
are concerned. When the animals were pre-treated with amphetamine, the administration of ethanol produced the same
change in the relationships between cortex and RF as did
ethanol alone. However, in this case, the inter-relationships between electrical activity in the hippocampus and
the cortex were also affected, as evidenced by a 3.5-fold
increase in the ratio of the cumulated variances.

Interestingly enough, amphetamine by itself, in the absence of any other treatment, was found to affect the

Tab. I: Ratios of the slopes of cumulated variances for amplitudes at cortical and subcortical structures in rabbits

Means ± S.D.

Condition	Reticular formation/ cortex	Hippocampus/ cortex
Baseline awake state	0.11 ± 0.04	0.42 ± 0.08
Following D-amphet- amine 2 mg/kg i.v.	0.33 ± 0.17[*]	1.15 ± 0.20[*]
Following ethanol 1 Gm/kg orally	0.48 ± 0.29[*]	0.47 ± 0.17
Ethanol 1 Gm/kg orally following pretreatment with D-amphetamine 2 mg/kg i.v.	0.47 ± 0.30[*]	1.46 ± 0.21[*]

Data from: GREENBERG, R.S. and GOLDSTEIN, L. (1969).

[*]Statistically significant difference from baseline awake state measurements. (p = 0.05 or less; 2-tailed t-test). Each variance value computed from 10 to 15 measurements, each including 10 individual numbers, on 5 to 10 animals.

relationships between hippocampus and cortex, as well as RF and cortex to almost the same extent as when its administration was followed by that of ethanol.

Thus, these data indicate that amphetamine, at the dose used, can affect the relationships of the electrical

activity for the two pairs of cerebral structures under
consideration, namely RF/cortex and hippocampus/cortex.

Effects of amphetamine on performance of an attention task in rats. The effects of amphetamine were investigated in rats well-trained on the continuous attention
task. The animals were first tested for 2 consecutive days
following intraperitoneal injections of saline (1 ml/kg).
Three days later, the same animals received on 3 consecutive days intraperitoneal injections of D-amphetamine
0.5 mg/kg (also, 1 ml/kg) immediately before the daily
test session.

The scores for omission and for commission errors
were averaged for the 2 days of saline and the 3 days of
amphetamine treatment. The data are summarized in Tab. II,
III and IV.

As can be seen in Tab. II, there were considerable
individual differences in the effects of amphetamine, especially in so far as the o.e.'s are concerned. On group
basis (Tab. III) amphetamine produced a statistically significant worsening of this behavioral feature (i.e., more
failures to make the appropriate response). The scores for
commission errors were not changed. It is interesting to
note that the effects of amphetamine on the omission and
commission error scores appeared to be independent (see
Tab. IV). A Chi-square analysis of the distribution of the
animals as regards their individual performances after
amphetamine (better, worse or not different than perfor-

Tab. II: Effects of D-amphetamine (comparison with saline) on two components of the attention task in rats

Animal		o.e./Reinf x 100	c.e./Reinf x 100
201	S* A**	9.0 3.3	14.3 15.3
202	S A	2.0 13.7	30.7 28.0
203	S A	85.6 163.3	11.7 12.9
204	S A	9.3 41.0	46.7 31.0
205	S A	10.0 54.9	15.3 13.8
206	S A	28.8 12.3	20.2 9.7
207	S A	6.3 6.0	11.8 8.5
208	S A	14.7 35.0	42.7 19.5
209	S A	51.6 29.5	29.2 63.0
210	S A	7.3 27.5	9.0 27.0
211	S A	30.3 169.4	26.3 32.9

* Control experiments. Intraperitoneal injections of saline during 2 consecutive days. The values are averages of the scores for the 2 days.

** Amphetamine experiments. Intraperitoneal injections of D-amphetamine, 0.5 ml/kg, during 3 days. The values given are averages of the scores for the 3 days.

See text for the definition of omission errors (o.e.) and commission errors (c.e.).

Tab. III: Overall means ± standard errors for o.e.'s and
          c.e.'s of animals after saline and after amphet-
          amine

Behavioral feature	Saline controls	Amphetamine
Omission errors	23.2 ± 7.6	50.5 ± 17.9*
Commission errors	23.4 ± 3.9	23.8 ± 4.7

*Statistically significant difference between saline and
 amphetamine (paired t-test) p = 0.05.

Tab. IV: Distributions representing the relative effect of
         D-amphetamine on o.e.'s and c.e.'s performance
         (The classification of effect is based on the com-
         parison of post-amphetamine performance with sa-
         line-control performance for each individual sep-
         arately).

Behavioral feature	Number of animals		
	Not affected	Improved	Worsened
Omission errors	1	3	7
Commission errors	5	3	3

mance after saline) yielded a value of 6.1, which indicates
a probability smaller than 0.05 that the distribution of
effects on the two components of behavior is the same.

When the scores were assigned to the 9 possible com-
binations of effects (e.g., worsened on both o.e.'s and
c.e.'s; improved on both scores; not affected on either
score; improved on o.e., but worsened on c.e.'s, etc.), a
random distribution became apparent, and was confirmed by
a Chi-square value of 6.24 (p = 0.40 for random assortment).

Thus, it appeared that even on a very well-learned
task, the ways animals were affected by moderate doses of
amphetamine were not only inconsistent and unpredictable,
varying from no effect to worsening to marked improvement,
but also different for the 2 components of the task.

Effects of amphetamine on the quantitated EEG in normal human volunteers. It has been found in a number of
studies (GOLDSTEIN et al., 1963a and 1963b; PFEIFFER et
al., 1964) that the 2 parameters of amplitude analysis,
the mean integrated amplitude (MIA) and the coefficient of
variation (CV) both decrease characteristically upon drug-
induced, or spontaneously occurring hyper-stimulation.
Anti-anxiety drugs, or spontaneous relaxation, produce, on
the contrary, an increase of the CV. Therefore, it was expected that amphetamine would produce typical decreases in
the MIA and in the CV when administered to normal volunteers. However, this expectation was not fulfilled. In a
study involving 10 normal volunteers (Tab. V), the overall
changes, up to 3 hours post-administration of an oral dose
of 15 mg, did not reveal any significant departure from
control, pre-drug, values. This was true both for the MIA
and for the CV. It was the case not only for the pre-drug,

Tab. V: <u>Effects of D-amphetamine (15 mg total oral dose)</u>
<u>and placebo on EEG parameters in normal subjects</u>

1. Averaged data on 10 subjects

Drug	EEG Parameter	Control recording	Hours post-administration		
			1	2	3
Amphetamine	M.I.A. [1]	79.7	75.3	77.7	75.5
	C.V. [2]	14.4	12.2	13.1	14.8
Placebo	M.I.A.	68.1	68.9	70.4	70.4
	C.V.	14.5	13.8	14.7	16.6

2. Averaged data on 5 subjects with high M.I.A. and high
C.V. during control recording

		Control	1	2	3
Amphetamine	M.I.A.	85.0	74.0*	82.0	87.3
	C.V.	18.9	11.3*	11.2*	14.0*
Placebo	M.I.A.	80.2	81.6	84.5	87.9
	C.V.	18.3	16.3	18.4	21.9

3. Averaged data on 5 subjects with low M.I.A. and low C.V.
during control recording

		Control	1	2	3
Amphetamine	M.I.A.	74.4	76.7	73.4	63.7*
	C.V.	9.8	13.0*	15.0*	15.6*
Placebo	M.I.A.	56.0	56.2	56.3	53.0
	C.V.	10.6	11.3	11.0	11.2

1) Mean Integrated Amplitudes
2) Coefficient of variation
* Statistically significant change from control levels
(p = 0.05 or less).

post-drug comparisons, but also for the within period comparisons between the amphetamine sessions, and placebo sessions which took place in the same subjects, either one
week before or one week after the amphetamine sessions according to a random assignment, double-blind design.

However, when these same data were rearranged, grouping separately the subjects with high MIA and high CV during the pre-drug runs (happening to number 5 in each group),
amphetamine was found to exert definite effects, but in-
versely directed in the 2 sub-groups (Tab. V). In the first
sub-group of subjects, the drug produced a decrease in the
MIA during the first hour post-drug recording run, and a
sustained progressive decrease of the CV during the 3 hourly post-drug runs. In the second sub-group of subjects,
with low initial MIA and CV, amphetamine did not change the
MIA, except for a decrease at the third post-drug hour, but
the drug produced a sustained and regular increase of the
CV. No changes were produced by placebo on either one of
the two sub-groups.

Thus, in normal human volunteers, when the brain waves
of the initial state were characterized by high amplitude
and high variability, amphetamine acted on the EEG as a
stimulant would, while when the initial state was characterized by low amplitude and low variability, the same drug
at the same dose acted as an anti-anxiety (i.e., tranquil-
lizer) drug would in changing electrical activity.

Effects of Nicotine on Electrophysiological Parameters
and on Performance

Effects of nicotine on cortical/subcortical relationships in rabbits. Our studies of nicotine, especially its
effects during chronic administration, resulted from an
interest in a possible neurophysiological basis for the
widespread voluntary self-administration of the drug by
human subjects by way of smoking tobacco.

The effects of nicotine were investigated in rabbits
after both acute and chronic administration. The BYFORD
procedure was used not only for the analysis of the relationships between electrical activity between RF and cortex, and hippocampus and cortex, but also for the pair,
amygdala and cortex. The results are summarized in Tab. VI.

In relation to the values prevailing during control
recordings, acute nicotine administration (10 µg/kg, i.v.)
produced marked, statistically significant changes for the
ratios of the slopes of cumulated variances for RF/cortex
and amygdala/cortex. There were no changes in the relationships hippocampus/cortex.

Upon chronic administration (200 µg/kg, subcutaneously,
5 times daily) over a 3 week period, there occurred, following an initial increase, a gradual decrease of the ratios for the slopes RF/cortex, with almost complete return
to pre-treatment control levels at the end of the third
week of chronic administration. The ratios of the slopes

Tab. VI: <u>Ratios of the slopes of cumulated variances for</u>
<u>amplitudes at cortical and subcortical structures</u>
<u>in rabbits</u>
Means $\pm$ S.D.

Condition	Reticular formation/ cortex	Hippocampus/ cortex	Amygdala/ cortex
Baseline awake state	0.16 ± 0.06	0.13 ± 0.05	0.12 ± 0.04
Acute nicotine 10 µg/kg i.v.	0.06 ± 0.001[*]	0.12 ± 0.05	0.52 ± 0.24[*]
Chronic nicotine 200 µg/kg subcutaneously 5 times daily.			
Week 1	0.50 ± 0.20[*]	0.25 ± 0.11	0.23 ± 0.13
Week 2	0.28 ± 0.05	0.61 ± 0.20[*]	0.19 ± 0.11
Week 3	0.29 ± 0.06	0.76 ± 0.20[*]	0.54 ± 0.39[*]
Chronic saline 5 times daily.			
Week 1	0.35 ± 0.09	0.54 ± 0.16[*]	-
Week 2	0.40 ± 0.13[*]	0.54 ± 0.14[*]	-
Week 3	0.35 ± 0.18	0.29 ± 0.08	-

Data from BHATTACHARYA, I.C. and L. GOLDSTEIN (1970).

[*]Statistically significant difference from baseline state
(p = 0.05 or less, 2-tailed t-test). Each variance point
computed from 10 to 15 groups of 10 measurements on each
of 5 to 10 animals.

for the pair cortex/amygdala were little affected during
the first 2 weeks of chronic administration. However, at
the end of the third week, a large scale change occurred.

Perhaps the most interesting change was seen in the
values of the ratios for the pair, hippocampus/cortex.
This change consisted of a gradual sustained increase in
the ratios, with indications of an inverse relationship
with the decrease occurring for the couple, RF/cortex. At
the end of the third week, a 6-fold increase in the ratios
of the slopes was detected.

Since the animals had to be handled great many times
in order to inject the drug, and thus were submitted to a
stress, control experiments were run on a different group
of animals, in which saline was substituted for the active
drug. As can be seen in Tab. VI, although some changes in
the ratios were found, these changes affected chiefly the
pair RF/cortex in which the ratios remained somewhat elevated. In so far as the pair hippocampus/cortex is concerned, the mean ratios at the end of the third week of
chronic administration were not statistically significantly
different from those existing during the pre-treatment recordings. The data of the effects of saline on the pair
amygdala/cortex were accidentally lost.

Thus, from these results it was concluded that chronic
nicotine produces a gradual shift in the predominant mechanism underlying cortical arousal from the RF to the hippocampus and also to the amygdala, the latter two structures

being part of the limbic system.

Effects of nicotine on the performance of a continuous performance task by rats. These just previously discussed results led us to speculate on the behavioral consequences of such a change. Our predictions were influenced by the schema set forth by ROUTTENBERG (1968). The experimental hypothesis was that chronic nicotine treatment would result in an improvement in performance on a difficult, yet well-learned task because of the shift from more general arousal to more incentive-related arousal.

To test the hypothesis, eleven adult, male Holtzman (Sprague-Dawley) rats were trained on the continuous attention task described in the "Material and Method" section. After more than two months of training, all rats were performing stably and at acceptable levels on both components of the task (o.e. and c.e.). The subjects (Ss) were then divided into two groups one of which received saline, subcutaneously 3 times daily, and the other received nicotine, 100 µg/kg, by the same route and with the same schedule of administration, for 4 weeks. After this first phase of treatment (denoted "saline treatment period" and "nicotine treatment period"), the groups were "crossed-over" so that the first went into a 4-week nicotine treatment period and the second into a saline treatment period. Following these periods, both groups received saline 3 times daily for 3 additional weeks ("saline recovery period"). All Ss were tested 5 days each week for the eleven week schedule just described with only rare exceptions.

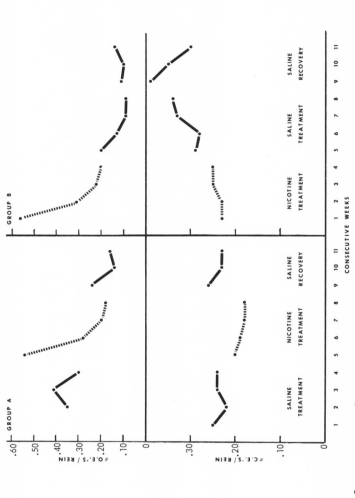

Fig. 1: Performance on the continuous attention task during the 11 week schedule of chronic nicotine and saline administration

Each point represents the group's weekly average of performance. The curves illustrate the decrease in commission errors during nicotine treatment periods and the apparently time-dependent decrease in omissions. Note the first week of nicotine treatment.

Fig. 1 represents a summary of the behavioral data
obtained during the course of the study. Much of these data
are described in a recent report (NELSEN and GOLDSTEIN,
1972). The Ss were divided into 2 groups at the outset in
order to better discriminate changes occurring between sessions which might be independent of drug treatment. Considering first omission error behavior (failure to make the
appropriate response); indeed it would appear that the o.
e.'s decrease over time in a manner independent of drug
treatment. Analysis of variance confirmed the visually indicated trends; i.e., time on the task and not treatment is
the critical factor in reducing o.e.'s.

It should be noted that there is one definite nicotine-
related phenomenon obvious in the o.e. data. The first several days of nicotine treatment result in gross disruption
of performance which is indicated on the figure by the high
o.e. values for the first week averages of the nicotine
treatment periods.

The lower part of Fig. 1 represents commission error
component of behavior (inappropriate responses) over eleven
weeks. Again, weekly averages for the two groups are shown
during the various phases of the study. An analysis of variance reveals that treatment order effects are not significant; i.e., the difference scores calculated by subtracting
the nicotine treatment scores from the saline treatment
scores did not differ statistically for the two groups. Since this was the case, the data from both groups were combined for a further analysis of variance which tested, among

other factors, the significance of the nicotine effect. Unlike omission error changes, the decrease in commission
errors during nicotine treatment was significant at less
than the 0.01 level, independent of any time effect.

Several interesting features of the c.e. performance
become evident when the daily average performance of all
Ss are considered during the 3 phases of the study (19
saline treatment sessions, 18 nicotine treatment sessions,
and 14 saline recovery sessions). First, these data show
that fewer c.e.'s were made during nicotine treatment than
during saline treatment and saline recovery periods (significant at the 0.001 and 0.02 levels, respectively). Also,
the saline treatment and saline recovery periods were not
significantly different. Thus, more evidence that the c.e.
effect is truly nicotine-induced is provided. Second, all
measures of session-to-session variability show that instability of daily performance is lowest under nicotine
conditions. For example, the coefficient of variation was
9.52 for the nicotine period and 18.97 and 18.38 for the
saline treatment and saline recovery periods, respectively.

Effects of nicotine on the acquisition of a continuous
performance task by rats. Having obtained the previously
described results which indicated that nicotine caused increased efficiency of performance on a well-learned task
by selectively reducing inappropriate responses (c.e.'s),
we set about to determine the effects of nicotine on the
acquisition of this same continuous performance task. These
studies are still going on. However, some preliminary data

are available and will be presented.

A group of 12 drug and task naive Holtzman adult, male
rats were randomly assigned to 2 groups. One group received
saline, subcutaneously, 3 times a day, and the other, nicotine, 100 μg/kg by the same route and with the same schedule
throughout the study. Injections were actually begun one
week before training on the task commenced in an attempt to
remove the disruptive effects caused by initial doses of
nicotine on the training. Also, to remove any experimenter
bias (which would be especially critical in acquisition or
training phase), the saline and nicotine solutions were
identified with a code so that the experimenter was "blind"
to which solution contained the active drug.

Training was carried out in a series of phases such
that initially the task was a very elementary one, i.e.,
continuous reinforcement in the presence of a constant C.S.
During successive training sessions, the I.T.I. and punishment contingency were introduced and the duration of the C.
S. was reduced in a step-wise fashion.

Fig. 2 represents the behavior of the 2 groups beginning at the point in training when all S̲s were first exposed
to the task with the final levels of the parameters set
(described earlier in the "Material and Method" section).
Each bar on the figure represents the group average of a
block of 4 successive sessions. Qualitatively, it is obvious
from the figure that both groups behaved similarly in so far
as the c.e. component of acquisition was concerned. Over

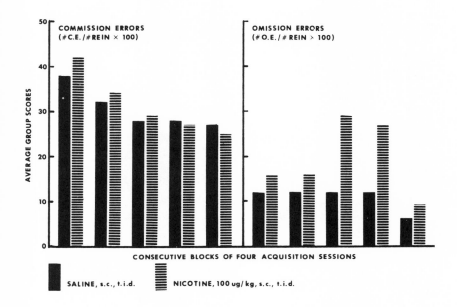

Fig. 2: <u>Preliminary results describing the acquisition of</u>
<u>the continuous attention task under conditions of</u>
<u>chronic nicotine and chronic saline treatment</u>
The bars illustrate the apparent impairment of
learning suffered by the nicotine-treated animals. (See omission error scores).

time, a regular improvement (decrease in inappropriate responding) occurred such that the successive bars in the
Figure resemble a classical "learning curve". Omission
error behavior (failure to make the appropriate response)
was consistently poorer for the animals treated with nicotine than for saline-treated animals.

This study is presently being continued into a "performance phase" (acquisition having been completed) with

the groups being "crossed-over" as regards their drug
treatments. A proposed interpretation of these results
(together with those found in the earlier performance study)
will be presented in the last part of this paper.

DISCUSSION

The review of the previously published data and the
presentation of more recent, unpublished results, are aimed
at supporting the concept that "arousal is not arousal",
more specifically that there exist in the brain two distinct systems. These produce (under normal conditions) a
similar end-product, namely cortical activation, but also
arousal which differs in nature depending on the mechanism
producing it. The existence of two systems for the control
of arousal, which was suggested by a number of findings
presented in the introduction, further can be assumed on
the basis of the pharmacological evidence presented in this
paper, concerning experimentation with 2 drugs known to induce arousal. Thus, in rabbits, direct electroencephalographic measures show that amphetamine acts, at the same
time, on the inter-relationships of electrical activity for
both the pair, RF/cortex and the pair, hippocampus/cortex.
Nicotine can also affect the same two pairs of structures,
but apparently to different relative degrees depending on
the condition of administration (acute vs. chronic). Upon
acute administration it acts primarily on the pair, RF/
cortex while following chronic administration, it gradually
shifts the predominant arousal system from RF/cortex to

hippocampus/cortex.

A proposal derived from the concept of a dual mechanism
for arousal is that the functional consequences of the RF
mediated arousal are different from those of hippocampal
mediated arousal. Because of our electroencephalographic
evidence concerning nicotine's initial and chronic effects
on brain activity relationships, we tested this proposal
behaviorally using nicotine. Indeed, our data indicate that
the initial or acute effects of nicotine on the performance
of a well-learned task (continuous attention) fit with what
would have been expected based on electrophysiological evidence. During initial phases of nicotine treatment (when
the influence of the RF is known to predominate leading to
a heightened drive state, perhaps even overdrive), our animals made many o.e.'s. We suggest that this behavioral deficit resulted from the predominance of drive-arousal (RF)
which in this situation interfered with performance: Ss
made more omissions because the control of response through
incentive-related stimuli (limbic system) had been reduced
via the inhibitory effects of RF predominance.

Within approximately one week of chronic nicotine
treatment, a shift toward limbic system control occurred.
The Ss no longer failed to make appropriate responses, but
continued to improve their o.e. scores apparently independently of nicotine treatment, perhaps as a function of
a continuous practice effect. However, our data show that
the reduction in inappropriate responding (c.e.'s) was
specifically related to chronic nicotine treatment, that

is, the period of predominance of limbic system influences. We suggest that this behavioral improvement is indeed related to the retrieval of stored information. The decrease in inappropriate responding (while, of course, appropriate responding is maintained) is indicative of effective retrieval of information concerning the task (memory traces concerning appropriate behavior within the test situation).

The interpretation of results from the acquisition study is much more difficult because of the complex interaction in our measures of behavior between learning or acquisition (storage) effects and performance (retrieval) effects. Our determination of the level of acquisition is based on the behavioral output of the animal (his performance). We would like to isolate completely the features of acquisition from those of performance, but this is not possible on such a task (or for that matter, on other simpler tasks such as one-trial avoidance learning which many investigators have applied to "memory" studies). However, if low levels of commission errors are indicative of successful retrieval, then our data show that nicotine treatment certainly did not interfere with such retrieval during the acquisition study.

An important question is: Are there differences in storage between saline-treated and nicotine-treated animals? Our o.e. data would indicate that there is, indeed, a difference, whereby more effective storage took place in animals treated with saline. The apparent interference with storage which was produced by nicotine should not have been an unexpected result since there is a body of direct and indirect evidence indicating that storage requires critical interaction and mutual modulation between the two arousal systems (see, e.g., ROUTTENBERG, 1968). Therefore, it is not startling that nicotine by inducing a shift in the relative states of the two systems might disrupt learning (storage). Thus, a pharmacological manipulation of the relationships between the 2 arousal systems which may be beneficial as regards its effects on behavior when the critical neurophysiological activity mediated by the systems involves retrieval of already stored information, may well be detrimental when the organism's neuronal activity is focused on the storage of information.

It is important to note that drugs which affect the relative states of the arousal system have the nature of their effects partially controlled by the pre-existing states of those systems. Thus, on the surface, the 10 subjects involved in the amphetamine experimentation were all similarly awake and responsive to normally occurring environmental stimuli. Yet, 5 of them reacted to amphetamine in an entirely different manner than the 5 other ones. If one postulates that the subjects with relatively higher amplitude and higher variability in their baseline EEG recordings were in a state of predominantly hippocampal-mediated arousal, one should expect the action of amphetamine to manifest itself on the RF (since it is essentially under-reactive when the hippocampal system is "over-driving"). This should result in a shift towards "drive-arousal", i. e., excitation and consequently a decrease in the CV. On

the other hand, if the baseline is one of drive arousal,
with low CV and low MIA, amphetamine should, by its action
on the hippocampus, shift arousal mechanisms to the limbic
system, and produce an increase in the CV.

The data presented in Tab. VI suggests very strongly
that such is the case. As a matter of fact, these data can
be interpreted as indicating that amphetamine transformed
the arousal of the subjects in the first group to that
prevailing in the second group, and vice versa.

One could wonder whether the "paradoxical" calming
effect exerted by amphetamine in hyperkinetic children is
not due to a similar type of shift in arousal mechanisms
from RF "over-drive" by a modulating effect via the hippocampus.

Such a dual arousal mechanism would also help account
for the results from experiments involving the effects of
amphetamine on information retrieval in rats. While some
animals were adversely affected, some performed more efficiently or were not affected in any measurable way. This
was true for omission errors, which can be conceived as
proceeding from "drive-arousal". But it was also true, and
independently, for commission errors, which would seem to
result from hippocampal-mediated "incentive-arousal". Thus,
one can conceive of amphetamine acting on both systems and
affecting them according to their particular state at the
time of administration.

The full significance of the concepts which we have attempted to introduce in describing and interpreting some of our data is only just beginning to be realized. For example, the phenomenon of state-dependent learning (the lack of transfer of trained responses from the state, often drug-induced, in which the responses have been learned to another physiological state) could be viewed quite parsimoniously within the framework of the dual arousal system. It is known that such learning cannot be explained on the basis of the simple discriminative cue properties of a drug (OVERTON, 1967). However, if the laying down of memory traces (and the subsequent retrieval of that information) is dependent on the existence of certain functional inter-relationships between the two arousal systems, then one might view state-dependent learning as a "special case" of the general phenomenon of learning. That is, although the physiological state of the organism (including the level of activation of the two arousal mechanisms) has been altered, the functional inter-relationships between the two systems remain intact, such that, memory traces can be laid down. It is quite likely that long-term memory is a result of the establishment of a permanent sensitivity of cell membranes or constituents thereof, to a particular array of neural impulses. It would follow that effective retrieval would depend on the re-establishment of the state existent during storage, thus allowing or making more probable the production of the required array of impulses.

At the very least, models of memory storage and retrieval, of learning and performance which recognize and incorporate the concepts of dual mechanisms for arousal have far greater heuristic value than those which are limited to either drive or incentive considerations, alone.

Finally, we are quite cognizant of the fact that biochemical or neurohumoral aspects of the processes we have discussed have not been considered in this paper. Presently, such considerations are beyond the scope of our experimental investigations. Of course, we do feel that just as the integration of neurophysiological and behavioral techniques has proven to be theoretically and experimentally fruitful, concern with the neurohumoral features involved would allow a more inclusive understanding of information storage and retrieval.

ACKNOWLEDGEMENTS

This work was supported by a Grant from the Council for Tobacco Research-USA, Contract No N00014-72-C-0203 from the Office of Naval Research, and allocations from Grant FR-05558, U.S. P.H.S.

LITERATURE

BATSEL, H.L.: Electroencephalographic synchronization and desynchronization in the chronic "cerveau isolé" of the dog.
Electroencephal. clin. Neurophysiol. 12, 421-430 (1960).

BHATTACHARYA, I.C., and L. GOLDSTEIN: Influence of acute
    and chronic nicotine on intra- and inter-structural
    relationships of the electrical activity in the rab-
    bit brain.
    Neuropharmacol. 10, 109-118 (1970).

BREMER, F.: Cerveau "isolé" et physiologie du sommeil.
    C.R. Soc. Biol. 118, 1235-1241 (1935).

BYFORD, G.H.: Signal variance and its application to con-
    tinuous measurements of EEG activity.
    Proc. R. Soc. B. 161, 421-437 (1965).

CARLI, G., V. ARMENGOL, and A. ZANCHETTI: Brain-stem-
    limbic connections and the electrographic aspects of
    deep sleep in the cat.
    Arch. Ital. Biol. 103, 725-750 (1965).

DROHOCKI, Z.: L'integrateur de l'électroproduction céré-
    brale pour l'électroencephalographie quantitative.
    Rev. Neurol. 80, 619-624 (1948).

GOLDSTEIN, L., H.B. MURPHREE, A.A. SUGERMAN, C.C. PFEIFFER,
    and E.H. JENNEY: Quantitative electroencephalographic
    analysis of naturally occurring (schizophrenic) and
    drug-induced psychotic states in human males.
    Clin. Pharmacol. Therap. 4, 10-21 (1963a).

GOLDSTEIN, L., H.B. MURPHREE, and C.C. PFEIFFER: Quantita-
    tive electroencephalography in man as a measure of
    CNS stimulation.
    Ann. N.Y. Acad. Sci. 107, 1045-1056 (1963b).

GREENBERG, R.S., and L. GOLDSTEIN: An EEG study of the
    relationships between brain structures in rabbits

under ethanol and D-amphetamine.

Quart. J. Studies Alcohol 30, 843-849 (1969).

JOUVET, M.: Telencephalic and rhombencephalic sleep in the
cat.
In: The Nature of Sleep (G.E.W. WOLSTENHOLME and M.
O'CONNOR, eds.), pp. 188-208, Ciba Foundation Symposium, Churchill, London (1961).

KORNETSKY, C., and M. ELIASSON: Reticular stimulation and
chlorpromazine: An animal model for schizophrenic
overarousal.
Science 165, 1273-1274 (1969).

MCGAUGH, J.L., and L.F. PETRINOVICH: Effects of drugs on
learning and memory.
Intern. Rev. Neurobiol. 8, 139-196 (1965).

MORUZZI, G., and H.W. MAGOUN: Brain stem reticular formation and activation of the EEG.
Electroencephal. clin. Neurophysiol. 1, 455-465 (1949).

NELSEN, J.M., and L. GOLDSTEIN: Improvement of performance
on an attention task with chronic nicotine treatment
in rats.
Psychopharmacologia (1972, in press).

OVERTON, D.A.: Differential responding in a three-choice
maze controlled by three drug states.
Psychopharmacologia 11, 376-378 (1967).

PFEIFFER, C.C., L. GOLDSTEIN, H.B. MURPHREE, and E.H.
JENNY: Electroencephalographic assay of anti-anxiety
drugs.
Arch. Gen. Psychiat. 10, 446-453 (1964).

ROUTTENBERG, A.: The two-arousal hypothesis: reticular
    formation and limbic system.
    Psychol. Rev. 75, 51-80 (1968).

SAWYER, C.H., J.W. EVERETT, and J.D. GREEN: The rabbit
    diencephalon in stereotaxic coordinates.
    J. comp. Neurol. 101, 801-824 (1954).

VILLABLANCA, J.: Electroencephalogram in the permanently
    isolated forebrain of the cat.
    Science 138, 44-46 (1962).

VILLABLANCA, J.: Electrocorticogram in the chronic cerveau
    isolé cat.
    Electroencephal. clin. Neurophysiol. 19, 576-586
    (1965).

    Present address: Leonide Goldstein, Dept. of
Psychiatry, Rutgers Medical School, New Brunswick, N.Y.,
USA.

MODE OF ACTION OF SOME DRUGS WHICH AFFECT LEARNING AND MEMORY

P.B. Bradley

Department of Pharmacology (Preclinical)
Medical School
Birmingham, B 15 2TJ, England

ABSTRACT

The effects of a number of centrally active drugs on the performance of a delayed discrimination task involving "recent memory" have been tested in primates. Whilst most of the drugs used disrupted performance, LSD 25 had a facilitatory effect 72 and 96 hours after injection. This drug was also found to facilitate reversal learning in rats, and this effect was accompanied by a significant increase of 5-HT levels in the brain. The results of investigations into the actions of LSD 25 on single neurones in the brain stem, using the technique of microiontophoresis have shown that the drug selectively antagonises the excitatory actions of 5-HT and in some cases also glutamate excitation. It is suggested that this action may be the basis for the psychotomimetic properties of LSD 25 and might also account for its effects on learning and memory.

193

We have studied the effects of some centrally acting
drugs on "recent memory" in primates (ROBERTS and BRADLEY,
1967). Monkeys were trained to perform a visual discrimination task which involved retention of stimulus parameters
during a variable time delay. The accuracy was recorded at
each delay together with the tendency to score errors of
omission or commission, the number of repetitive errors,
the total number of responses and their latency. The effects
of D-lysergic acid diethylamide (LSD 25), psilocybin, Brom-
LSD (BOL-148), 5-hydroxytryptophan (5-HTP), chlorpromazine,
pentobarbitone, atropine, atropine methonitrate and physostigmine on these performance measures were compared with
the effects of changes in levels of motivation and of distraction. The drugs and treatments used, together with the
measures of performance are shown in Tab. I. LSD 25 disrupted
accuracy at the longer delays, its effects on the other
measures being similar to that of distraction. The effects
of psilocybin were similar to those of LSD 25. BOL-148 and
5-HTP had no effect on performance. Chlorpromazine, pentobarbitone and atropine all depressed performance, as did
physostigmine. The disruptive effects of atropine appeared
to be related to its peripheral actions. The most striking
result obtained in these experiments was an increase in the
accuracy of performance produced by LSD 25, in a dose of
15 µg/kg, at 72 and 96 hours after injection (Fig. 1). This
effect was also shown by psilocybin but to a lesser extent
by BOL-148. It is possible therefore, that this facilitatory
effect on performance by LSD 25 might be related to its
psychotomimetic properties. Although the disruptive effect
of LSD 25 did not appear to be related to an antagonism of
5-hydroxytryptamine (5-HT), we were unable to demonstrate

Tab. I:

Treatments		Measures
LSD 25	5 and 15 µg/kg	Accuracy.
Chlorpromazine	2.5 and 5.0 mg/kg	Accuracy at each delay.
2 Brom LSD	15 µg/kg	Repetitive errors.
Psilocybin	1.5 mg/kg	Ratio of Go to No-Go errors.
5 Hydroxytryptophan	0.5 mg/kg	Total respones.
Pentobarbitone	5.0 mg/kg	Distribution of errors.
Atropine	0.1 mg/kg	Distribution of repetitive errors.
Atropine Methonitrate	0.01 mg/kg	Response latency.
Saline	—	
Greater food deprivation	—	
Less food deprivation	—	
Distraction	—	

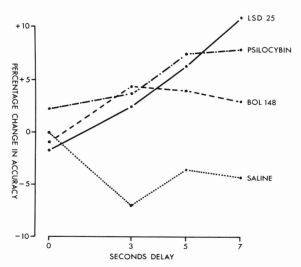

Fig. 1: <u>The effects of psychotomimetic drugs on the accuracy</u>
<u>of performance of a delayed visual discrimination</u>
<u>in primates</u>
The accuracy, expressed as a percentage change of
the pre-drug control value, is plotted against the
delay time, 3 and 4 days after the injection of
LSD 25 (15 µg/kg), psilocybin (1.5 mg/kg) or BOL 148
(15 µg/kg).
(From: ROBERTS, M.H.T. and BRADLEY, P.B., 1967).

whether or not this was the case with the facilitatory
effects of LSD 25.

Recently, my colleagues, Dr. KING, Mr. MARTIN and Mrs.
SEYMOUR have studied the effects of LSD 25 on reversal

learning in rats. They found that reversal learning was facilitated by LSD 25 in doses of 12.5, 25 or 50 µg/kg (i.p.) and that this effect was not present when either saline or BOL-148 was injected (KING et al., 1972). The levels of amines (noradrenaline, dopamine and 5-HT) in the brains of the animals used in these experiments were determined (MARTIN and ANSELL, 1972) and it was found that the dose of LSD 25 which caused facilitation of learning also produced a significant increase in 5-HT levels. No significant changes were observed in the levels of noradrenaline or dopamine.

The results of early investigations in this laboratory (BRADLEY and ELKES, 1953, 1957) suggested that a possible site of action for LSD 25 in the brain might be the reticular formation of the brain stem. It was also shown that this drug increases sensitivity to sensory stimuli (BRADLEY and KEY, 1958) and generalisation of auditory stimuli (KEY, 1961). More recently we have re-investigated the original hypothesis of GADDUM (1953), that the central effects of LSD 25 might be related to an antagonism to the actions of 5-HT, as was demonstrated peripherally. Using the technique of microiontophoresis the effects of LSD 25 and its interactions with certain neurotransmitter substances were studied (BOAKES et al., 1970). Acetylcholine (ACh), noradrenaline (NA) and 5-HT, applied iontophoretically excite some neurones in the brain stem and inhibit others. When the effects of iontophoretically applied LSD 25 were examined, it was found that this substance antagonised only the excitatory actions of 5-HT (Figs. 2 and 3A) but not its inhibitory actions (Fig. 3B), nor any actions, either excitatory or inhibitory,

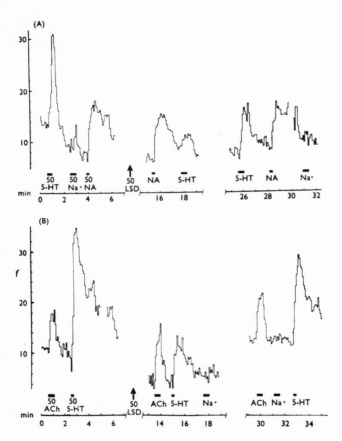

Fig. 2: <u>The effects of iontophoretically applied LSD 25 on</u>
        <u>excitatory responses of single brain stem neurones</u>

        <u>to 5-HT, NA and ACh</u>
        The firing rate in impulses per sec (f) is plotted
        against time (min). Iontophoretic applications are
        shown by the bars, at the currents indicated (in nA).
        A: The excitatory response to 5-HT, but not that to
        NA, was blocked after applying LSD 25 for 5.5 min.
        A further application of 5-HT 13 min later showed
        partial recovery of the excitatory resonse.
        B: Application of LSD 25 for 10 min blocked the ex-
        citatory response to 5-HT but not that to ACh.
        Partial recovery of the 5-HT response was present
        18 min later. (From: BOAKES et al., 1970).

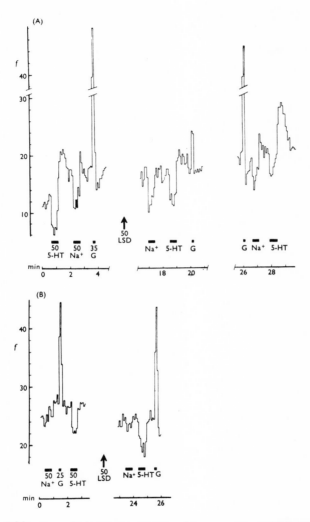

Fig. 3: <u>Effects of iontophoretically applied LSD 25 on
responses to 5-HT and glutamate</u>

A: Excitation of neurone firing by both 5-HT and
glutamate (G) blocked by application of LSD 25 for
10 min. Both responses show partial recovery.
B: A neurone inhibited by 5-HT and excited by
glutamate; LSD 25 was without effect. Other details
as in Fig. 2. (From: BOAKES et al., 1970).

Tab. II: Antagonism by LSD 25 to effects of excitatory and
         inhibitory agents

		blocked	not blocked
5-HT	+	32	2
	-	0	8
ACh	+	0	15
	-	0	2
1-NA	+	1	9
	-	0	10
GLUT	+	17	6
DLH	+	0	7
GLY	-	0	9
GABA	-	0	8

+ = excitation; - = inhibition. The numbers refer to numbers
of neurones.

of ACh or NA. Glutamate excitation, where it was associated
with excitation by 5-HT was antagonised (Fig. 3A), however,
but excitation by other amino acids (e.g. DL-homocysteic
acid) or inhibition by glycine or GABA was unaffected (Tab.
II). BOL-148 was rarely effective as an antagonist to 5-HT
but similar antagonistic effects were observed in experiments
where LSD was administered intravenously. Thus, it was concluded that the antagonism to 5-HT (and glutamate) excitation
of brain stem neurones by LSD 25 may be the basis of the
psychotomimetic action of this drug. It is further suggested

that this may be related to the mechanism by which the drug
produces facilitation of learning.

LITERATURE

BOAKES, R.J., P.B. BRADLEY, I. BRIGGS, and A. DRAY:
    Antagonism of 5-hydroxytryptamine by LSD 25 in the
    central nervous system: a possible neuronal basis for
    the actions of LSD 25.
    Br. J. Pharmac. 40, 202-218 (1970).

BRADLEY, P.B., and J. ELKES: The effects of amphetamine and
    D-lysergic acid diethylamide (LSD 25) on the electrical
    activity of the brain of the conscious cat.
    J. Physiol. 120, 13P (1953).

BRADLEY, P.B., and J. ELKES: The effects of some drugs on
    the electrical activity of the brain.
    Brain 80, 77-117 (1957).

BRADLEY, P.B., and B.J. KEY: The effect of drugs on arousal
    responses produced by electrical stimulation of the
    reticular formation of the brain.
    Electroenceph. clin. Neurophysiol. 10, 97-110 (1958).

GADDUM, J.H.: Antagonism between lysergic acid diethylamide
    and 5-hydroxytryptamine.
    J. Physiol. 121, 15P (1953).

KEY, B.J.: The effect of drugs on discrimination and sensory
    generalisation of auditory stimuli in cats.
    Psychopharmacologia 2, 352-363 (1961).

KING, A.R., I.L. MARTIN, and K.A. SEYMOUR: Reversal learning
    facilitated by a single injection of lysergic acid
    diethylamide (LSD 25) in the rat.
    Br. J. Pharmac. 45, 161-162 (1972).

MARTIN, I.L., and G.B. ANSELL: A sensitive gas chromato-
    graphic procedure for the estimation of noradrenaline,
    dopamine and 5-hydroxytryptamine in rat brain.
    Biochem. Pharmacol., in press (1972).

ROBERTS, M.H.T., and P.B. BRADLEY: Studies on the effects
    of drugs on performance of a delayed discrimination.
    Physiol. and Behav. 2, 389-397 (1967).

COMPENSATORY MECHANISMS FOLLOWING LABYRINTHINE LESIONS IN
THE GUINEA-PIG. A SIMPLE MODEL OF LEARNING.

K.-P. Schaefer and D.L. Meyer

Psychiatrische Klinik der Universität Göttingen

34 Göttingen, v.-Siebold-Straße 5, W.-Germany

ABSTRACT

Compensation of symptoms resulting from labyrinthine
lesions was investigated under normal conditions, after
hemispherectomy, cerebellectomy, transection of the spinal
cord and under the influence of drugs. The symptoms considered were head-deviation, permanent eye-nystagmus and
rotation of head and body about the animal's longitudinal
axis. The labyrinthine lesions were unilateral in most
animals, in other animals the second labyrinth was subsequently destroyed after the first in order to study the
"Bechterew-compensation". In these cases the intervals between the two lesions varied from 3 days to 6 weeks.

The compensatory functions described here are interpreted in terms of simple learning processes.

INTRODUCTION

The study of adaptation to afferent stimulus patterns,
posttetanic potentiation and conditioned reflexes has made
it possible to investigate and comprehend simple learning
processes and to test how these can be influenced pharma-
cologically. The experimental conditions should be as
simple as possible since the molecular biological and electrophysiological bases of such processes are still largely
hypothetical.

The mechanisms of compensation resulting from labyrinthine lesions can also be employed as a model for simple learning processes. These mechanisms have already been
studied by several authors in animals as well as in humans
(BECHTEREW, 1883; MAGNUS, 1924; SPIEGEL and DEMETRIADES,
1925; SPIEGEL and SATO, 1926; LORENTE DE NO, 1928; DOW,
1938; von HOLST, 1948; MITTERMAIER, 1950; SCHOEN, 1950,
van EYCK, 1954; KOLB, 1955; FLUUR, 1960; LANGE and KORN-
HUBER, 1962; PRECHT et al., 1966; STENGER, 1955 and others).
In our laboratory we have likewise concentrated on this
phenomenon for several years, but with the additional aim
of a quantitative analysis (DAL RI and SCHAEFER, 1956;
1957; SCHAEFER and WEHNER, 1966; DIETL, 1966; MATTHAEI,
1966; ALTKÄMPFER, 1966; HOLZBACH, 1967; KRAUS, 1970; AKIL,
1972). Some of our results will be presented here.

The vestibular system seems to be particularly well-
suited for experiments of this type. It responds to stimuli
which can be exactly defined in terms of postural changes

and acceleration, and the responses take the form of
quantitatively measurable reactions, such as vestibular
nystagmus, which permits an objective analysis of the
most complicated control processes (see JUNG, 1953; KORN-
HUBER, 1966). Another useful feature is the high level of
spontaneous activity among the neurons of the vestibular
nuclei (LOEWENSTEIN and SAND, 1940; GERNANDT, 1949;
DUENSING and SCHAEFER, 1958; PRECHT and SHIMAZU, 1965;
SHIMAZU and PRECHT, 1965). This means that destruction of
one labyrinth leads to a marked asymmetry of tone within
the vestibular system. In the course of several hours after
the operation the central nervous system is able to eliminate partially this asymmetry. The quantitative analysis
of this process of compensation is further facilitated by
the fact that it is relatively constant in time from one
animal to another.

The general problem was well defined by GERARD (1963)
when he wrote the following: "Any system must have a certain structure which is reasonably constant in time. In
response to stimuli, this system interacts with its environment, yields and restores itself. Under certain conditions the interaction of the system and its environment
leads to irreversible changes, and a different system than
exists. This progressive specification or alteration with
time, is, in the broadest sense, fixation of experience".
GERARD touches on a point of particular interest for us,
when he says: "There is good electrical evidence that many
neurons or collectivities of neurons are involved in each
engram, and equally, that every such neuron is involved in
many engrams". The following experiments will show that

the process of compensation after destruction of the mem-
branous labyrinth in guinea-pigs is not an isolated function of the vestibular system, but of the central nervous
system as a whole.

METHODS

We needed a rapid and technically simple method for
eliminating the labyrinth, particularly for the pharmacological investigations, in which large numbers of animals
were to be screened. We therefore used a chemical rather
than a surgical method. Under light ether anaesthesia
0.1 - 0.2 cc. of chloroform was injected into the middle
ear. This method had previously been described by MAGNUS
(1924) and was modified by us. The symptoms of unilateral
destruction of the labyrinth appear after a few minutes
and are permanent unless compensated. If seven days later
the other labyrinth is destroyed, the same symptoms appear,
but in the opposite direction (BECHTEREW-compensation).

The symptoms suitable for quantitative analysis are
head-deviation, permanent eye-nystagmus and rotation of
head and body about the animal's longitudinal axis. Head
nystagmus, rolling movements, circular walking movements
and asymmetry of limb tonus were not subjected to quantitative analysis. The experimental procedure is summed up
in the following table:

1. Compensatory behavior in normal animals (ca. 300 g body
   weight).

    a) unilateral labyrinthine lesion.

    b) bilateral lesion (BECHTEREW-compensation).

2. Correlation with age of animal (4, 10, 20 days, 9 months).

3. Influence of light (compensation in light and darkness).

4. BECHTEREW-compensation with various intervals between the first and second lesions (3, 7 days, 3, 6 weeks).

5. Labyrinthectomy after cerebral hemispherectomy.

6. Labyrinthectomy after cerebellectomy.

7. Compensatory behaviour with animals suspended in harness (elimination of proprioceptive influence from limbs).

8. Labyrinthectomy after transection of spinal cord at level of first thoracic vertebra.

9. Pharmacological influence on compensatory behaviour:

    a) sedative drugs (phenobarbital, chlorpromazine and others).

    b) stimulatory drugs (amphetamine, strychnine, cholinesterase inhibitor E 600, anabolic substances and others).

RESULTS

Unilabyrinthal Lesion (Short-Term Compensation)

As mentioned above compensatory processes after labyrinthine lesions have already been investigated by several other authors. In order to investigate this phenomenon under

quantitative conditions we first examined the normal compensatory processes after unilateral labyrinthine lesions
in groups of 6 - 10 experimental animals. Head-deviation,
permanent eye-nystagmus and body rotation due to a labyrinthine lesion were compensated as follows.

After elimination of one labyrinth a <u>head-deviation</u>
of ca. 160^{0} occurred to the side of the labyrinthine lesion. Three to four hours after the operation the degree
of head-deviation had decreased to half of its initial
value, and after 8 - 12 hours it had asymptotically reached
its terminal value. In a few animals there was still a
head-deviation of 30 degrees after 12 hours. The decline
in head-deviation did run linearly on a semilogarithmic
scale. The rate of decline was independent of the animals'
age, but ran at a higher level in older animals. Fig. 1
demonstrates the compensation of head-deviation after uni-
labyrinthal lesion in 4-day old and 8-month old guinea-pigs.

The experimentally induced <u>permanent eye-nystagmus</u>
towards the side of the lesion had an initial frequency of
150 - 180 beats/min after a unilateral labyrinthine lesion.
This rate declined to a half after 3 - 4 hours, and was fully compensated after about 9 hours. The frequency plot did
also run linearly on a semilogarithmic scale, though the
standard deviations were rather large. The decline of the
eye-nystagmus was also independent of the animals' age,
but again the curve obtained from older animals did run at
a higher level.

There was a different course of events with the

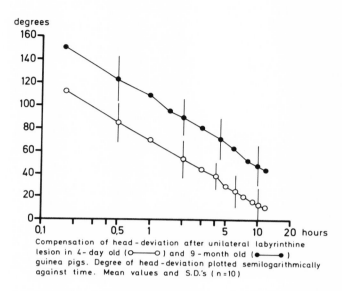

Compensation of head-deviation after unilateral labyrinthine
lesion in 4-day old (O——O) and 9-month old (●——●)
guinea pigs. Degree of head-deviation plotted semilogarithmically
against time. Mean values and S.D.'s (n=10)

Fig. 1: <u>Compensation of head-deviation after unilateral
        labyrinthine lesion in 4-day old and 9-month old
        guinea-pigs</u>

<u>longitudinal body rotation</u>. As already described by MAGNUS
(1924), this rotation was towards the side of the lesion,
amounted to about 100° of turning and was permanent. That
is, it was not compensated after irreversible destruction
of the labyrinth. After a few days the turning can even
increase in magnitude, but the turning may not be detectable if the animal is sitting upright with all four ex-
tremities in contact with the ground.

Bilateral Labyrinthine Lesion (Long-Term Compensation)

The experiments described so far give some insight into compensatory processes in the acute situation. We may
now ask what mechanisms are responsible for consolidation
of the equilibrium achieved. To answer this question we
examined in further experiments (MATTHAEI, 1966) the so-
called BECHTEREW-compensation after varying intervals between the two operations. In four groups of 6 - 10 experimental animals the second labyrinth was destroyed 3 days,
7 days, 3 weeks or 6 weeks after the first.

After 3 days the newly achieved equilibrium was still
very unstable and the asymmetry of tone within the vestibular system was not yet completely levelled out. This is
evident from the fact that the second operation produced
only slight symptoms. After a period of three days head-
deviation and body rotation were still directed towards
the side of the first labyrinthine lesion. The eyes however showed a BECHTEREW nystagmus, that is, the permanent
eye-nystagmus after lesion of the second labyrinth which
was directed towards the second lesion.

These results might be proof of the fact that after
vestibular defects the above mentioned symptoms such as
eye-nystagmus and head-deviation are compensated at a different rate. The same phenomenon can be observed in the
recovery processes of humans when one labyrinth is temporarily defect, as for instance in vascular diseases (see
JUNG, 1953; KORNHUBER, 1966).

In animals in which the second labyrinth was destroyed 7 days after the first, all the symptoms such as head-deviation, eye-nystagmus and body rotation were now directed towards the other side, that is, to the side of the second lesion. With this interval of 7 days between the operations, the compensation proceeded rather faster after the second operation - it was complete in about 6 hours. In this case the body rotation was also compensated - a consequence of the fact that both labyrinths were eliminated, so that the asymmetry of tone was purely a function of the central nervous system.

With longer intervals the symptoms of the second operation became even more marked, the compensation took longer, and more or less constant terminal values were reached which were not compensated. The question therefore arises, whether the relatively unstable degree of compensation reached at 3 and 7 days after the first operation may be purely functional in nature, while after 6 weeks the remaining symptoms may be anchored in structural changes. The present findings show that the late phase of compensation is amenable to experimental investigation and is governed by predictable conditions.

Fig. 2 shows the BECHTEREW-compensation of the eye-nystagmus with intervals of 7 days and 6 weeks between the two labyrinthine lesions. It is obvious that longer intervals between the two operations increase the basic frequency.

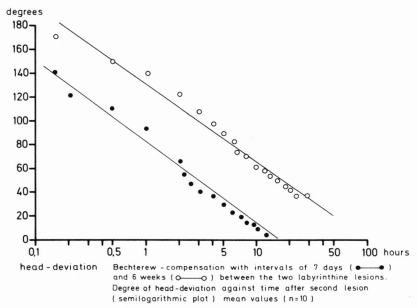

Fig. 2: <u>BECHTEREW-compensation with intervals of 7 days</u>
        <u>and 6 weeks between the two labyrinthine lesions</u>

Influence of the Cerebral Cortex (Hemispherectomy)

Of the nervous pathways arising from the vestibular
nuclei only a few ascend as far as the cerebral cortex. In
monkeys the vestibular projection in the cortex has recently been localized by KORNHUBER (1966) to area 2. In its
turn the cortex exerts an influence on the vestibular system, mainly via the reticular formation and the cerebellum.
Movements of the eyes and the head can be elicited by electrical stimulation of almost any region of the cortex. Corresponding to these observations asymmetrical phenomena of
the limb and trunk muscles can be induced by unilateral

lesions of the cerebral cortex. As is also known, the
cerebral cortex is able to compensate the effects of a
hemicerebellectomy (LUCIANI, 1891; MANNI and DOW, 1963).
On the other hand the cerebellum as well as the spinal
cord seem to be involved in the compensation of the deficiency phenomena induced by the unilateral cerebral lesions (GIRETTI, 1971).

In order to investigate the part played by the cerebral cortex in the process of compensation HOLZBACH (1967)
tested in our laboratory the asymmetrical position of head
and body after elimination of one hemisphere in otherwise
intact guinea-pigs. Destruction of the left hemisphere resulted in a turning of the head to the right. In the course
of a few days compensation occurred and the head was held
straight again, although the head-deviation could still be
elicited if the animals were lifted off the ground or by
transection of the spinal cord. If on the other hand the
destruction of the left hemisphere was preceded by the
elimination of both labyrinths, the initial direction of
head-deviation was to the left, that is, ipsilateral. Not
until three days later did the animal turn its head to the
right. In this latter experiment an eye-nystagmus was observed, but there is not the space to go into the reasons
for this here.

The essential part of our experiments was the examination of the case in which hemispherectomy was followed by
the destruction of a labyrinth. If the left hemisphere was
removed and subsequently the right labyrinth, that is, the
contralateral one, was destroyed, the tendency of each op-

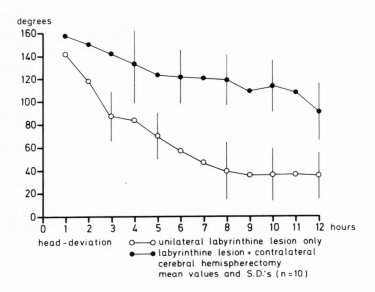

Fig. 3: <u>Head-deviation after hemispherectomy followed by</u>
<u>the destruction of a labyrinth</u>

eration to induce a deviation of the head to the right was
summated, and the course of compensation slowed down as is
demonstrated in Fig. 3. If the destruction of the labyrinth
was performed on the same side as the hemispherectomy, compensation took place at the normal rate.

The results suggest therefore that the vestibular
nuclei, deprived of vestibular afferents, can be brought
back to a more or less normal level of excitation by afferents from the contralateral cerebral cortex as has been
discussed by MENZIO (1949). This pathway probably runs via
the cerebellum and from there to the Deiter's Nucleus.

On the other hand the permanent eye-nystagmus did not
last longer than when the labyrinthectomy was performed
alone. Under these conditions too a difference can be observed between the compensatory processes of head-deviation and eye-nystagmus.

Influence of the Cerebellum (Cerebellectomy)

It is well known that embryologically the cerebellum
is closely connected with the vestibular apparatus. Vestibular fibres have a strong projection onto the so-called
vestibulo-cerebellum and the cerebellar Purkinje cells have
a direct inhibitory effect on the large neurons of Deiter's
Nucleus. After removal of the cerebellum the vestibular
nystagmus is disinhibited,but there is a longer-lasting
impairment of body-position reflexes (LUCIANI, 1891; MAG-
NUS, 1924; RADEMAKER, 1931). It may therefore be assumed
that removal of the cerebellum may also disturb the compensatory process following destruction of a labyrinth.

The first effect of cerebellectomy was that, when a
labyrinth was eliminated, the symptoms were more marked
than they were in otherwise intact animals. The turning
of the head was more pronounced, the animals carried out
more long-lasting rolling movements, and the nystagmus
rate could rise from 150 to as much as 180 - 200 beats
per minute (HOLZBACH, 1967; KRAUS, 1970).

The subsequent compensation was much delayed and, as
in the previous experiments,there was a difference between

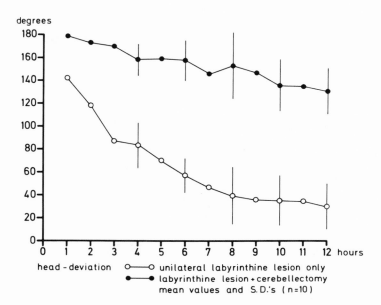

Fig. 4: <u>The influence of cerebellectomy on the compensation</u>
       <u>of head-deviation due to unilateral labyrinthine</u>
       <u>lesion</u>

the compensation of head-deviation and eye-nystagmus. The
compensation of the eye-nystagmus was only affected during
the first 5 days after cerebellectomy, what may well be a
direct result of operative trauma.

    The compensation of the head-turning, on the other
hand, was delayed for weeks, even as long as as nine months.
Fig. 4 demonstrates the influence of cerebellectomy on the
compensation of head-deviation due to unilateral labyrinthine lesion. In this case the cerebellectomy preceded
the labyrinthectomy by 6 days. Under these circumstances
the compensation is significantly impaired - even up to
12 hours.

In view of the fact that the compensation of head-deviation was delayed for this length of time we conclude that the cerebellum is in some way necessary for the recovery of the normal head position. Direct afferents from the spinal cord (see below) as well as from the cortex and other cerebral regions, seem to be ineffective. In other words, the intact spino-reticular or the cortico-reticular pathways were not able to replace the somatosensory or cortical afferents arriving via the cerebellum. According to these findings the question arises whether the cerebellum is able to organize any long-term processes such as the compensatory mechanisms investigated here. Nowadays it is generally believed that the cerebellum is only capable of short-term storage (FREEMAN, 1965 and others). If this hypothesis is correct it is possible that the cerebellum is not the agent but only the mediator of the processes in question.

Influence of the Spinal Cord (Cordotomy)

As is well known vestibular impulses reach the cord by mono- and polysynaptic pathways and also exert an influence on the gamma spindle loops. In the reverse direction, proprioceptive and somatosensory afferents reach the vestibular nuclei either directly or via the cerebellum. The close functional link between the vestibular apparatus and the body-supporting mechanisms helps to understand why asymmetry in the position of back and limbs is found after destruction of one labyrinth. The existence of this link may also explain why proprioceptive and somatosensory af-

ferents from the spinal cord have an influence on the process of compensation after destruction of a labyrinth (KOLB,
1955; AZZENA, 1969). Likewise the spinal cord is known to
compensate the effects of the complete removal of the
cerebellum (BATINI et al., 1957) or the cerebral release
phenomena after hemispherectomy (GIRETTI, 1971). In our
own experiments (MATTHAEI, 1966; HOLZBACH, 1967; KRAUS,
1970) we were also able to demonstrate a certain influence
of the spinal cord on the compensation after labyrinthectomy. Delay in compensation was produced by suspending the
animals in the air so that they had no contact with the
ground. If after 8 hours the animals were put down on the
ground, there was a strong turning of the head which was
compensated from this point on at the usual rate.

However, a delay in compensation could also be observed after transection of the spinal cord at the level
of the first thoracic vertebra. Under these conditions too
the compensation of head-deviation was much more severely
affected than that of the eye-nystagmus. By means of statistical investigations into a large number of experiments
we were however able to show that the compensation of the
eye-nystagmus was significantly affected too.

In further experiments the influence of the spinal
cord could still be demonstrated after compensation had
occurred. In this case lifting the animals off the ground,
transecting the spinal cord, or applying light ether anaesthesia led to a recurrence of the positional asymmetry.
These findings can most probably be explained by the inter-
ruption of the spinocerebellar pathways, for the spino-

reticular pathways are not capable of organising compensation in the usual way after cerebellectomy, as has already been mentioned.

Pharmacological Investigations

Pharmacological investigations after labyrinthine lesions have only taken into account the acute effects of individual drugs on the vestibular symptoms mentioned above (MAGNUS, 1924; DAL RI and SCHAEFER, 1956, 1957). The influence of these drugs on the described long-term compensatory processes have not been investigated so far. Under the experimental conditions we applied it is however possible to reach a quantitative analysis of such pharmacological influences.

Drugs delaying compensation: In our laboratory experiments with phenobarbital (DIETL, 1966) showed that a single intraperitoneal injection of the drug at a dose of 50 mg/kg, performed shortly after destruction of the labyrinth, produced a marked inhibition of the compensatory process. When this dose was given, the animals lay at first on their sides and it took three times as long for them to return to the normal position - thirty-six hours instead of twelve. If the injection was repeated at the same dosage, the process of compensation could be delayed for three days or longer.

Chlorpromazine too had a strong delaying effect (ALT-KÄMPFER, 1966; AKIL, 1972). This drug inhibited compensa-

Tab. I: Equieffective doses of neuroleptic drugs on com-
        pensation in guinea-pigs after unilateral laby-
        rinthine lesion

	Equieffective doses	Compensatory factor
1. Levomepromazine (Neurocil)	40 mg/kg	5.088
2. Chlorprothixen (Taractan)	30 mg/kg	3.816
3. Perazine (Taxilan)	20 mg/kg	3.09
4. Chlorpromazine (Megaphen)	10 mg/kg	1
5. Clopenthixol (Ciatyl)	10 mg/kg	0.64
6. Perphenazine (Decentan)	20 mg/kg	o.254
7. Butyrylperazine (Randolectil)	10 mg/kg	0.065

tion in doses which had only slight sedative effect and
very little effect on the guinea-pig's vestibular functions. The inhibitory effect was slight at a dose of 5 mg/
kg injected intraperitoneally, more marked at 10 mg/kg.
Again the compensatory process could be blocked for many
hours by repeated application of the drug.

In relation to chlorpromazine Tab. I gives the equieffective doses of several neuroleptic drugs tested in the
same manner. The so-called compensatory factor listed at
the right side of Tab. I represents the product of these
equieffective doses and the average clinical dosage in

humans.

$$\text{Compensatory factor} = \frac{\substack{\text{average clinical} \\ \text{dosage of drug} \\ \text{applied}}}{\substack{\text{average clinical} \\ \text{dosage of chlor-} \\ \text{promazine}}} \times \frac{\substack{\text{equieffective} \\ \text{dosage of drug} \\ \text{applied}}}{\substack{\text{10 mg/kg} \\ \text{chlorpromazine}}}$$

Butyrylperazine,e.g.,is characterised by the same equieffective doses as chlorpromazine.

In clinical use butyrylperazine, however, is given in
significantly smaller doses which lead to the low compensatory factor of this drug. In general neuroleptic drugs,
such as butyrylperazine, clopenthixol and perphenazine
which show a marked neuroleptic potency combined with strong
antipsychotic effects exerted little influence on the compensatory behaviour. The high compensatory factor of chlorpromazine, chlorprothixen and perazine however shows that
neuroleptic drugs with a high sedative component have a
strong delaying effect on compensatory processes.

Drugs accelerating the compensation: The reverse effect is shown by drugs which excite the CNS. These substances accelerate the process of compensation. In these
experiments too we found that the head-deviation was more
strongly affected than the vestibular eye-nystagmus. Among
the drugs with marked accelerative effects were metamphet-
amine, pentetrazole, strychnine, paraoxon (E 600) and
risatarun, a glutamate-containing substance (SCHAEFER and

Tab. II: <u>Influence of stimulatory drugs on compensatory behaviour in guinea-pigs after unilateral labyrinthine lesion</u>

<u>Drug</u>	<u>Dose</u>	<u>Effect</u>
1. Nortestosterone (Primobolan)	100 mg/kg	Ø
2. Caffeine	100 mg/kg	+
3. Pentetrazole (Cardiazol)	20 mg/kg	+
4. Metamphetamine (Pervitin)	3 mg/kg	++
5. Strychnine	0.3 mg/kg	++
6. Risatarun[R]	100 mg/kg	+++
7. Paraoxon (E 600) (+ PAM 50 mg/kg)	1 mg/kg	+++

Ø negative   + slight   ++ moderate   +++ strong

WEHNER, 1966). Tab. II summarises the substances tested.

Of particular interest is the influence of some of
these drugs on body-rotation about the longitudinal axis,
a symptom which the nervous system does not compensate
spontaneously. Caffeine had a slight effect on this symptom, strychnine and cardiazol did have a rather stronger
one. But E 600, a cholinesterase inhibitor completely eliminated the body-rotation after a few hours, and the rotation
could not be elicited by lifting the animal off the ground.
The compensatory effect of the drug disappeared after one
to two days. Risatarun, an anabolic substance, had a particularly strong and long-lasting effect, the body-rotation did not reappear until eight to fourteen days after
the application of the drug was discontinued. In Fig. 5

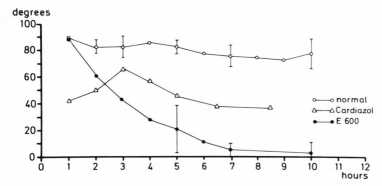

Fig. 5: Effects of several drugs on the body-rotation

the effect of pentetrazole (cardiazol) and E 600 is demonstrated.

We assume that these drugs have two basic effects:
firstly a non-specific generalized arousal effect, accelerating learning processes especially compensatory mechanisms, and secondly a pharmacological activation of inactive synapses and synaptic pathways. From a consideration of the time course of these processes it seems im-
probable that the formation of new synapses plays a role
in compensatory mechanisms. It is indeed generally considered unlikely that the formation of synapses de novo
is an important factor in learning processes (GUTMANN,
1957; LASSEK and EMERY, 1959; FISCHER, 1963; KÄLLEN, 1965
and others).

CONCLUDING REMARKS

Symptoms suitable for quantitative analysis after labyrinthectomy in guinea-pigs proved to be head-deviation, permanent eye-nystagmus and rotation of head and body about the animal's longitudinal axis. As described by MAGNUS (1924) and several other authors head-deviation and eye-nystagmus were compensated spontaneously and declined asymptotically to their terminal value within 8 to 12 hours. The rate of decline was independent of the animals' ages.

The mechanisms involved to consolidate the equilibrium achieved, could be investigated by making use of the so-called BECHTEREW-compensation after successive lesion of the two labyrinths. In order to gain insight into the long-term processes of compensation we varied in other experiments the intervals between the two operations. It could also be demonstrated that the late phase of compensation, that is the period between about 6 weeks and 9 months, was governed by predictable conditions. In this connection the question arose whether the remaining permanent symptoms might possibly be anchored in structural changes.

Further experiments were performed in order to investigate the influence of the cerebral cortex, the cerebellum and the spinal cord in the process of compensation. After hemispherectomy, cerebellectomy as well as after transection of the spinal cord a marked delay in compensation could be observed. Under these conditions too the compensation of head-deviation was more severely affected than that of the eye-nystagmus, a phenomenon which could also be demonstrated in pharmacological experiments with drugs delaying or accelerating the compensatory processes. This might be proof of the fact that the compensation of the eye-nystagmus is partly a function of the vestibular system itself. Recovery of the normal position of head and body is, on the other hand, more strongly dependent on cortical, cerebellar and somatosensory afferents. A similar phenomenon, that is, a dissociation of vestibular induced eye- and body-movements, has been observed in humans (see JUNG, 1959; KORNHUBER, 1966).

In this paper it is not possible to present all our relevant findings in detail. In spite of this we hope to have demonstrated that the process of compensation, which follows a labyrinthine lesion and the subsequent asymmetry of muscle tone, is not merely a function of the vestibular apparatus, but of the central nervous system as a whole. Other neurophysiological investigators have also suggested this conclusion, which should be taken into consideration when investigating the cellular and molecular basis of learning, retention and recall. We believe that the biological model for the investigation of compensation, described here, is well suited for investigations at the cellular and molecular levels as has already been demonstrated (HYDEN and EGYHAZI, 1963 and others).

From a clinical point of view the pharmacological experiments presented here suggest that it would be worth-

while to search for substances having, in even smaller
dosage, and with the least possible side-effects, a maximal accelerative effect on compensation. Investigations of
this kind can be of particular value in the clinical treatment of side-effects after labyrinthine lesions as well as
in the rehabilitation of perinatal and other traumatic brain
lesions.

LITERATURE

AKIL, R.: unpublished

ALTKÄMPFER, R.: Zur Wirkung von Phenothiazinen und ande-
    ren Psychopharmaka auf zentralnervöse Kompensations-
    vorgänge nach Labyrinthausschaltung.
    Diss. Göttingen 1966.

AZZENA, G.B.: Role of the spinal cord in compensating the
    effects of hemilabyrinthectomy.
    Arch. Ital. Biol. 107, 43-53 (1969).

BATINI, G., G. MORUZZI, and O. POMPEIANO: Cerebellar re-
    lease phenomena.
    Arch. Ital. Biol. 95, 71-95 (1957).

BECHTEREW, W. von: Ergebnisse der Durchschneidung des N.
    acusticus, nebst Erörterung der Bedeutung der semi-
    circularen Kanäle für das Gleichgewicht.
    Arch. ges. Phys. 30, 312-347 (1883).

DAL RI, H., and K.-P. SCHAEFER: Pharmakologische Untersu-
    chungen am artifiziellen Dauernystagmus des Meer-
    schweinchens.

Naunyn-Schmiedebergs Arch. exp. Path. Pharmak. 234,
79-90 (1956).

DAL RI, H., and K.-P. SCHAEFER: Beeinflussung des Nystag-
mus durch Stell- und Haltereflexe am nichtfixierten
Meerschweinchen.
Pflügers Arch. 265, 125-137 (1957).

DIETL, G.: Zur Wirkung von Luminal, Megaphen und Pervitin
auf kompensatorische Vorgänge nach Labyrinthausschaltung.
Diss. Göttingen 1966

DOW, R.S.: The effects of unilateral and bilateral labyrinthectomy in monkey, boboon and chimpanzee.
Am. J. Physiol. 121, 392-399 (1938).

DOW, R.S., and E. MANNI: The relationship of the cerebellum to extraocular movements.
In: The Oculomotor System (B. MORRIS and M.D. BENDER,
eds.), pp. 280-292, Harper and Row, New York, Evanston
and London (1964).

DOW, R.S., and G. MORUZZI: The physiology and pathology of
the cerebellum.
The University of Minnesota Press., Minneapolis (1958).

DUENSING, F., and K.-P. SCHAEFER: Die Aktivität einzelner
Neurone im Bereich der Vestibulariskerne mit Horizon-
talbeschleunigung unter besonderer Berücksichtigung
des vestibulären Nystagmus.
Arch. Psychiat. Nervenkr. 198, 225-252 (1958).

DUENSING, F., and K.-P. SCHAEFER: Die Aktivität einzelner
Neurone der Formatio reticularis des nicht gefessel-

ten Kaninchens bei Kopfwendungen und vestibulären
Reizen.
Arch. Psychiat. Nervenkr. 201, 97-122 (1960).

EYCK, M., van: Recherches sur le pheñomène de compensation
après labyrinthectomie.
Acta Oto-laryng., (Stockholm) 44, 456-461 (1954).

FISCHER, E.: The relative importance of central nervous
system, peripheral nerve and muscle cell in use, disuse and denervation phenomena of muscular function.
In: The Effect of Use and Disuse on Neuromuscular
Functions (E. GUTMANN and P. HNIK, eds.), pp. 359-
367, Elsevier, Amsterdam, London, New York (1963).

FLUUR, E.: Vestibular compensation after labyrinthine destruction.
Acta Oto-laryng., (Stockholm) 52, 367-375 (1960).

GERNANDT, B.E.: Response of mammalian vestibular neurons
to horizontal rotation and caloric stimulation.
J. Neurophysiol. 12, 173-184 (1949).

GERARD, R.W.: The material basis of memory.
Journal of verbal learning and verbal behavior 2,
22-33 (1963).

GIRETTI, M.L.: Spinal compensation of the cerebral release
phenomena.
Exp. Neurol. 30, 459-466 (1971).

GUTMANN, E.: Neurotrophic relations in the regeneration
process.
In: Mechanisms of Neural Regeneration (M. SINGER and
I.P. SCHADE, eds.), pp. 72-112, Elsevier, Amsterdam,

London, New York (1957).

HOLST, E. von: Die Gleichgewichtssinne der Fische.
Verh. dtsch. zool. Ges. 37, 108-114 (1935).

HOLST, E. von: Quantitative Untersuchungen über Umstim-
mungsvorgänge im Zentralnervensystem.
Z. vgl. Physiol. 31, 134-148 (1948).

HOLZBACH, E.: Kompensatorische Vorgänge nach Labyrinth-
ausschaltung unter besonderer Berücksichtigung der
Einflüsse von Großhirn und Kleinhirn.
Diss. Göttingen 1967.

HYDEN, H., and E. EGYHAZI: Nuclear RNA changes of nerve
cells during a learning experiment in rats.
In: The Effect of Use and Disuse on Neuromuscular
Functions (E. GUTMANN and P. HNIK, eds.), pp. 219-
228, Elsevier, Amsterdam, London, New York (1963).

JUNG, R.: Nystagmographie:Zur Physiologie und Pathologie
des optisch-vestibulären System beim Menschen.
In: Handbuch Innere Medizin V/1 (G.V. BERGMANN et
al., eds.), pp. 1325, Springer, Berlin, Göttingen,
Heidelberg (1953).

KÄLLEN, B.: Degeneration and regeneration in the vertebra
central nervous system during embryogenesis.
In: Degeneration Patterns in the Nervous System
(M. SINGER and J.P. SCHADE, eds.), pp. 77-96,
Elsevier, Amsterdam, London, New York (1965).

KOLB, E.: Untersuchungen über zentrale Kompensation und
Kompensationsbewegungen einseitig entstateter Frösche.
Z. vgl. Physiol. 37, 136-160 (1955).

230   K.-P. SCHAEFER AND D.L. MEYER

KORNHUBER, H.: Physiologie und Klinik des zentral-vesti-
    bulären Systems (Blick- und Stützmotorik).
    In: Hals-Nasen-Ohren-heilk., Bd. III, Teil 3,
    2151-2351, Stuttgart (1966).

KRAUS, U.: Kompensationsvorgänge nach Labyrinthausschal-
    tung beim Meerschweinchen und ihre Beeinflussung
    durch Cerebellektomie und Despinalisierung.
    Diss. Göttingen 1970.

LANGE, G., and H. KORNHUBER: Zur Bedeutung peripher- und
    zentralvestibulärer Störungen nach Kopftraumen.
    Arch. Ohr-, Nas- u. Kehlk.-Heilk. 179, 366-378 (1962).

LASSEK, A.M., and S.L. EMERY: Hidden neurologie mechanism
    in motor function restitution. Experiments with cor-
    tical lesions in monkeys.
    Neurol. 9, 107 (1959).

LORENTE DE NO, R.: Die Labyrinthreflexe auf die Augenmus-
    keln nach einseitiger Labyrinthexstirpation nebst
    einer kurzen Angabe über den Nervenmechanismus der
    vestibulären Augenbewegungen.
    Urban und Schwarzenberg, Wien (1928).

LOEWENSTEIN, O., and A. SAND: The individual and inte-
    grated activity of the semicircular canals of the
    elasmobranch labyrinth.
    J. Physiol. 99, 89-101 (1940).

LUCIANI, L.: Il Cerveletto. Le Monnier, Firenze (1891).

MAGNUS, R.: Körperstellung. Springer, Berlin (1924)

MATTHAEI, H.G.: Untersuchungen über Kompensationsvorgänge
    einseitiger und doppelseitiger, zweizeitiger Laby-

rinthausschaltung.

Diss. Göttingen 1966.

MANNI, E., and R.S. DOW: Some observations of the effect
of cerebellectomy in the rat.

J. comp. Neurol. 121, 189-194 (1963).

MENZIO, P.: Rapporti fra la corteccia cerebral ed i feno-
meni di emislabirintazione.

Arch. Fisiol. 49, 97-104 (1949).

MITTERMAIER, R.: Über die Ausgleichsvorgänge im Vestibu-
larapparat.

Z. Laryng. Rhinol. 29, 487-585 (1950).

PRECHT, W., and H. SHIMAZU: Functional connections of
tonic and kinetic vestibular neurons with primary
vestibular afferents.

J. Neurophysiol. 28, 1014-1028 (1965).

PRECHT, W., H. SHIMAZU, and C.H. MARKHAM: A mechanism of
central compensation of vestibular function following
hemilabyrinthectomy.

J. Neurophysiol. 29, 996-1010 (1966).

RADEMAKER, G.G.J.: Das Stehen: Statische Reaktionen.
Gleichgewichtsreaktionen und Muskeltonus unter besonderer Berücksichtigung ihres Verhaltens bei klein-
hirnlosen Tieren.

J. Springer, Berlin (1931).

SCHAEFER, K.-P., and H. WEHNER: Zur pharmakologischen Beeinflussung zentralnervöser Kompensationsvorgänge
nach einseitiger Labyrinthausschaltung durch Krampf-
gifte und andere erregende Substanzen.

Naunyn-Schmiedebergs Arch. Pharmak. u. exp. Path. 254, 1-17 (1966).

SCHOEN, L.: Quantitative Unterschungen über die zentrale Kompensation nach einseitiger Utriculusausschaltung bei Fischen.
Z. vgl. Physiol. 32, 121-150 (1950).

SHIMAZU, H., and W. PRECHT: Tonic and kinetic respones of cat's vestibular neurons to horizontal angular acceleration.
J. Neurophysiol. 28, 991-1013 (1965).

SPIEGEL, E.A., and T.D. DEMETRIADES: Die zentrale Kompensation des Labyrinthverlustes.
Pflügers Arch. ges. Physiol. 210, 215-222 (1925).

SPIEGEL, E.A., and G. SATO: Experimentalstudien am Nervensystem. V. Über den Erregungszustand der medullären Zentren nach doppelseitiger Labyrinthausschaltung.
Pflügers Arch. ges. Physiol. 215, 106-119 (1926).

STENGER, H.H.: "Erholungsnystagmus" nach einseitigem Vestibularisausfall, ein dem Bechterew-Nystagmus verwandter Vorgang.
Arch. Ohr-, Nas-, u. Kehlk.-Heilk. 175, 545-548 (1959).

Supported by a grant from "Deutsche Forschungsgemeinschaft" (SFB 33).

ACTION OF VARIOUS DRUGS ON THE FORMATION AND FIXATION OF LONG TERM INFORMATION IN THE GOLDFISH (CARASSIUS AURATUS)

H.P. Zippel, I. Gremmels, and H. Abdel Ghaffar

Physiologisches Institut
Lehrstuhl II der Universität
34 Göttingen, Humboldtallee 7, W-Germany

ABSTRACT

Pyrithioxine (40 mg/kg; 2 mg/kg) not only causes an increase in general activity, but also has a marked positive effect on the learning behavior of goldfish. Amphetamine (2 mg/kg) causes a similar enhancement of learning but has less pronounced effects on the overall activity. With Fencamphamine (2 mg/kg) there is no increase in activation, although some improvement in learning is manifested.

INTRODUCTION

A large number of experiments have been published in the literature describing the effects of stimulant drugs on general activity and learning in both animal and human subjects.

The present investigations were conducted using three drugs which have previously been reported to cause an increase in activity and an enhancement of learning; Pyrithioxine (Encephabol[R], Merck, Darmstadt, Germany), Amphetamine (Amphetamin sulfuricum = d,1 Benzidrinsulfat, Merck Darmstadt, Germany) and Fencamphamine (Fencamfamin - HCl = H610, Merck, Darmstadt, Germany). The majority of these previous experiments have involved not only the use of widely differing training techniques and drug dosages, but also a variety of experimental subjects (PYRITHIOXINE: HOTOVY et al., 1964: CAT, 150 mg/kg, p.os, general activity: decrease; RHESUS MONKEY, 40 mg/kg, p.os, aggressive behavior: no influence; RAT, 50 mg/kg, p.os, activity during training (conditioned avoidance): positive effect; MAN, 200 mg/kg, p.os, psychomotoric tests: positive effect. - A great number of results have been published by many authors almost exclusively on human patients: positive effects, e.g., on learning, general activity, motivation. AMPHETAMINE: BIANCHI and MARAZZI-UBERTI, 1969: MICE, 2 mg/kg, i.p., avoidance training: no effect; DOTY and DOTY, 1966: RAT, 2 mg/kg, i.p., avoidance behavior: positive effect; ILYUCHENOK, 1970: RAT, i.p. improvement of learning; KRIVANEK and MCGAUGH, 1969: MICE, 0.25, 0.5, 1, 1.5, 2, 2.5 mg/kg, i.p., Y-maze, shock-free: maximum effect (positive) 2.0 mg/kg; KULKARNI, 1968: RAT, 1 mg/kg, i.p., shock avoidance: increase of general activity and learning; LAL and BROWN, 1968: RAT, 0.3 mg/kg, i.p., shock-free: increase of activity; LAL, 1969: MICE, 2 mg/kg, i.p. shock avoidance: positive effect on learning. OWEN, 1963: RAT, 1.0, 2.0 mg/kg, s.c., shock avoidance: positive effect.

FENCAMPHAMINE: HOTOVY et al., 1961: MICE, RAT, CAT, 10 mg/
kg, p.os: increase of general activity; SOMMER and HOTOVY,
1961: RAT, 10 mg/kg, p.os: increase of general activity.
An increase of general activity in human patients has been
reported by several authors). Because of these differences
it is very difficult to make a ready comparison of the ef-
ficacy of the various drugs.

The experiments reported here, using a shock-free
training procedure, were initiated to investigate the effects of these three drugs in the same animal. The goldfish (Carassius auratus) was chosen as the experimental
subject for two reasons: firstly, the behavior of these
animals during shock-free color discrimination training
has already been intensively studied (ZIPPEL et al., in
prep.) and secondly, the results of the above investigations indicated that a color discrimination ability is
not easily acquired by goldfish and thus, any positive
effects of the drugs on the animals' behavior should be
more readily apparent.

                            METHODS

Goldfish 12-15 cm long were first adapted to laboratory conditions for at least 6 weeks. Groups of three fish
were then habituated for a further 4 weeks to individual
training tanks.

The experimental set-up (Fig. 1) was similar to that

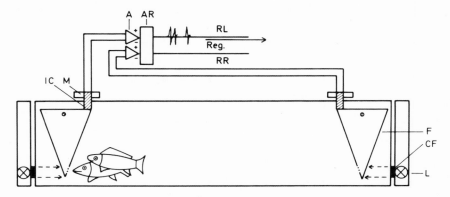

Fig. 1: <u>The training apparatus</u>

A = amplifier; AR = automatic recorder; CF = color
filters; F = feeding funnel; IC = induction coil,
L = white light source; M = magnet; Reg = registration from the left (RL) and right (RR) funnels.

used in previous investigations (ZIPPEL et al., in prep.).
Two feeding funnels are suspended in the water at opposite
ends of the tank, so that they are free to oscillate on
contact. During each contact made by a fish at the funnels,
an electric current is induced which is then amplified and
directly registered on a two-channel automatic recorder
(Varioscript 443, Schwarzer, Munich). At the tip of each
funnel are a series of holes through which the fish are
fed Tubifex worms. The color stimuli are presented through
frosted glass at each end of the tank, at the level of the
feeding areas. These stimuli are produced by means of a
white light source which passes through circular (Ø 40 mm)
gelatine color filters (Göttinger Farbfilter, yellow No.
2701; orange No. 255; red No. 114/1; green No. 682; blue-

green No. 442; blue No. 473).

Initially, the spontaneous behavior to the various
color stimuli is observed and the animals' general activity is assessed in terms of the number of reactions per
minute made to the feeding funnels. On the basis of these
observations the animals are divided into two groups, one
of which is classified as active (high general activity)
and the other as less active (low general activity). Each
of these groups is then further divided into control and
test animals. Subsequently all fish are trained to differentiate between a spontaneously disliked color quality
(positive stimulus) and a spontaneously preferred color
quality (negative stimulus).

The drugs are administered orally, in powder form, in
the center of small balls of minced meat, which are dropped
into the water and are immediately eaten by the fish. Control groups are fed balls of meat which do not contain the
drug. The animals are trained daily and the drugs are administered at varying intervals of time following training.
The behavior of the animals is evaluated in terms of biting
the funnel (food expectation) or swimming against it. The
total number of recorded contacts on the left plus right
sides is counted as 100 %.

RESULTS

Spontaneous Behavior

Prior to training, the spontaneous behavior in all groups of fish was in accordance with unpublished findings (ZIPPEL et al., in prep.). The natural responses to the various color stimuli can be arranged in hierarchical order (preference, positive signs; avoidance, negative signs) as follows: blue (+++), blue-green (++), red (+), green (-), orange (--) and yellow (---).

Effects of Pyrithioxine (40 mg/kg) on the General Activity

The effects of Pyrithioxine on the general activity was examined in 6 groups of animals. The results obtained from two of these groups are shown in Fig. 2A and B. In both instances, the level of overall activity, prior to the administration of the drug, was recorded on a number of occasions (2-5 times). The mean activity level was then estimated and this "control activity" (C, open histograms) was taken as 100 %. The level of activity one day after administration of Pyrithioxine (shaded histograms) is represented as the increase over the basic "control activity" (100 %). As can be seen, after the fish have received the drug, there is usually a marked increase in activity. In some cases, with repeated exposure to the drug, the increase is less pronounced (Fig. 2A), or the maximum effect

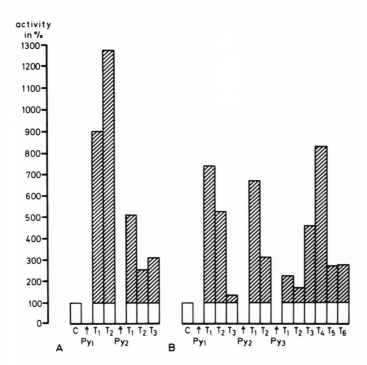

Fig. 2: <u>Effects of Pyrithioxine on the general activity in</u>
<u>two groups of animals (A and B)</u>

C = control activity prior to drug administration.
Py_{1-3} = administrations of Pyrithioxine.

is only realised after a latency period of 1-3 days (Fig.
2B). In all cases, following the maximum activity, there
is a gradual return to the control level as the effect of
the drug wears off.

However, the results shown here were obtained with
animals in which the basic activity was at an extremely
low level. In groups with a higher level of general activ-

ity the observed increases are not as great. And in animals in which the activity is already at a very high level,
the effects of the drug are not readily apparent.

Differentiation Training Following Administration of Pyrithioxine (40 mg/kg)

In this series of experiments, 5 groups of animals
were treated with Pyrithioxine and a further 5 groups were
used as controls. All animals were trained to differentiate
between yellow (positive) and blue (negative) color stimuli.
In Fig. 3A, a comparison between one of these groups and its
control is represented. As can be seen from the test session prior to training (Fig. 3A), the general activity is
at a very low level. As one would expect from the results
described above, daily administration of the drug causes
a distinct increase in activity, with the maximum effects
being manifested during the acquisition period. The control
group, on the other hand, shows very little change in behavior. In addition, the treated animals reach a very high
level of correct responses (80-100 %) after only 5 days of
training (Fig. 3B), and even during these early stages of
testing the learning curve is remarkably stable. Conversely, the learning curve for the control group shows a considerable amount of fluctuation and the animals, even when
trained over a long period of time (40-50 days), do not
reach the level attained by the treated group.

Fig. 4 shows the reults obtained from 4 groups of ani-

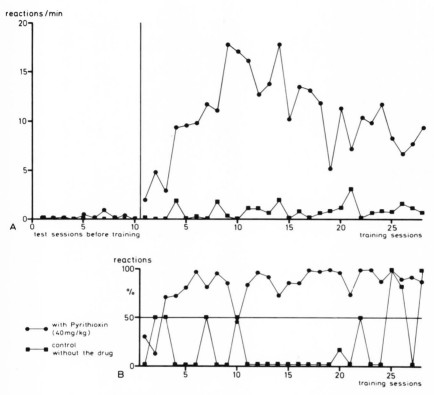

Fig. 3: General activity (A) and differentiation training
        (B) in two groups of inactive animals

mals. Two of these groups were given Pyrithioxine during
the first 20 training sessions and then the final 18 sessions were conducted without drug administration (Fig. 4A).
In the other two groups the reverse procedure was applied
(Fig. 4B). The stabilisation effect of the drug, as described above, is again quite obvious. In animals which
receive the drug during the inital part of the testing

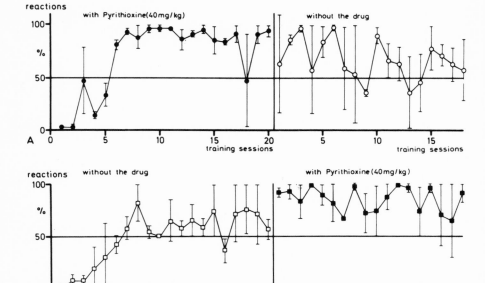

Fig. 4: <u>Differentiation training in 4 groups of animals
with or without administration of Pyrithioxine</u>

Each of the symbols represents the mean value obtained from two groups; the range is shown by the
vertical lines.

period (Fig. 4A), not only is the learning curve at a very
constant level, but the range of responses is also very
small. This effect is also apparent in animals treated
during the latter part of the testing period (Fig. 4B),
but to a lesser degree. The results obtained with untreated animals can be interpreted in a similar manner.

A statistical analysis of the results (Tab. I), using
the Wilcoxon test, was made for both treated and control
animals over the whole of the experimental period (20 days

Tab. I: Statistical evaluation of the results obtained from 5 Pyrithioxine-treated groups and 5 control groups, by the Wilcoxon test

5 groups (plus controls) received 20 training sessions, 3 of these groups (plus controls) received a further 30 sessions.

Training Sessions	1 - 5	6 - 10	11 - 15	16 - 20	21 - 25	26 - 30	31 - 35	36 - 40	41 - 45	46 - 50
Pyrithioxine (40 mg/kg) p - values	> 0.075	< 0.00005	< 0.00005	< 0.0025	< 0.00005	< 0.00005	< 0.00005	< 0.00005	< 0.00005	< 0.00005
Number of tests =	25	25	25	25	15	15	15	15	15	15
Control animals p - values	> 0.075	< 0.010	< 0.025	< 0.00005	< 0.0005	< 0.00005	< 0.0005	< 0.0005	< 0.005	< 0.005
Number of tests =	25	25	25	25	15	15	15	15	15	15

for all 5 treated groups, wit!: _ontrols, and a further 30
days for 3 of these treated g oups, with controls). As can
be seen, the preference for the training stimulus is significantly greater in animals which have received the drug
than in controls.

Differentiation Training Following Administration of
Amphetamine (2 mg/kg) and Second Order Differentiation
Following Administration of Pyrithioxine (2 mg/kg)

In these experiments 6 groups of animals were treated
with the drugs and 6 groups were used as controls. 3 of the
experimental groups, and their controls, were acclimated
to the laboratory conditions in the usual manner (see methods), whilst the remaining 3 groups (plus controls) received no period of habituation. In the initial series of experiments, the 6 test groups, given daily dosages of Amphetamine, were trained to differentiate between yellow (positive) and blue (negative) color stimuli. Controls were
trained in the same manner without receiving the drug.

The results obtained are shown in Figs. 5 and 6. As
can be seen from Fig. 5, in the non-adapted animals, the
learning curves for both treated and control groups are
not essentially different. In fact there is very little
evidence of learning with or without the drug.

By complete contrast, fully adapted animals show
significant learning effects following administration of
the drug (Fig. 6); these effects are absent in the control

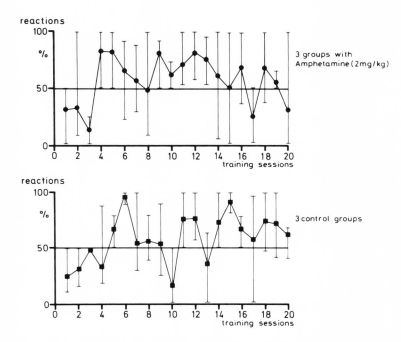

Fig. 5: Differentiation training in non-acclimated ani-
         mals with or without administration of Amphetamine
         For details see legend to Fig. 4.

animals. The learning curves for the Amphetamine-treated
animals are directly comparable with those obtained from
Pyrithioxine-treated fish in the previous experiment (see
Fig. 4). The stabilisation effect, mentioned earlier, can
also clearly be seen. The only difference arising from the
two treatments is in the level of general activity shown
by the animals: the activity being at a much higher level
in those animals which were given Pyrithioxine. The difference in the dosage levels of the two drugs is not per

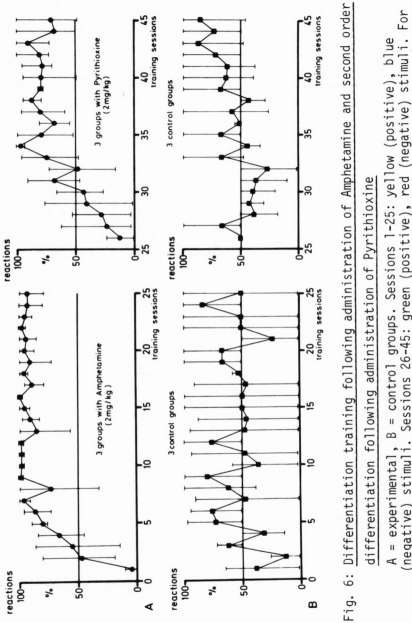

Fig. 6: Differentiation training following administration of Amphetamine and second order differentiation following administration of Pyrithioxine

A = experimental, B = control groups. Sessions 1-25: yellow (positive), blue (negative) stimuli. Sessions 26-45: green (positive), red (negative) stimuli. For details see legend to Fig. 4.

se a sufficient explanation of the observed results (see
below).

After 25 training sessions, the spontaneous behavior
of the fully acclimated animals to red and green lights
was tested, without either drug treatment or reinforcement.
In accordance with the results of previous investigations
(ZIPPEL and DOMAGK), there is a preference for red and a
dislike for green. This preference for red was most marked
in those animals which had previously shown a consistently
high level of stable behavior to the yellow stimulus.

Subsequently these animals were trained, using green
light as the positive stimulus and red light as the negative, in a further 20 training sessions (Fig. 6). The experimental animals which had previously received Amphetamine were now administered daily with the same dosage of
Pyrithioxine. The control animals were untreated as before.
As one would expect from the results of the spontaneous behavior, the learning curves of the treated animals start
initially at a very low level, but after only 10 trials a
significant learning effect is manifested. By comparison,
the control curve starts at a higher level, but no real
evidence of learning is shown before the 15th training session.

The stabilising effect of Pyrithioxine on the learned
behavior is very similar to that found earlier when the
drug was given after an initial untreated training period
(see Fig. 4B). Furthermore, Pyrithioxine causes a much more
pronounced increase in the overall activity than Amphet-

Tab. IIA: Statistical evaluation of the results obtained with adapted animals, by the Wilcoxon test

The treated animals were initially administered with Amphetamine (training sessions 1-25) and then later with Pyrithioxine (sessions 26-45).

Training Sessions	1 - 5	6 - 10	11 - 15	16 - 20	21 - 25
Amphetamine (2 mg/kg) (3 groups) p - values	> 0.075	< 0.0005	< 0.0025	< 0.0005	< 0.0005
Number of tests =	15	15	15	15	15
Control animals (3 groups) p - values	> 0.075	< 0.0025	< 0.025	< 0.050	< 0.025
Number of tests =	15	15	15	15	15

Training Sessions	26 - 30	31 - 35	36 - 40	41 - 45
Pyrithioxine (2 mg/kg) (3 groups) p - values	< 0.015	< 0.015	> 0.075	< 0.0005
Number of tests =	15	15	15	15
Control animals (3 groups) p - values	> 0.075	> 0.075	> 0.074	> 0.075
Number of tests =	15	15	15	15

Tab. IIB: Statistical evaluation of the results obtained
          with non-adapted animals, by the Wilcoxon test

Training sessions	1 - 5	6 - 10	11 - 15	16 - 20
Amphetamine (2 mg/kg) (3 groups) p - values	> 0.075	> 0.075	> 0.075	> 0.075
Number of tests =	15	15	15	15
Control Animals (3 groups) p - values	> 0.075	> 0.075	< 0.025	< 0.005
Number of tests =	15	15	15	15

amine.

A statistical evaluation of the above data by the
Wilcoxon test is presented in Tab. IIA and B.

Differentiation Training Following Administration
of Fencamphamine (2 mg/kg)

6 groups of animals were again used in this series of
experiments, 3 groups being treated with the drug and the
remaining 3 groups being used as controls. The animals were
trained to differentiate between green (positive) and red
(negative) color stimuli and the results obtained are shown
in Fig. 7. In two of the groups the behavior is practically

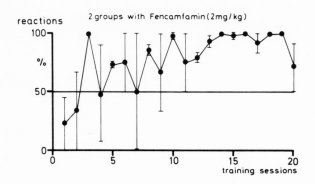

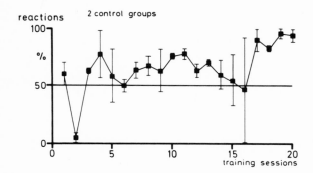

Fig. 7: <u>Differentiation training following administration</u>
<u>of Fencamphamine</u>
For details see legend to Fig. 4.

identical to that obtained with Amphetamine (see Fig. 6).
In the third group, however, the drug has essentially no
effect, but these animals were again not fully adapted to
the laboratory conditions.

Unlike Amphetamine and Pyrithioxine, Fencamphamine, in
this low dosage, has no effect at all on the general activity.

Tab. III: <u>Statistical evaluation of the results obtained</u>
          <u>from 3 Fencamphamin-treated groups and 3 control</u>
          <u>groups, by the Wilcoxon test</u>

Training sessions	1 - 5	6 - 10	11 - 15	16 - 20
Fencamphamin (2mg/kg) p - values	> 0.075	< 0.025	< 0.005	< 0.0005
Number of tests =	10	10	10	10
Control animals p - values	> 0.075	> 0.075	> 0.075	< 0.005
Number of tests =	10	10	10	10

Tab. III gives a statistical summary of the results
(Wilcoxon test).

DISCUSSION

The above results show that the three drugs have different effects on the general activation and the learning
ability of the goldfish.

Fencamphamine, unlike Amphetamine and Pyrithioxine,
has no observable effect on the general actitivity in the
low dosage (2 mg/kg) used in these experiments. The maximum
increase in activity was obtained with high dosages of Pyrithioxine (40 mg/kg), although, even in the low dosage (2 mg/
kg), this drug still produced a marked increase in overall

activity. Amphetamine was only administered in a low dosage (2 mg/kg) and although the level of activity showed some increase, the effect was much less marked than with the same dosage of Pyrithioxine. However, it should be pointed out that the effects of all the drugs on the general activity depend to a considerable extent on the normal level of activity manifested by the animals prior to drug administration. Thus, in animals which show a low basic activity the resulting increments caused by the drugs are far more pronounced than in animals where the basic activity is already at a high level.

In contrast to their effects on the general activity, all three drugs were found to have a positive effect on the learning behavior of the animals. Furthermore, the different concentrations used did not appear to have such a substantial effect on the learning ability, as they did on activity levels. However, the period of adaptation to which the animals are subjected prior to training appears to be very important. This point is clearly shown in the results obtained with adapted and non-adapted animals following administration of Amphetamine. Similarly, it is possible that the effects observed with Fencamphamine would have been more pronounced in better adapted animals.

ACKNOWLEDGEMENTS

This manuscript has been carefully revised by A. BURT, DAAD fellow.

LITERATURE

BIANCHI, C., and E. MARAZZI-UBERTI: Acquisition and re-
    tention of a conditioned avoidance response in mice
    as influenced by pemoline, by some of its derivatives
    and by some CNS stimulants.
    Psychopharmacologia 15, No. 1, 9-18 (1969).

DOTY, B.A., and L.A. DOTY: Facilitative effects of amphet-
    amine on avoidance conditioning in relation to age
    and problem difficulty.
    Psychopharmacologia 9, 3, 234-241 (1966).

HOTOVY, R., H.J. ENENKEL, J. GILLISSEN, U. JAHN, H.-G.
    KRAFT, H. MÜLLER-CALGAN, P. MÜRMANN, S. SOMMER, and
    R. STRULLER: Zur Pharmakologie des Vitamin B_6 und sei-
    ner Derivate.
    Arzneimittel-Forsch. (Drug-Res.) 14, 26-29 (1964).

HOTOVY, R., H.J. ENENKEL, J. GILLISSEN, A. HOFFMANN, U.
    JAHN, H.-G. KRAFT, H. MÜLLER-CALGAN, and R. STRULLER:
    Pharmakologische Eigenschaften des 2-Äthylamino-3-
    phenyl-norcamphan.
    Arzneimittel-Forsch. 11, 20-24 (1961).

ILYUCHENOK, R.Y.: The action of drugs on memory and learning.
    Farmakol. Toksikol. (Novosibirsk, USSR) 33, No. 2,
    237-246 (1970).

KRIVANEK, J.A., and J.L. MCGAUCH: Facilitating effects of
    pre- and posttrial amphetamine administration on dis-
    crimination learning in mice.
    Agents and Actions 1, 36-42 (1969).

KULKARNI, A.S.: Facilitation of instrumental avoidance
    learning by amphetamine: an analysis.
    Psychopharmacologia 13, 418-125 (1968).

LAL, H.: Control of learned conditioned-avoidance responses
    (CAR) by amphetamine and chlorpromazine.
    Psychopharmacologia 14, 33-37 (1969).

LAL, H., and R.M. BROWN: Interaction of amphetamine with
    social and other environmental stimuli in increasing
    inter-trial operant responding.
    Proc. Int. Union Physiol. Sci. XXIV Congress, Washing-
    ton, Vol. VII (1968).

SOMMER, S., and R. HOTOVY: Zur Differenzierung der zentral-
    erregenden Wirkung von 2-Äthylamino-3-phenyl-norcamphan.
    Arzneimittel-Forsch. 11, 969-972 (1961).

ZIPPEL, H.P., and G.F. DOMAGK: Spontaneous behavior and co-
    lor differentiation in the goldfish (Carassius aura-
    tus) after a shock-free training procedure.
    J. biol. psychol. 12, 2, 3-7 (1971).

ZIPPEL, H.P., A. BURT, R. SCHÄFER, I. GREMMELS and H.
    ABDEL GHAFFAR: Spontanverhalten und straffreie Dres-
    sur beim Goldfisch (Carassius auratus) bei verschie-
    denen Farbreizen.
    in prep.

COMMUNICATION BETWEEN NERVES AND MUSCLES: POSTNATAL DEVELOPMENT IN KITTEN HINDLIMB FAST AND SLOW TWITCH MUSCLE

R.A. Westerman[*], D.M. Lewis, J. Bagust[**],

G.D. Edjtehadi and D. Pallot[***]

Department of Physiology
University of Bristol
Bristol BS8 1TD, England

ABSTRACT

Some properties of hindlimb fast- and slow-twitch muscles (F.H.L.[+] and Soleus) and their innervation are described for two age groups of kittens (2 and 6 weeks). The

---

[+] The long "flexor" on the toes has two heads in the cat and separate tendons fuse in the foot before dividing to serve all digits. The medial head has been called "flexor hallucis longus" by some workers (BULLER et al., 1960). This nomenclature has been retained in this paper although some other workers have used the anatomically more correct name "flexor digitorum longus" (OLSON and SWETT, 1966).

[*]Current address: Department of Physiology, Monash University, Victoria, Australia.
[**]Supported by the Muscle Dystrophy Group of Great Britain.
[***]Supported by the Science Research Council.

length of single muscle fibres and their sarcomeres at
three ages (1, 2 and 6 weeks) are given, together with
the diameters of the muscle fibres at these ages. Isometric length - tension characteristics of the whole muscle
have been normalised in terms of individual muscle fibre
length. Isometric tetanic tension of at least 50 single
motor units from both muscles at each age are given, and
their mean tension relative to that of the whole muscle
was found to be approximately twice as large as that in
the adult. Total nerve axon counts and alpha motor axon
counts are presented and show a progressive increase of
approximately 15 % from 2 to 6 weeks, and a further equal
increment thereafter. Evidence for the presence of a significant degree of polyneuronal innervation of these twitch
muscles in the young kitten is presented and the functional implications are discussed.

  Skeletal muscles of the cat have been divided into
two groups, fast twitch and slow twitch, on the basis of
their isometric characteristics (DENNY-BROWN, 1929; BULLER
et al., 1960a) but the functional motor units in any one
adult muscle, defined as a large motor nerve cell and the
group of muscle fibres innervated by its axon, vary in their
speed of isometric contraction and in the tension that they
develop in tetanic contraction (WUERKER et al., 1965). Experiments in which the nerves of a fast and of a slow twitch
muscle are crossed surgically and allowed to regenerate
have shown that these characteristics are at least partly
dependent on the specific type of nerve fibre innervating
the muscle (BULLER et al., 1960a; 1960b).

In the newborn kitten the hindlimb muscles do not show this clear difference in contraction times and BULLER et al. (1960a) have demonstrated that a muscle such as flexor hallucis longus (F.H.L.) which has a fast twitch in the adult is slow contracting at birth and progressively speeds in the first 8 weeks of postnatal life. During this period of differentiation the muscle size and tension also increase, although the differences between the alpha motor nerve fibres to muscles destined to become fast and slow twitch are complete at birth (RIDGE, 1967). Furthermore it was shown by intracellular recording (ECCLES et al., 1963) that in kittens 3 weeks of age the adult pattern of moto-neurone monosynaptic innervation was established, and that antidromic spikes were generally similar to those in adult motoneurones.

This paper reports changes in the muscle fibres and their innervation during the first six weeks of postnatal development in one fast (F.H.L.) and one slow (soleus) twitch muscle in the hindlimb of the kitten and evidence for polyneuronal innervation of these twitch muscles in the young kitten is presented. It also describes the iso-metric properties of single motor units in these two mus-cles, the lengths of teased single muscle fibres and their sarcomere counts, and the fibre diameter spectra of these muscles and their nerves.

In order to simplify analysis kittens were selected to fall into groups by age: 2 days and 1, 2 and 6 weeks. In the last two groups a range of $\pm$ 2 and $\pm$ 4 days re-

spectively was allowed, and motor units were only studied
in these. The adult cats weighed between 1.75-2.9 kg. The
animals were anaesthetised with pentobarbitone sodium intraperitoneally, supplemented intravenously when necessary.
One hindlimb was denervated except for either the nerve to
F.H.L. or the nerve to soleus. A lumbar laminectomy allowed
ventral roots segments lumbar 5 - sacral 2 to be cut centrally and these were split for isolation of single motor
units (WUERKER et al., 1965). A diagram of the recording
arrangement and sample records illustrating the method of
identifying single motor units is given in Fig. 1. Muscles
were maintained at $37.5^{0}C$ and tension recorded isometrical-
ly (CLOSE and HOH, 1967; LEWIS et al., 1972). The output
of the isometric dynamometer (Ether UF2) was amplified and
analysed by a digital computer (Modular I, C.T.L.) which
also controlled stimulation and length changes of the muscle.

Muscle fibres were dissected from whole muscles which
had been fixed in situ by perfusion while immobilised at
the optimum length for tetani. The subsequent fixation,
maceration and teasing is described by AL-AMOOD and POPE
(1972). The number of sarcomeres in each fibre was counted
and the length of each of at least 8 fibres per muscle was
measured.

For both Soleus and F.H.L. there is a progressive increase in the length of the muscle fibres at each age examined from 1 week to adult. By contrast, there are no
consistent differences in the lengths of sarcomeres in

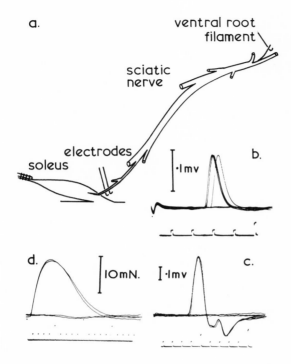

Fig. 1: The recording arrangement and criteria for motor
unit identification

a. shows the muscle tendon cut and attached to a
dynamometer; the motor nerve was stimulated to
elicit isometric twitches and tetani. In all animals single motor units were isolated by splitting
ventral roots (a) and were identified as single by
the all - or - none behaviour of the antidromic
action potential (b) and confirmed by the muscle
electromyogram (c) and twitch (d). (From J. BAGUST
1971, with thanks).

either muscle at the various ages, but there is a small
trend towards a decrease in sarcomere length with increas-

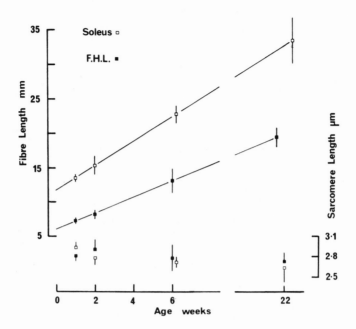

Fig. 2: <u>Muscle fibre length and sarcomere length at various</u>
<u>ages</u>

Open squares - soleus, filled squares - F.H.L.;
abscissa - age of animals (weeks), ordinate: left -
fibre length (mm), right - sarcomere length (µm).
With the exception of the 1 week data obtained
from one animal each point is the mean of at least
24 fibres for each muscle from at least 3 animals
and vertical bars on each point show (+ and or -)
standard deviation.

ing postnatal age which is not statistically significant
with the numbers of animals tested at each age.

The length - tension curves from a 2 day kitten (Fig.
3A) show that the change in length (L) on either side of

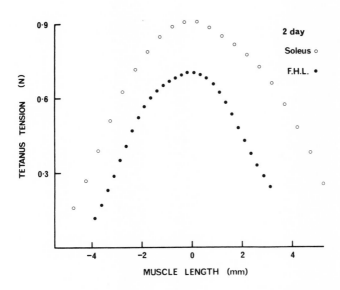

Fig. 3A: <u>Length - tension curves obtained from the F.H.L.</u>
<u>and soleus of a 2-day old kitten</u>
A tetanus was elicited at short length, and during lengthening by equal increments; one tetanus
was measured at each length after allowing a period of 60 sec during which twitches were elicited
at 1/15 sec.
The whole of this sequence was controlled by the
computer and so standardised. Ordinate - tetanic
tension (Newtons). Abscissa - change of muscle
length (mm).

the tetanus optimum (Lo) which reduces tetanic tension (P)
by 50 % relative to the maximum (Po) is only $\pm$ 0.25 mm
greater for soleus than for F.H.L. This is in contrast
with the findings of BULLER and LEWIS (1963) using older
animals, but may be explained by reference to muscle fibre
length. Thus it is of interest to compare animals of different ages and also fast and slow muscles in the same

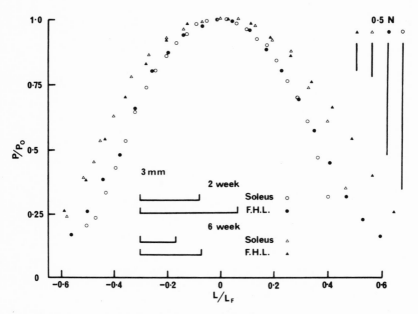

Fig. 3B: <u>Normalised length - tension curves</u>

In these curves tetanic tension (P) is expressed
as a fraction of that at muscle optimum (Po) on
the ordinate and plotted against the change of
muscle length (L) relative to the optimum length
for maximum tetanic tension (Lo). The horizontal
calibration bars show 3 mm scaled by the fibre
length of 2 and 6 week F.H.L. and soleus. The
vertical ones show 0.5N also scaled.

animal and to do this such curves are replotted with change
of muscle length expressed as a fraction of mean fibre
length calculated from comparable muscles (Fig. 3B). The
resulting curve for fast muscle closely corresponds to that
for slow muscle in any one animal, and both fit with less
than 15 % overall variation between the 2 week, 6 week and
adult animals. This is consistent with the observed differ-

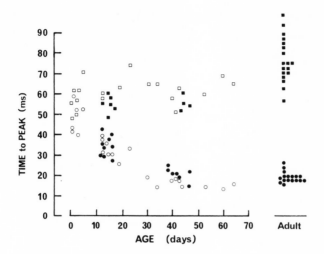

Fig. 4: <u>Isometric twitch times to peak for whole muscles</u>

Ordinate - time to peak (msec); abscissa - age of animal (days); squares - Soleus, circles - F.H.L.

Solid black symbols are values from the three age groups in the present experiments; open symbols represent data kindly supplied by BULLER and LEWIS (1965) for comparison.

ences of up to 18 % of fibre length between animals of the same age. The data of adult fibre and sarcomere lengths were taken from AL-AMOOD and POPE (1972).

Fig. 4 shows the times to peak for both the F.H.L. and soleus muscles at various ages (includes data of BULLER and LEWIS, 1965b) and emphasizes the considerable speeding of the F.H.L. isometric twitch during differentiation from birth to adult described by BULLER et al. (1960a).

Differences between the fast and slow twitch muscles are
obvious at 2 weeks and the contraction time of the F.H.L.
isometric twitch at 6 weeks approaches that for the adult.
Soleus contraction times are slowing during this time to
approach the adult value.

In order to compare motor units from muscles widely
differing in size (cf., BULLER and LEWIS, 1965a,b) unit
maximum tensions (at tetanus optimum) have been expressed
as a percentage of whole muscle tetanic tension (Tab. I,
II). The mean motor unit tensions were significantly different in each group of kittens for F.H.L. (P < 0.002) and
soleus (P < 0.01) from those in adults (Tab. I, II).

The histograms in Fig. 5 demonstrate the distribution
of tetanic tension of F.H.L. motor units, which is not immediately seen in Tab. I and II. The distribution of motor
unit tensions was significantly different in each group of
kittens for F.H.L. (P < 0.002) from that in adults. Some
of the kitten motor units had tensions either considerably
smaller (0.027 %) or larger (2.4 %) than those found in
adult F.H.L. (Fig. 5). Similarly the largest of the 2 week
soleus motor units had a relative tetanic tension twice
that of the largest unit found in adult soleus (BAGUST et
al., 1972a). These measurements of isometric contractions of
motor units in cats at various ages indicate that the mean
number of muscle fibres innervated by one motor axon is
significantly greater in kitten skeletal muscle than that
of the adult cat (BAGUST et al., 1972b). One possible explanation of this is that the number of functioning motor

Tab. I: Some characteristics of motor units of cat F.H.L. at various ages

Age of Cat	12 - 16 days	38 - 45 days	Adult
Motor units			
Number	47	53	68
Tension (%Po)	1.19 (0.005 - 4.3)	1.05 (0.022 - 4.2)	0.65 (0.027 - 2.4)
Whole muscle			
Number	10	6	15
Tension (N)	1.19 ($\pm$ 0.13)	3.79 ($\pm$ 0.89)	14.2 ($\pm$ 1.6)

Tab. II: Some characteristics of motor units of cat Soleus at various ages

Age of Cat	12 - 16 days	38 - 45 days	Adult**
Motor units			
Number	55	52	100
Tension (%Po)	1.39 ($\pm$ 0.62)	0.85 ($\pm$ 0.30)	0.69 ($\pm$ 0.27)
Whole muscle			
Number	6	4	17
Tension (N)	1.5 ($\pm$ 0.3)	4.5 ($\pm$ 0.5)	14.5 ($\pm$ 3.3)

Tetanic tensions of motor units are expressed as a percentage of whole muscle tension (Po).
The mean and (in brackets) range are shown for motor units in Tab. I since the distribution was asymmetrical; means ($\pm$ standard deviation) are quoted for the data from whole muscles, and for motor units in Tab. II.
**The adult soleus motor unit data was taken from BAGUST, 1971.

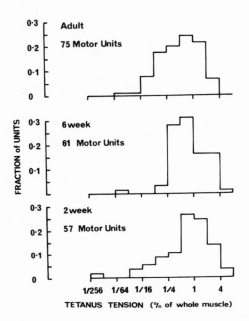

Fig. 5: <u>Times to peak and tetanic tension for F.H.L. motor</u>
<u>units at various ages</u>
Ordinate - number of units (n) expressed as a
fraction of the total population (n - inset at top
of the histogram age group); abscissa - tetanus
tension (% of whole muscle tetanus tension) on a
logarithmic scale.

axons increases during development. There may also be
changes in the number of muscle fibres or their contractile
properties during early postnatal development. As an initial investigation of these we have measured and counted
the nerve and muscle fibres of F.H.L. and Soleus. Muscles

fixed in 3 % glutaraldehyde were embedded in paraffin,
sectioned at 15 μm and stained with phosphotungstic acid.
In four representative regions of each mid-section approximately 10 % of the total number of muscle fibres were
measured and counted using the Zeiss TGZ3 particle size
analyser (NYSTROM, 1968a) and from the cross sectional
area of the muscle section total fibre counts were calculated.

The photomicrograph sections in Fig. 6B and the data
in Fig. 6A and Fig. 7 show the progressive increase in muscle fibre diameter with age and the consistent tendency
for the soleus fibres to be larger and less variable in
diameter than F.H.L. The variance ratios between pairs
of muscles at each age range from 1.35 to 1.52 and are
significantly higher for F.H.L. in every case (P < 0.01).
The larger mean diameter and greater homogeneity of soleus
fibres is in good agreement with the findings of SISSONS
(1963) in man, and GUTMANN and ZELENA (1962) in rat. The
total muscle fibre count shows no consistent variation
with age, but in each instance except the 6 week group
the total fibre count is greater in Soleus than in the
corresponding F.H.L. In his study of innervation ratios
of motor units CLARK (1931) divided the muscle into three
portions, sectioned these at their centre and added the
three counts in order to correct for the soleus fibre
length being only one third of muscle length. He obtained
a mean total count of 26,600 muscle fibres in the solei
from 6 adult animals. Because the muscle fibre length remains a constant fraction of the muscle length at all ages

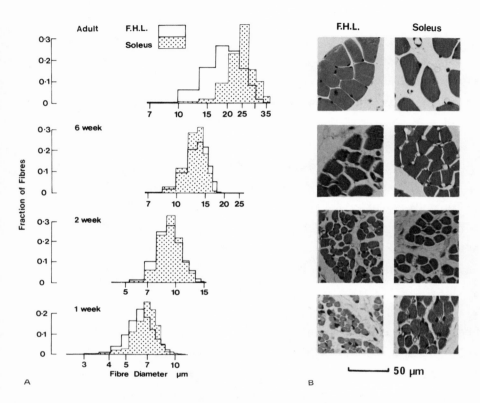

Fig. 6: <u>Total muscle fibres in F.H.L. and soleus at various</u>
<u>ages</u>

Fig. 6A: Histogram showing the distribution of
fibre diameters obtained from pairs of muscles of
kittens aged 1 week, 2 weeks, 6 weeks and adult.
Ordinate - number of fibres in each class as a
fraction of the total; abscissa - fibre diameter
(μm) on a logarithmic scale.

Fig. 6B: Ordinate - age (weeks). Representative
sections of F.H.L. and soleus from animals aged
1,2,6 weeks and adult, from below upwards.

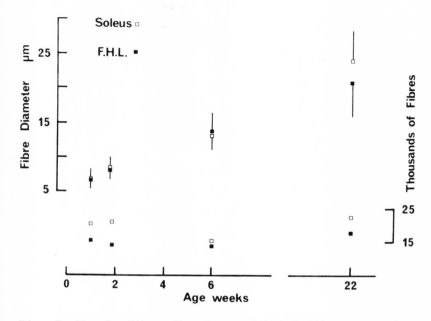

Fig. 7: <u>Muscle fibre diameter and total fibre count in</u>
<u>F.H.L. and soleus at various ages</u>
Ordinate: left - fibre diameter (µm), right -
fibre count (thousands) at middle of muscle belly.
Abscissa - age (weeks).

(cf., Fig. 2) if our total counts based on a single trans-
verse section in the widest diameter of the muscle do not
change with age, then any correction which is applied may
reasonably be assumed to not alter the proportion of the
totals at the various ages. Our uncorrected mean total
count in adult soleus is 22,600 and the figure corrected
for comparison with CLARK's data assuming a uniform soleus
fibre length 50 % of muscle length (AL-AMOOD and POPE,
1972) is 28,200. Both these figures agree quite well with
that of CLARK (1931).

From these figures it may be calculated that the smallest motor unit in soleus contains at least 75 average sized fibres (or 300 small ones). The estimate of total fibre count in F.H.L. must be less certain in view of our technique but a reasonable estimate from fibre arrangement (AL-AMOOD and POPE, personal communication) would put it at between 1 and 1.2 times the maximum diameter count of 16,000, i.e., a maximum of 20,000. The smallest motor unit (0.005 % tension) then might only consist of one muscle fibre, but if it consisted of only the smallest fibres (one half the diameter, one quarter the cross-sectional area) the figure would rise to four fibres but still well below the corresponding value for soleus.

The motor nerve axons were counted and their diameters were measured by an indirect method (cf., BOYD and DAVEY, 1969) in two kittens de-efferented unilaterally over segments S2-L6 inclusive when aged 2 weeks and then killed and perfused when 43 days old, one side being retained as control. Both F.H.L. and soleus nerve axon diameters were measured by the method of NYSTROM (1968a), modified in that nerves were fixed for 2 hours in 3 % glutaraldehyde then post-fixed for 2 hours in 1 % osmium tetroxide, both fluids were buffered with cacodylate 0.1 M. Tissues were dehydrated and embedded in Epon and 0.1 µm (silver-gold) sections were examined and photographed with a Philips EM 300 electron microscope. Enlarged photographic prints and a Zeiss TGZ3 particle size analyser were employed for counting and measuring axon diameters (NYSTROM, 1968a). Total axon counts and diameter spectra were measured in

nerves to both F.H.L. and soleus in animals aged 1, 2 and
6 weeks for comparison with the kitten data of NYSTROM.

The electronmicrographic appearance of one de-efferented F.H.L. nerve is shown in Fig. 8C and the diameters
of the remaining afferent nerve axons (at right) are seen
to range between almost the smallest and largest found in
the nerve from the unoperated limb (at left). The lower
histogram in Fig. 8A shows the motor nerve axon spectrum
obtained by subtraction from the sections displayed in
part in Fig. 8C. The axon diameters corresponding to the
alpha and gamma ranges were selected arbitrarily but with
reference to the separation in conduction velocities measured on 109 F.H.L. motor units in one 2 week animal and 71
soleus units in another (Fig. 8B).

A comparison of the total axon counts for F.H.L. and
soleus nerves is provided in Tab. III. The total nerve
axon counts and mean α axon diameters for both nerves from
2 and 6 week kittens are in close agreement with the values
obtained by NYSTROM (1968a) at equivalent ages. The mean
total axon count for soleus nerves is 16 % less in the 2
week animal compared with ones of 6 weeks (P < 0.01), but
this is sufficient to account only in part for the decrease in motor unit size during this period of postnatal
development.

ECCLES and SHERRINGTON (1930) demonstrated that considerable branching of motor nerve axons occurs in adult
cat hindlimb at levels as distant as 60 mm from the muscle.

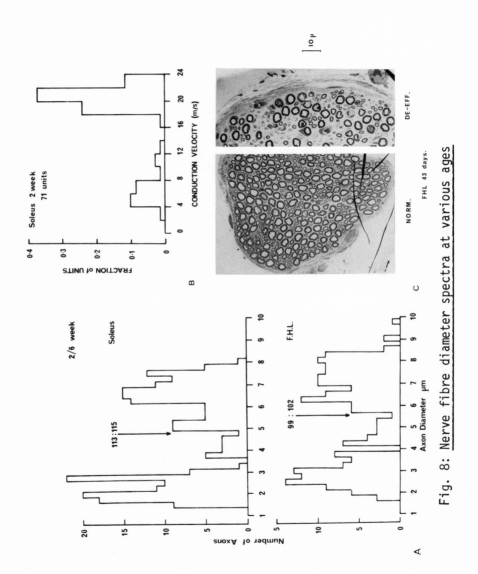

Fig. 8: Nerve fibre diameter spectra at various ages

Fig. 8A: Histogram of fibre diameters for soleus and F.H.L. nerves obtained by substraction of remaining afferent fibres on the operated side, from the total count at the same level on the control side. Arrows indicate the diameter which corresponds to the division between smaller (γ) fibres and large (α) fibres. The proportions $\alpha:\gamma$ and totals are shown for each muscle.

Fig. 8B: Histogram of conduction velocities of 71 axons isolated at random from nerve to soleus in a 2 week kitten. Ordinate - fraction of total units in each bin; abscissa - conduction velocity (m/sec).

Fig. 8C: Electronmicrographs of normal (left) and de-efferented (right) nerve to F.H.L. muscle aged 43 days (i.e., 29 days after ventral root section of segments S2 - L6 inclusive).

Tab. III: Count of myelinated axons in intact muscle nerves

Age	Axon Count				
	F.H.L.			Soleus	
1 week	Total	304	(1)	294	(1)
2 week	Total	329 ± 26	(6)	313 ± 23	(3)
6 week	Total	355 ± 10	(3)	374 ± 17	(2)
	α-motor	88 ± 11	(2)	109 ± 7	(2)
	γ-motor	72 ± 22	(2)	105 ± 8	(2)
*Adult	Total	478		422	
	α-motor	116		118	
	γ-motor	107		91	

Only myelinated axons have been included in these counts. Figures are means ± standard error of the mean; the number of cats is in parenthesis. *Adult figures are taken from page 39 of BOYD and DAVEY, 1966. The axon diameter spectra for soleus closely resemble that shown in the histograms of NYSTROM, 1968a, and this author gave no data for F.H.L.

In our preliminary nerve axon counts we have not examined
the degree or location of branching at different ages but
further experiments relating to this question are in progress.

Because polyneuronal innervation in kitten muscles
would provide an alternative explanation of the disparity
in motor unit size we used a method similar to that of
BROWN and MATTHEWS (1960) to investigate this possibility.
The ventral roots were prepared and split to provide two
subdivisions which on stimulation produced about equal
tensions in the muscles. The tension produced by tetanic
stimulation of the two roots simultaneously was compared
with the sum of the tensions from the two roots stimulated
individually. If polyneuronal innervation is present, then
the sum of the tension produced by stimulating the two
root fractions individually (added tension) compared with
the tension produced by simultaneous stimulation of the
two fractions (combined tension) should show an excess of
tension. Fig. 9 shows the records of one such experiment
in which the roots innervating the soleus muscle of a 2
week old animal were tested, and 20 % tension excess of
added tension relative to the tension from combined stimulation is seen.

Our data concerning adults indicates that there is
probably no polyneuronal innervation and is in accord with
other reports (BROWN and MATTHEWS, 1960; BULLER and LEWIS,
unpublished observations). It can be seen that in all animals in the 2 week group for both muscles tested there was

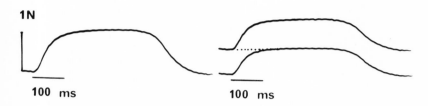

Fig. 9: Isometric tetanic contractions of 2 week old
        soleus from stimulation of two ventral root sub-
        divisions either simultaneously (at left) or sep-
        arately (at right). Ordinate - tension (Newtons);
        abscissa - time (calibration bars = 100 msec).

Tab. IV: Ratios of tension from vr subdivisions

Muscle	12-16 days	38-46 days	Adult*
FHL	117 % (1.0)	104 % (2.0)	100 % (2.6)
No. of animals	5	8	5
Soleus	121 % (3.2)	101 % (0.3)	99.9 % (0.7)
No. of animals	6	5	9

$$^{added}/_{combined} \quad i.e. \quad \frac{Tn \ (VR1) \ + \ Tn(VR2)}{Tn(VR1 \ + \ VR2)} \ \%$$

Figures are mean % ratios and the standard error of the
mean (+) is given in parenthesis. Three age groups of animals were used.
*The data for adults in part kindly supplied by AL-AMOOD
and POPE, unpublished observations.

such an excess of tension from the added fractions compared with the combined tension.

This has been assumed to be related to the degree of polyneuronal innervation which progressively changes towards the adult pattern over the first few weeks of postnatal development to reach the adult state in which there is no significant polyneuronal innervation (Tab. IV, above; see also BROWN and MATTHEWS, 1960) by 6 weeks or shortly thereafter. Further studies on the amount of polyneuronal innervation at ages less than 2 weeks and the time course of its disappearance are in progress.

Preliminary histological examination of F.H.L. and soleus muscles in the 2 week kitten has been carried out using Loewit's Gold Chloride technique (CARLETON and DRURY, 1957) in an attempt to corroborate the electrophysiological findings and an example is shown in Fig. 10. The results can only be described as suggesting the presence of multiple innervation in the soleus and F.H.L. muscle of the 2 week old kitten, but do not distinguish between polyneuronal innervation or two endplates from separate preterminal branches of the same axon.

In discussing the results presented in the paper we wish to reaffirm that,although their relevance to the study of memory is tenuous, they do suggest that some rather more extensive synaptic plasticity is occurring at the neuromuscular junctions of skeletal fast and slow twitch muscle in the young kitten than that which occurs

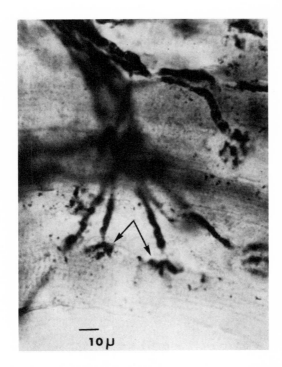

Fig. 10: Photomicrograph of a gold chloride preparation of
         F.H.L. from a 2 week old kitten. The end plates
         have just begun to differentiate from their more
         primitive disc-like stage seen in the newborn
         kitten (NYSTROM, 1968c; COERS and WOOLF, 1959).
         The arrows indicate two end plates apparently on
         the same muscle fibre.

in the adult (BARKER and IP, 1965). The teased muscle fibre
data on fibre length and sarcomere length of F.H.L. and
soleus at the ages studied indicate that during their postnatal development these muscle fibres increase in length
by the addition of standard-sized sarcomeres as has been
shown by CLOSE (1964) to occur in the rat. The results obtained from both whole muscle and motor units accord well

with many published reports concerning the differentiation
of fast muscles and slow muscle twitch characteristics
(BANU, 1922; DENNY-BROWN, 1929; BULLER et al., 1960; CLOSE,
1964; BULLER and LEWIS, 1965).

The alteration in both force-velocity characteristics
and active state properties of skeletal muscle during early postnatal development may be determined by a chemical
influence from their motor nerve (BULLER et al., 1960).
We have found that the isometric length tension characteristic of both fast and slow twitch muscle correspond
very well in kittens aged 1, 2 and 6 weeks with those of
adults (BULLER and KEAN, unpublished observations) if the
results are normalised with respect to fibre length. The
functional motor units of both types of twitch muscle in
young kitten show a striking disparity at the upper end
of the range of tensions they produce relative to that of
the whole muscle. This tension disparity suggests that the
motor axon to each such large motor unit is able to activate more muscle fibres than nerve axons supplying large
motor units in the adult (BAGUST et al., 1972c). Although
there is a progressive increase in both the total and the
α motor nerve axon count with age, the observed differences
in α motor count seem only adequate to partly explain the
amount of reduction which occurs in the mean motor unit
size over the postnatal period and this might anyway be
explained partly by differences in the rostro-caudal level
of axon branching.

The absence of significant change in the number of

muscle fibres during postnatal muscle histogenesis supports
previous reports denying any mitotic activity in normal
muscle after birth (RUBINSTEIN, 1960), and precludes any
explanation of the motor unit tension differences which
involves an actual change in the number of muscle fibres.

The results of comparing tetanic tension produced by
stimulation of ventral root subdivisions separately and
simultaneously strongly suggest that polyneuronal innervation is present to a significant extent in the 2 week
animals, but progressive reduction in the amount of polyneuronal innervation occurs over the next 2 - 3 weeks to
approach the adult state where no polyneuronal innervation
can be demonstrated by these techniques (BROWN and MATTHEWS,
1960). Other explanations including stimulus spread between root fractions or the ephaptic excitation of intra-
muscular nerve branches by the muscle action potential were
tested and largely excluded using the methods described by
BROWN and MATTHEWS. The possibility of some ephaptic excitation of preterminal nerve branches must be considered
because of the behaviour of developing neuromuscular synapses in tissue culture to electrical stimuli (ROBBINS
and YONEZAWA, 1971) and the 'immaturity' of end plates
found in newborn kitten by BULLER and LEWIS (1965b). When
testing with paired stimuli at different intervals no evidence for ephaptic transmission has yet been found in either the 2 or 6 week kitten.

There are a number of abnormal situations in which
mammalian skeletal muscle fibres receive more than one

endplate per fibre (WEISS and HOAG, 1946; GUTH, 1962; cf., MILEDI, 1960; MARK and MAROTTE, 1971) but amphibian and teleost skeletal muscle fibres do so normally (MILEDI, 1960; MARK and MAROTTE, 1971). We have no evidence concerning any specificity in the formation of connections in the kitten twitch muscles and at present it is simplest to assume that initial innervation is a random process. It is possible that the degree and type of usage of particular motor neurones with consequent alteration of their discharge pattern may modify the muscle fibres with which synaptic contact has been made (BULLER et al., 1960a; BURKE et al., 1971; SALMONS and VRBOVA, 1969) but a change in innervation ratio is required to explain the present motor unit findings. The fact that little or no polyneuronal innervation is found in the 6 week old kitten and none in adults implies that some neuromuscular junctions are lost or rendered non-functional during early postnatal development.

Although the mechanism by which such a change can be brought about is unknown, there is indirect evidence relevant to this question. First, innervation has obvious trophic effects on some sensory cells such as taste buds (TORREY, 1934), muscle spindles (HNIK and ZELENA, 1961) and on muscle (GUTH, 1968). In the latter case innervation restricts receptor sensitivity to acetylcholine in the region of the endplate (AXELLSON and THESLEFF, 1959; MILEDI and POTTER, 1971; cf., VRBOVA and WAREHAM, 1972). It also stops muscle fasciculation by some other means (since fasciculation is not blocked by curare, ROSEN-

BLUETH and LUCO, 1937) and has a trophic influence on muscle fibre size (GUTMANN, 1964). Secondly, a two-way movement of particles and vacuoles up and down living axons
has been described (LUBINSKA, 1964; WATSON, 1968) and
there is evidence that some substances (e.g., glutamate,
mitochondria) may pass up the axon towards the cell body
(KERKUT et al., 1967; WEISS and PILLAI, 1965). In addition
there is the recent evidence that protein and A.T.P. are
released together with the transmitter acetylcholine at
mammalian neuromuscular junctions (MUSICK and HUBBARD,
1972). Thus, mechanisms exist whereby nerves may transport
and release trophic substances. Such postnatal changes in
motor unit innervation as can be inferred from the present
results could be achieved by the repression of supernumary
endplates with some redistribution of territory between
functioning axons. The concept of repressor substances is
not new and the suggestion of CRAGG (1970) that neurones
produce a substance that represses RNA production and
loses some of this repressor when the axon is injured or
sprouts comes nearest to explaining all the experimental
findings concerning chromatolysis. A more relevant situation in which release of a repressor substance is suggested is in the teleost extraocular muscle where suppression of synaptic transmission occurs at previously functioning neuromuscular synapses following reinnervation by
the original nerve (MARK and MAROTTE, 1971). A very recent
report concerns neuromuscular junction in kittens where as
well as inducing and maintaining the high sensitivity of
the endplate to depolarizing drugs, the nerve by its activity can increase this sensitivity above a 'resting'

level (VRBOVA and WAREHAM, 1972). There is no evidence
from which to speculate further about the nature of the
information transfer between nerve endings and muscle
fibres.

Histological methods of demonstrating polyneuronal
innervation are notoriously uncertain, and although RED-
FERN (1970) obtained good evidence for multiple innervation in the young rat, he also could not with certainty
distinguish preterminal branching of the same axon from
polyneuronal innervation. It is unlikely that even cholinesterase histochemical techniques (COERS and WOOLF, 1959)
would demonstrate polyneuronal innervation with certainty.
In kittens NYSTROM (1968b,c) found no systematic variation
in the structure of the developing motor nerve terminals
on small nerve fibres and only 'en plaque' endings were
seen in both slow and fast twitch muscle. At 2 weeks the
motor nerve terminals had begun to differentiate from their
primitive disc-like stage and by 1 month a meshwork of
terminal twigs was beginning to form but no polyneuronal
innervation was observed.

Micro-electrode studies probably offer the best hope
of establishing without question the extent of polyneuronal
innervation in these twitch muscles of the young kitten.
Meanwhile, on the present evidence this degree of polyneuronal innervation alone explains in part the discrep-
ancies between motor unit size in kittens and adults together with the apparent increase in the number of nerve
fibres during this postnatal period probably accounts for

most of the quantitative differences in soleus and F.H.L.
motor units of the young kitten compared with those of the
adult.

One functional implication of such 'macromotor units'
in the young kitten is that motor control would be less
precise, but this would seem no serious disadvantage at
the stage when only crawling movements are required. It
may be advantageous if the pattern and quantitative specification of motor unit innervation is developed at the
time when skilled movements are being learned and motor
neurone firing patterns established.

ACKNOWLEDGEMENTS

Much of the work reported here was supported by a
grant from the Muscular Dystrophy Group of Great Britain
and the histological work by the Science Research Council.
The Modular I computer which was used in these studies was
purchased by the Medical Research Council. The EM 300 was
purchased by a grant from the Wellcome Trust to T.J. Biscoe.

LITERATURE

AL-AMOOD, W., and R. POPE: A comparison of the structural
        features of muscle fibres from a fast- and a slow-
        twitch muscle of the pelvic limb of the cat.
        J. Anat., in press (1972).

AXELLSON, J., and S. THESLEFF: A study of supersensitivity
    in denervated mammalian skeletal muscle.
    J. Physiol. (Lond.) 147, 178-193 (1959).

BAGUST, J.: Motor unit studies in cat and rabbit solei.
    Ph.D. Thesis, University of Bristol, Briston (1971).

BAGUST, J., D.M. LEWIS, and R.A. WESTERMAN: Development
    of motor units in a slow twitch muscle of the cat
    hindlimb.
    Proc. Regional Meeting Int. Union Physiol. Sciences,
    Sydney, Australia 21-25 August (1972a).

BAGUST, J., D.M. LEWIS, J.C. LUCK, and R.A. WESTERMAN:
    Development of motor units in a fast twitch muscle
    of the cat hindlimb.
    Communication to the Physiological Society, March
    (1972b).

BAGUST, J., D.M. LEWIS, D. PALLOT, and R.A. WESTERMAN:
    Polyneuronal innervation of twitch muscle in the
    kitten hindlimb.
    Communication to the Physiological Society, July
    (1972c).

BANU, G.: Recherches physiologiques sur le development
    neuromusculaire chez l'homme et l'animal.
    Paris: Maretheux (1922).

BARKER, D., and M.C. IP: The probable existence of a proc-
    ess of motor end-plate replacement.
    J. Physiol. (Lond.) 176, 11P (1965).

BOYD, I.A., and M.R. DAVEY: The composition of peripheral
    nerves.
    In: Control and Innervation of Skeletal Muscles (B.L.
    ANDREW), pp. 35-52, University of St. Andrews,
    Dundee (1966).

BROWN, M.C., and P.B.C. MATTHEWS: An investigation into
    the possible existence of polyneuronal innervation
    of individual skeletal muscle fibres in certain hind-
    limb muscles of the cat.
    J. Physiol. (Lond.) 151, 436-457 (1960).

BULLER, A.J., J.C. ECCLES, and R.M. ECCLES: Differentia-
    tion of fast and slow muscles in the cat hind-limb.
    J. Physiol. (Lond.) 150, 399-416 (1960a).

BULLER, A.J., J.C. ECCLES, and R.M. ECCLES: Interactions
    between motoneurones and muscles in respect of the
    characteristic speeds of their responses.
    J. Physiol. (Lond.) 150, 417-439 (1960b).

BULLER, A.J., and D.M. LEWIS: Factors affecting the dif-
    ferentiation of mammalian fast and slow muscle fibres.
    In: The Effect of Use and Disuse on Neuromuscular
    Functions (E. GUTMANN and P. HNIK, eds.), Czechoslovak
    Academy of Sciences, Prague (1963).

BULLER, A.J., and D.M. LEWIS: The rate of tension develop-
    ment in isometric tetanic contractions of mammalian
    fast and slow skeletal muscle.
    J. Physiol. 176, 337-354 (1965a).

BULLER, A.J., and D.M. LEWIS: Further observations on the
    differentiation of skeletal muscles in the kitten

hind-limb.

J. Physiol. (Lond.) 176, 355-370 (1965b).

BURKE, R.E., D.N. LEVINE, F.R. ZAJAC III, P. TSAIRIS, and
    W.K. ENGEL: Mammalian motor units: physico-histochemi-
    cal correlation in three types in cat gastrocnemius.
    Science 174, 709-712 (1971).

CARLETON, H.M., and R.A.B. DRURY: Histological technique,
    pp. 251-252, 3rd Ed. Oxford University Press, London
    (1957).

CLARK, D.A.: Muscle counts of motor units: a study in in-
    nervation ratios.
    Amer. J. Physiol. 96, 296-304 (1931).

CLOSE, R.: Dynamic properties of fast and slow skeletal
    muscles of the rat during development.
    J. Physiol. (Lond.) 173, 74-95 (1964).

CLOSE, R., and J.F.Y. HOH: Force: velocity properties of
    kitten muscles.
    J. Physiol. (Lond.) 192, 815-822 (1967).

COERS, C., and A.L. WOOLF: The innervation of muscle.
    Blackwell, Oxford (1959).

CRAGG, B.G.: What is the signal for chromatolysis?
    Brain Research 23, 1-21 (1970).

DENNY-BROWN, D.: The histological features of striped
    muscle in relation to its functional activity.
    Proc. Roy. Soc. B. 104, 371-411 (1929).

ECCLES, R.M., C.N. SHEALY, and E.D. WILLIS: Patterns of
    innervation of kitten motoneurones.
    J. Physiol. (Lond.) 165, 392-402 (1963).

ECCLES, J.C., and C.S. SHERRINGTON: Numbers and contraction
    values of individual motor units examined in some
    muscles of the limb.
    Proc. Roy. Soc. B. 106, 326-357 (1930).

GUTH, L.: Neuromuscular function after regeneration of
    interrupted nerve fibers into partially denervated
    muscles.
    Exper. Neurol. 6, 129-141 (1962).

GUTH, L.: 'Trophic' influence of nerve on muscle.
    Physiol. Rev. 48, 645-680 (1968).

GUTMANN, E.: Neurotrophic relations in the regenerating
    process.
    In: Mechanisms of Neural Regeneration (M. SINGER and
    J.P. SCHADE, eds.), Progress in Brain Research,
    vol. 13, pp. 72-112, Elsevier, Amsterdam (1964).

GUTMANN, E., and J. ZELENA: Morphological changes in the
    denervated muscle.
    In: The Denervated Muscle (E. GUTMANN, ed.), pp. 57-
    102, Czechoslovak Academy of Sciences, Prague (1962).

HNIK, P., and J. ZELENA: Atypical spindles in reinnervated
    rat muscles.
    J. Embryol. Exptl. Morphol. 9, 456-467 (1961).

KERKUT, G.A., A. SHAPIRA, and R.J. WALKER: The transport
    of labelled material between CNS and muscles along a
    nerve trunk.
    Comp. Biochem. Physiol. 23, 729-748 (1967).

LEWIS, D.M., J.C. LUCK, and S. KNOTT: A comparison of
    isometric contractions of the whole muscle with those
    of motor units in a fast-twitch muscle of the cat.
    Exp. Neurol. In press (1972).

LUBINSKA, L.: Axoplasmic streaming in regenerating and in
    normal nerve fibres.
    In: Mechanisms of Neural Regeneration (M. SINGER and
    J.P. SCHADE, eds.), Progress in Brain Research, vol.
    13, pp. 1-71, Elsevier, Amsterdam (1964).

MARK, R.F., and Lauren R. MAROTTE: Suppression of synaptic
    transmission in reinnervated muscle.
    Proc. XXV Int. Union. Physiol. Sciences Congress
    (Munich), 1095, p 369 (1971).

MCPHEDRAN, A., R.B. WUERKER, and E. HENNEMAN: Properties
    of motor units in a homogenous red muscle (Soleus)
    of the cat.
    J. Neurophysiol. 28, 71-84 (1965).

MILEDI, R.: Formation of extra nerve muscle junctions in
    innervated muscle.
    Nature (Lond.) 199, 1191-1192 (1960).

MILEDI, R., L.T. POTTER: Acetylcholine receptors in muscle
    fibres.
    Nature 233, 599-603 (1971).

MUSICK, J., and J.I. HUBBARD: Release of protein from
    mouse motor nerve terminals.
    Nature, in press (1972).

NYSTROM, B.: Fibre diameter increase in nerves to "slow-
    red" and "fast-white" cat muscles during postnatal

development.
Acta neurol. scand. 44, 265-294 (1968a).

NYSTROM, B.: Histochemical studies of end-plate bound
esterases in "slow-red" and "fast-white" cat muscles
during postnatal development.
Acta neurol. scand. 44, 295-318 (1968b).

NYSTROM, B.: Postnatal development of motor nerves terminals in "slow-red" and "fast-white" cat muscles.
Acta neurol. scand. 44, 363-383 (1968c).

OLSON, C.B., and C.P. SWETT, Jr.: A functional and histochemical characterization of motor units in a heterogeneous muscle (flexor digitorum longus) of the cat.
J. Comp. Neurol. 128, 475-498 (1966).

REDFERN, P.A. Neuromuscular transmission in newborn rats.
J. Physiol. (Lond.) 209, 701-709 (1970).

RIDGE, R.M.A.P.: The differentiation of conduction velocities of slow twitch and fast twitch muscle motor in-
nervations in kittens and cats.
Quart. J. Exp. Physiol. 52, 293-304 (1967).

ROSENBLUETH, A., and J.V. LUCO: A study of denervated mammalian skeletal muscle.
Amer. J. Physiol. 126, 39-57 (1937).

ROBBINS, N., and T. TANEZAWA: Developing neuromuscular
junctions: First signs of chemical transmission during
formation in tissue culture.
Acta Neurol. scand. 44, 363-383 (1968c).

RUBINSTEIN, L.J.:
    In: Structure and function of Muscle (G. BOURNE,ed.),
    3, 209, Academic Press, New York (1960).

SALMONS, S., and G. VRBOVA: The influence of activity on
    some contractile characteristics of mammalian fast
    and slow muscles.
    J. Physiol. (Lond.) 201, 535-549 (1969).

SISSONS, H.A.: Investigations of muscle fibre size.
    In: Research in Muscular Dystrophy. Proc. 2nd Symp.
    on Current Research in Muscular Dystrophy, Pitman
    Medical, London, P 89-98 (1963).

TORREY, T.W.: The relation of taste buds to their nerve
    fibres.
    J. comp. Neurol. 59, 203-220 (1934).

VRBOVA, G., and A.C. WAREHAM: The effect of activity on
    some properties of neuromuscular junction of mam-
    malian skeletal muscle.
    Communication C33 to Physiological Society, April
    (1972).

WATSON, W.E.: Centripetal passage of labelled molecules
    along mammalian motor axons.
    J. Physiol. (Lond.) 196, 122-123P (1968).

WEISS, P., and Ann HOGG: Competitive reinnervation of rat
    muscles by their own and foreign nerves.
    J. Neurophysiol. 9, 413-418 (1946).

WEISS, P., P.A. PILLAI: Convection and fate of mitochondria
    in nerve fibres: axonal flow as a vehicle.
    Proc. nat. Acad. Sci. (Wash.) 54, 48-56 (1965).

WUERKER, R.B., A.M. MCPHEDRAN, and E. HENNEMAN: Properties
of motor units in a heterogeneous pale muscle (M.
gastrocnemius) of the cat.
J. Neurophysiol. 28, 85-99 (1965).

SOME NEUROPHYSIOLOGICAL CONSIDERATIONS CONCERNING "MEMORY"

O.D. Creutzfeldt

Department of Neurobiology, Max-Planck-
Institute of Biophysical Chemistry
34 Göttingen, Germany

ABSTRACT

It is proposed that memory is an alteration of filter
functions of cortical networks. Such alterations may take
place at all levels of sensory (input) and motor (output)
systems, and may also involve intrinsic systems such as the
hypothalamic and limbic structures. As a simple model for
such functional alterations of an analysing network, sensory
after-effects are suggested. The McCOLLOUGH-effect is demonstrated and interpreted as temporary alteration of inhibitory connections between colour sensitive cells and
orientation sensitive cells in a cortical column. Such an
interpretation suggests that an alteration of inhibitory
connections may play an essential role in "learning". Local
changes in minute neuronal circuitries such as certain
cortical "columns" in an otherwise homogeneous neuronal
structure may be sufficient to induce an alteration of the
filter function of the whole cortical area. The hypothesis

that specific memory substances may be responsible for de-
position and recall of complex perceptual or behavioural
patterns is not necessary for a neurophysiological memory
model, and is actually incompatible with many neurophysiological observations.

Neurophysiologists have as yet searched in vain for an
electrophysiological correlate of learning. Nor can any of
the present psychological or biochemical hypotheses be
reasonably fitted into our present "model of neuronal
functioning". Some of the main elements of this model are:

The individual units of the circuitry of the brain
are neurones.

Neurones can be excited by other neurones via excitatory and inhibitory synapses. The electrophysiological and
chemical time course of inhibitory actions is longer than
that of excitatory actions.

Neurones have a clearly defined threshold of excitation.

In the different levels of the nervous system (thalamus, cortex), neurones are connected with each other by inhibitory and/or excitatory synapses resulting in "neuronal
networks".

The physical environment of the body is recorded by
sense organs and projected into the brain in topographical

maps. The topography of these maps is distorted by a scale which takes into account the importance of certain sensory informations (weighted topographical representation).

Each relay network has the function of a spatio-temporal filter. Several such filters are connected in series. In addition, simultaneous processing is performed in parallel systems which receive, in principle, the same or similar spatio-temporal input.

The output of the system, which finally results in motor activity, is a function of its input as well as of its internal state. The internal state of the whole body is recorded in the brain by intrinsic sub-systems (intrinsic sensors) and reported to the rest of the brain. The intrinsic sensors are sensitive to internal biochemical states of the body, such as metabolic and neuroendocrine, but also report on physical states such as posture etc.

The aim of the nervous system is to keep the whole organism alive until it has reproduced itself. In order to achieve this goal, the system has an inbuilt mechanism which keeps it in a state of instability until the conditions for the fulfilment of the above purpose are satisfied. The purpose itself is not consciously known by the system, but induces, when not fulfilled, a state of restlessness, which may be called "drive".

One of the essential properties of higher nervous systems is that they can adapt themselves to the conditions

of an environment. This adaptation may be considered a
learning process in the broadest sense of the word, com-
prehending various forms of learning such as imprinting,
Pavlovian procedures, learning, one time experience,
systematic training and even various forms of habituation
or adaptation.

These processes change the system in such a manner
that it will respond differently to a stimulus than before
the learning. Such changes may take place on the efferent
side (e.g., the motor output), on the afferent side (e.g.,
the cortical sensory relay), or somewhere in between where
also more complex systems such as integrated activities
from different sense modalities and/or "internal state
sensors" may be involved. In the latter case, the hypothalamic and/or "limbic" systems come into play. Probably
only rarely the function of only one input or output system
will be altered by learning, nor can it be assumed, that
only one level of a sensory analysis or motor output system
will be involved.

Let the assumption be that a complex sensory stimulus
leaves a mark in the corresponding cortical filter networks.
The idea corresponds in some way to the old idea of the
"memory trace" of some Gestalt-psychologists (KÖHLER and
WALLACH, 1944). In terms of the above considerations it
would mean that the filtering properties of the system
for an afferent stimulus have been temporarily altered:
the same stimulus or new stimuli will not be transformed
in exactly the same way as they were before the exposure
to the learned stimulus. Or, in more neurophysiological

terms, the spatio-temporal excitation profile in the primary sensory and in related secondary, tertiary etc. areas
will not be the same. Or the story which the sensory organ
tells the brain after "learning" has taken place, is different than before the learning procedure. Repeated exposure
may lead to a more enduring change; combination with other
stimuli or with internal responses may lead to the establishment of combined, linked response patterns of various
systems. In such a broad concept, also Pavlovian and
Skinnerian learning may find their place.

Such alterations of filter functions of cerebral network are daily experience. Long, monotonous exposure to a
visual or auditory stimulus will lead to perceptual after-
effects which everyone has experienced: the young electro-
encephalographer who has looked at EEG-curves for hours,
and afterwards sees EEG-elements in various natural forms
such as the silhouette of mountains; or the physiologist
who has sat for hours in front of an oscilloscope and sees
the traces before his eyes when he goes to bed; the perceptual distortions of complex nature after having been in
a movie etc.

A fine example for such a perceptual after-effect is
the McCOLLOUGH-effect (McCOLLOUGH, 1965). Demonstration:
look at a pattern of black-red vertical and black-green
horizontal stripes in repeated temporal succession for
some 10 min, and then look at black-white vertical and
black-white horizontal stripes. The vertical white stripes
will look greenish and the horizontal white stripes reddish
for some time, up to hours and even days if the experimen-

tal conditions are appropriate. Experimental evidence indicates that the mechanisms for this perceptual distortion
must be located in the visual cortex, probably the primary
visual cortex.

        In the primary visual cortex, single neurones are
wired up with their afferent input and with each other in
such a way that they preferentially respond to moving
edges of a certain orientation(HUBEL and WIESEL, 1962).
Cells sensitive to one orientation are clustered in cortical
columns. Some of these neurones receive their primary input from single cones and are thus specifically sensitive
to monochromatic light (GOURAS, 1970). From these findings
in animal experiments we can extrapolate and may say that
the vertical black-red stripe pattern will have excited
mainly neurones in columns, whose cells are sensitive to
vertical stripes. Of the monochromatically sensitive cells
in these columns only the red sensitive cells were activated.
This, we may assume, has resulted in the establishment of
a selective functional connection between the red sensitive
cells and the non-colour sensitive vertical "edge detectors"
of these columns. If, in the next step, only the vertical
black-white stripes are shown, the same columns with cells
sensitive to vertical stripes will be excited. But the red-
cells in these columns will be less excited (if at all) than
before. The brain sees the complementary colour, similar to
a negative afterimage following monochromatic bleaching of
the retina. But the McCOLLOUGH effect is not due to a
mechanism based on colour adaptation of the retinal receptors.

It is now experimentally well established that an
edge stimulus optimally oriented to excite a given cortical cell will not only excite this cell but will also
cause strong inhibition in this cell as well as in other
cells of the same column (BENEVENTO et al., 1972; CREUTZ-
FELDT et al., in prep.). Based on such findings, we may
specify the functional connection postulated above by
saying that the strong excitation of red sensitive cells
in the vertical orientation column might have established
a strong and lasting inhibitory connection between these
cells and all orientation sensitive cells of the same
column. When, later on, only the black-white stripes are
shown, this inhibitory connection will lead to a suppression of the red-cells, so that they will be relatively less
excited by the white light than cells which are sensitive
to other spectral components and which had not been predominantly active during the preceding red-black pattern.

It is understood that such an interpretation is still
purely speculative, and other interpretations have been
proposed before (McCOLLOUGH, 1965; MURCH and HIRSCH, 1972).
An experimental test of any of these explanations is still
lacking. But I choose this interpretation in order to indicate that network properties may not only be altered by
strengthening or weakening excitatory connections between
neuronal elements, but also by variation of the inhibitory
connections.

Inhibitory pathways are themselves particularly good
for such temporary changes in neuronal connections, since
inhibitory mechanisms are longer lasting. After local

electrophoretic application of an inhibitory substance
such as 5-HT or nor-adrenaline, the inhibitory action of
this substance may outlast the time of application considerably, in contrast to the immediate termination of the
excitatory effect of acetylcholine or glutamic acid after
the end of the application (STEINER, 1971). In Aplysia
neurones, inhibitory actions of up to half an hour or more
have been described (TAUC, 1969). Such findings and their
possible application to long lasting functional changes
of groups of neurones, such as we have tried in this assay,
may stimulate the search not only for lasting facilitation
of excitation but also for lasting facilitation of inhibition as a possible mechanism underlying long lasting
changes of functional properties of neuronal networks.

It is understood that the mentioning of an effect
like the McCOLLOUGH effect may evoke the protest of many
memory specialists. But what seems paradigmatic in this
experiment is the lasting alteration of a simple perceptual
function, which can be tested psycho-physically and can be
subject to direct experimentation in animals. There is no
serious reason to assume that similar mechanisms, taking
place in input or output systems of the brain or in combined systems, may not be the basis of complex and longer
lasting functional alterations of these systems, the result
of which we may call "memory". There are, in fact, many
experimental findings in favour of this assumption.

This example was also chosen in order to stress the
local factors for the establishment of "memory traces".

Functional changes of connectivity, such as proposed, can be effective if they take place only in local circuits (in our case only in certain columns of the visual cortex), which are imbedded in an otherwise homogeneous cortical network. It is difficult for a neurophysiologist to imagine the action of a specific chemical memory substance in such an experimental situation.

LITERATURE

BENEVENTO, L.A., O.D. CREUTZFELDT, and U. KUHNT: Significance of intracortical inhibition in the visual cortex.
Nature New Biology 238, 124-126 (1972).

CREUTZFELDT, O.D., U. KUHNT, and L. BENEVENTO: An intracellular analysis of excitation and inhibition in visual cortex neurones.
In prep.

GOURAS, P.: Trichromatic mechanisms in single cortical neurons.
Science 168, 489-492 (1970).

HUBEL, D.H., and T.N. WIESEL: Receptive fields, binocular interaction and functional architecture in the cat's visual cortex.
J. Physiol. (Lond.) 160, 106-154 (1962).

KÖHLER, W., and H. WALLACH: Figural aftereffects: an investigation of visual processes.
Proc. Amer. Phil. Soc. 88, 269-357 (1944).

McCOLLOUGH, C.: Color adaptation of edge-detectors in the
    human visual system.
    Science 149, 1115-1116 (1965).

MURCH, G.M., and J. HIRSCH: The McCollough effect created
    by complementary afterimages.
    Am. J. Psychol. 85, 241-248 (1972).

STEINER, F.A.: Neurotransmitter und Neuromodulatoren.
    Technik und Resultate der Mikroelektrophorese im
    Nervensystem.
    Thieme, Stuttgart (Sammlung psychiatrischer und
    neurologischer Einzeldarstellungen) (1971).

TAUC, L.: Polyphasic synaptic activity.
    Progr. in Brain Res. 31, 247-257 (1969).

TRANSFER OF ACQUIRED

INFORMATION

THE STRUCTURE OF THE "MEMORY-CODE-WORD" SCOTOPHOBIN

Wolfgang Parr and Gunther Holzer

Chemistry Department
University of Houston
Houston, Texas 77004

ABSTRACT

The tentative structure of scotophobin, a pentadeca-peptide amide isolated from brain of rats which had been trained to have a fear of the dark, was confirmed by a synthetic approach. The synthetic material was prepared via the solid phase method and was cleaved from the solid carrier by direct ammonolysis. The purified product showed 100 % biological activity and an undistinguishable dose-response-curve as the natural scotophobin. The synthetic peptide and its tryptic fragments showed identical pro-perties by thin layer chromatography. Furthermore, micro-dansylation of the synthetic and natural materials gave the same results. According to these findings scotophobin has the following primary structure: Ser-Asp-Asn-Asn-Gln-Gln-Gly-Lys-Ser-Ala-Gln-Gln-Gly-Gly-TyrNH$_2$
1 2 3 4 5 6 7 8 9 10 11 12 13 14 15

UNGAR and coworkers reported in 1968 experiments in
which the innate dark preference of rodents was reversed
by appropriate training so that the animals showed an
avoidance of the dark. Within two years UNGAR and his
group (1970, 1972) trained some 4000 rats, and collected
their brains. The isolation of the specific behavior inducing substance was done by various techniques of chromatography. Finally 300 µg of a peptide with an unknown
structure was isolated. A quantitative amino acid analysis
of the peptide gave the following ratios 3 Asx, 2 Ser,
4 Glx, 3 Gly, 1 Ala, 1 Tyr, 1 Lys and ammonia in an unde-
termined amount. Enzymatic cleavage with trypsin yielded
two fragments T_1 and T_2. Both of them had Ser as N-termini
according to the results obtained from microdansylation.
The C-terminus of the whole peptide was not free because
no reaction with carboxypeptidase was achieved. Since the
overall yield of scotophobin was only 300 µg and part of
the material already had been used for amino acid analysis,
tryptic digests and biological tests, the conventional
sequential analysis was no longer possible. With samples of
20 µg of the whole pentadecapeptide and the corresponding
octa- and heptapeptides which were obtained by tryptic
digestion we attempted to determine the structure by high
resolution mass spectrometry. The details of this technique
are described elsewhere (DESIDERIO et al., 1971). Fortu-
nately sufficient smaller fragments in the form of tri-
and dipeptides were formed which after overlapping led to
a tentative sequence, which is given in Fig. 1. The mass
spectrometric data clearly showed that Asp, Glu and Glu
were in the uncertain positions 2, 5 and 11 but it could

1	2	3	4	5	6	7	8	9	10	11	12	13	14	15
Ser	Asx	Asn	Asn	Glx	Gln	Gly	Lys	Ser	Ala	Glx	Gln	Gly	Gly	TyrNH₂
		Asn	Asn	Glx	Gln	Gly	Lys	Ser	Ala	Glx	Gln	Gly	Gly	
						Gly					Gln	Gly		

(1) Ser Asx Asn Asn Glx Gln Gly Lys Ser Ala Glx Gln Gly Gly TyrHN₂

Fig. 1: Sequence Assignment:
Determination of the tentative amino acid sequence of scotophobin, deduced by overlapping of di- and tripeptides obtained by mass spectrometry.

not be excluded that amide groups could have been originally in the side chain of the dicarboxylic acids and could
have been lost during the ionization process.

In view of this fact and the lack of further natural
material is was decided to undertake a synthetic approach
in order to establish the correct structure of scotophobin.
Since three amino acids were uncertain, eight possible
structural analogs could be expected. The structure having
three free carboxylic groups in the uncertain positions
(peptide IV, Fig. 2) has been synthesized by ALI and coworkers (1971) and was found to have only 2,5 % activity
(75 U/mg) (Tab. I) of the natural product. Furthermore,
the analytical data did not agree with those of the natural
material, isolated from rat brain. Our first attempt to
establish the structure of scotophobin resulted in a fully
amidated peptide (PARR and HOLZER, 1971) (Peramidoscotophobin, peptide III, Fig. 2). This material already showed
10 % biological activity (300 U/mg) (Tab. I). However the
analytical data of peramidoscotophobin was different from
the natural peptide and this structure was also rejected.

Conclusions were drawn that possibly only one of the
three acids were free and therefore the synthesis of the
peptide (II) (Fig. 2) was attempted by the solid phase method because it has been shown that this technique is very
suitable for the preparation of smaller peptides. However,
longer reaction times were used and after each cleavage
step the free NH_2-groups were titrated in order to have a
control over the percentage - yield in every coupling step.

	1	2	3	4	5	6	7	8	9	10	11	12	13	14	15
I	Ser	Asx	Asn	Asn	Glx	Gln	Gly	Lys	Ser	Ala	Glx	Gln	Gly	Gly	TyrNH$_2$
II	Ser	Asp	Asn	Asn	Gln	Gln	Gly	Lys	Ser	Ala	Gln	Gln	Gly	Gly	TyrNH$_2$
III	Ser	Asn	Asn	Asn	Gln	Gln	Gly	Lys	Ser	Ala	Gln	Gln	Gly	Gly	TyrNH$_2$
IV	Ser	Asp	Asn	Asn	Glu	Gln	Gly	Lys	Ser	Ala	Glu	Gln	Gly	Gly	TyrNH$_2$

Fig. 2: Synthetic analogs of scotophobin

I represents the tentative amino acid sequence of scotophobin with un-certainties of portions 2, 5, and 11. The peptides II, III, and IV have been synthesized. III and IV showed only portions of activity (300 U/mg and 75 U/mg, respectively); however the biological and chemical pro-perties of II are identical with those of the natural.

Tab. I: <u>Comparison of biological activities of natural and</u>
        <u>synthetic scotophobin</u>

Natural Scotophobin   3000 U*/mg

Synthetic Scotophobin (II)   3000 U/mg

  Sequence (III)   300 U/mg

  Sequence (IV)   75 U/mg

*One unit of activity is the amount of material that re-
 duces the mean time spent in the dark from 130 to 60 sec.

Dicyclohexylcarbodiimide was used as coupling agent for
Boc-L-alanine, Boc-O-benzyl-L-serine, N-E-Cbo-N-α-t-Boc-L-ly-
sine, Boc-glycine and N-Dpoc-L-aspartic acid-β-tert. butyl
ester. All other Boc amino acids were used as their p-nitrophenyl esters. The OH-group of the C-terminal tyrosine was
protected by a benzyl group.

     The synthesis was designed so that the cleavage of the
fully protected pentadecapeptide (Fig. 3) from the polymer
could be achieved by direct ammonolysis (BAYER et al., 1971).
This technique can be carried out even if the side chains
in aspartic acid or glutamic acid are protected by tert.
butyl esters since no ammonolysis of these groups takes
place. For the temporary protection of aspartic acid β-tert.
butyl ester the N-2-(p-diphenyl)-2-isopropyloxy-carbonyl
group was used since this group can be cleaved in acetic
acid in presence of tert. butyl esters. The synthesis was
carried out with 800 mg of chloromethylated resin which
was esterified with Boc-O-benzyl-L-tyrosine in absolute

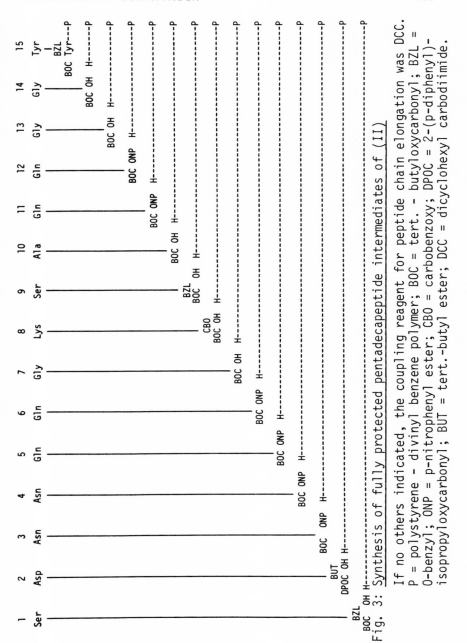

Fig. 3: <u>Synthesis of fully protected pentadecapeptide intermediates of (II)</u>

If no others indicated, the coupling reagent for peptide chain elongation was DCC.
P = polystyrene - divinyl benzene polymer; BOC = tert. - butyloxycarbonyl; BZL =
O-benzyl; ONP = p-nitrophenyl ester; CBO = carbobenzoxy; DPOC = 2-(p-diphenyl)-
isopropyloxycarbonyl; BUT = tert.-butyl ester; DCC = dicyclohexyl carbodiimide.

ethanol for 48-hrs. under reflux using triethylamine as
acid binding reagent. Amino acid analysis of a dried sample gave 0.1-mmole of L-tyrosine per 1 g of resin. The following reaction cycles were used to build the amino acid
sequence 15 to 3, cleavage of the Boc-groups by addition
of 20 ml of N HCL in glacial acetic acid and shaking for
30 minutes, washing three times with 20 ml portions of
glacial acetic acid, absolute ethanol, N,N-dimethylforma-
mide and chloroform for 10 minutes, washing three times
with 20 ml portions of chloroform and methylene chloride,
coupling of the new amino acid to the free amino groups on
the polymer in DMF and methylene chloride. Aspartic acid
in position 2 was introduced as Dpoc-β-tert.-butyl asparate, which was obtained by treating the corresponding
DCHA-salt with citric acid in methylene chloride. The Dpoc-
group was split off by shaking the resin for 8 hours with
30 ml acetic acid, and finally the N-terminal amino acid
serine was attached as described above. Boc-amino acids,
Boc-amino acid p-nitrophenyl esters and Dpoc-β-tert.-
butyl asparate were used in a 6-fold excess. The reaction
time was three hours in the case of the DCC method and 8
hours in the case of the p-nitrophenyl esters. The dried
fully protected pentadecapeptide-polymer (Fig. 3) was
transferred into a pressure bottle, suspended in 50 ml DMF
and 50 ml liquid NH_3 was added at -72°C. The bottle was
closed and the reaction mixture was shaken for 7 days. After filtration and washing with DMF, the combined filt-
rate was evaporated in vacuo and yielded about 100 mg.

The above amide was suspended in trifluoroacetic acid

and treated with anhydrous HBr for 2 hrs, using anisole
as a scavenger. After removal of the HBr and evaporation
of the trifluoroacetic acid in vacuo, the crude product
was dissolved in glacial acetic acid and precipitated
with ether. The yield at this point was about 80 mg.

   A portion of the crude product was dissolved in methanol and the alcoholic solution was submitted to gel filtration on Sephadex LH-20 using methanol for elution. The
active material, as determined by the dark avoidance test,
was eluted. It was further purified by preparative thin
layer chromatography on silica gel sheets with n-butanol,
ethanol, acetic acid, and water (80:20:10:30) as solvents.
Three ninhydrin-positive spots at R_F 0.07, 0.38 and 0.57
were obtained. The spot at R_F = 0.57 (R_F of the natural
scotophobin is 0.58) (Tab. II) was eluted with pyridine
acetic acid buffer, pH 3.7. Using this technique repeatedly, 39.6 mg were obtained. Amino acid analysis (6N HCl, 36
hours, 115°C) gave the following molar ratios, Gly being
taken as 3.00:

      Ala 1.05 Gly 3.00 Asp 2.80 Glu 3.80

      Tyr 1.15 Lys 0.95 Ser 1.90 NH_3 7.15

After incubation of the synthetic material with trypsin
two fragments were obtained which showed about identical
R_F values with those of the products obtained by the hydro-
lysis of the natural substance (Tab. II). Microdansylation
of the natural and synthetic product with 5-dimethyl amino-
naphthaline-sulfonyl chloride gave identical results in
both dimensions (Tab. II).

Tab. II: Comparison of chromatographic properties of natural and synthetic scotophobin

R_F - Values	Natural (I) Scotophobin	Synthetic (II) Scotophobin	Solvent - System
Pentadecapeptide	0.58	0.57	n-Butanol/Ethanol/Acetic Acid/Water 80:20:10:30
Dansyl-derivatives			
1. Dimension	0.16	0.16	Formic Acid/Water 1.5:100
2. Dimension	0.16	0.18	Benzene/Acetic Acid 9:1
Tryptic Fragments			
T_1	0.26	0.30	n-Butanol/Ethanol/Acetic Acid/Water 80:20:10:30
T_2	0.40	0.37	

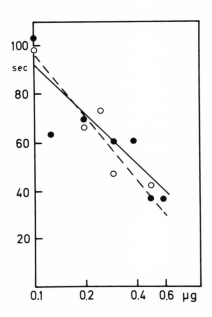

Fig. 4: Dose-response curve of natural (0) and synthetic
(0) scotophobin
Abscissa: dose in µg per 20-25 g mouse (log scale);
ordinate: seconds in dark box out of total of 180 s
(DBT). Before injection DBT was 130-135 s. Solid
line, natural scotophobin; broken line, synthetic
scotophobin. Regression lines were calculated by
the least squares method. Natural scotophobin was
tested in groups of twelve mice at each dose; the
synthetic peptide was tested in groups of twenty-
four mice, except for 150 and 200 ng which were
tested in twenty mice each and 400 ng tested in
twelve.

The purified pentadecapeptide was tested in mice for
biological activity in comparison with natural scotophobin.
Before injection the mice spent an average of 130 to 140 sec
in the dark out of the 180 sec duration of the test. Fig. 4
shows a dose-response curve of the natural in comparison

with the synthetic product. The lines were calculated by
the least squares method. The dose reducing the dark-box-
time   to 60 sec was 0.27 mg (95 % confidence interval
0.186-0.396) for the natural and 0.3 mg (0.238-377) for
the synthetic. The slope functions were 2.34 (1.44-3.78)
and 2.55 (1.50-430) respectively. The potency ratio 1.11
(0.72-1.71) and the slope ratio 1.09 (0.54-2.22) indicated
that there is no significant difference between the behavioral activity of the substances. All values were calculated by the method of LICHTFIELD and WILCOXON (1949).
Other laboratories have confirmed the activity of the synthetic material (GUTTMAN, 1972; DOMAGK, MALIN).

Control animals injected with saline or extracts of
normal rat brain remained above 120 sec. The tests were
done by automatic recording under "double blind" conditions.
For a detailed description of the test see reference.

ACKNOWLEDGEMENT

The authors are indebted to Dr. G. UNGAR, Baylor
School of Medicine, Houston, Texas, for providing the
bioassay. This work was supported by the Robert A. Welch
Foundation, Houston, Texas through Grant Number E-404.

LITERATURE

ALI, A., J.H.R. FAESEL, D. SARANTAKIS, D. STEVENSON, and
    B. WEINSTEIN: Synthesis of a structure proposed for

scotophobin.

Experientia <u>27</u>, 1138-1139 (1971).

BAYER, E., E. BREITMAIER, G. JUNG, and W. PARR: Synthese
des C-terminalen Hexapeptidamids an einem neuen fe-
sten Träger.
Hoppe-Seyler's Z. Physiol. Chem. <u>352</u>, 759-760 (1971).

DESIDERIO, D.M., G. UNGAR, and P.A. WHITE: The use of mass
spectrometry in the structural elucidation of scotophobin - a specific behavior-inducing brain peptide.
Chem. Comm. <u>9</u>, 432-433 (1971).

GUTTMAN, H:N., G. NATWYSHYN, and G.H. WARRINER, III:
Synthetic scotophobin-mediated passive transfer of
dark-avoidance.
Nature (New Biology) <u>235</u>, 26-27 (1972).

LICHTFIELD, J.T., Jr., and F. WILCOXON: A simplified method
of evaluating dose-effect experiments.
J. Pharmacol. Exp. Ther. <u>96</u>, 99 (1949).

PARR, W., and G. HOLZER: Synthesen von Scotophobin-Analoga. Peptide mit gedächtnisübertragender Wirkung.
Hoppe-Seyler's Z. Physiol. Chem. <u>352</u>, 1043-1048
(1971).

UNGAR, G., L. GALVAN, and R.H. CLARK: Chemical transfer
of learned fear.
Nature (London) <u>217</u>, 1259-1261 (1968).

UNGAR, G., I.K. HO, and L. GALVAN: Isolation of a dark
avoidance inducing brain peptide.
Fed. Proc. <u>29</u>, 658 (1970).

UNGAR, G., D.M. DESIDERIO, and W. PARR: Isolation, iden-
    tification, and synthesis of a specific-behavior in-
    duced brain peptide.
    Nature (London) <u>238</u>, 198-202 (1972)

EVIDENCE FOR MOLECULAR CODING OF NEURAL INFORMATION

Georges Ungar

Baylor College of Medicine

Houston, Texas, USA

ABSTRACT

The essential significance of the chemical transfer of
learning experiments is that they provide bioassay methods
for the isolation of behavior-inducing substances. The results obtained by these methods have provided evidence for
a molecular code of neural information. Once the specific
behavior-inducing substances are chemically defined they can
be further investigated by appropriate microanalytical techniques. It is hoped that this general approach will allow
within the next few years the identification of several in-
formation-carrying molecules and thus open the way to a
systematic study of the molecular code.

The main purpose of this communication is to place the
experiments of chemical transfer of information in what I
believe to be their proper perspective. The attempts to

317

transfer acquired information from one individual to another by chemical means, called "memory transfer", "inter-
animal transfer of learned behavior", "chemical transfer
of learning", etc., have been regarded, at best, as a cu-
riosity and, at worst, as an "absurd and chimerical" en-
terprise. I believe that the controversy over these experiments is largely due to the fact that they have been taken
out of the context of a broader problem to which they belong, the problem of the mechanism of information processing in the nervous system.

The processing of information is generally recognized
to be the main function of the nervous system. Its study
has been almost entirely taken over by neurophysiologists
and cyberneticists (GERARD and DUYFF, 1962). In spite of
the now universally recognized chemical nature of the processing of genetic information, the possibility of a molecular mechanism for the coding of acquired information has
been largely disregarded (PERKEL and BULLOCK, 1968). The
early hypotheses, formulated by MONNE (1948), FOERSTER
(1948) and KATZ and HALSTEAD (1950) attracted little attention until the first experimental approach to the problem
by HYDEN in 1959.

Since then, evidence has been accumulating for the
existence of chemical correlates of learning (see review
by BOOTH, 1970). I shall not go into the details of this
approach, which will undoubtedly be discussed by other participants to this conference. The interpretation of these
experiments is still hotly disputed but it is reasonable to

conclude that acquisition of information is accompanied by increased synthesis of RNA and protein.

Whether this augmented rate of synthesis is an essential condition of learning or only an index of heightened functional activity has been answered by the use of metabolic inhibitors capable of blocking RNA or protein synthesis (review by COHEN, 1970). There again, the interpretation of the results is far from unanimous but there is good evidence to suggest that inhibition of increased synthesis of RNA and, especially, of protein can impair the acquisition of new information without interfering with the utilization of established memory traces.

The question that remains to be answered concerns the significance of the chemical changes taking place in the brain in correlation with the learning process. Do the newly synthesized molecules actually contain information in the form of a molecular code or do they simply fulfill some nonspecific metabolic or functional need?

THE BIOASSAY APPROACH

It is unlikely that any of the chemical or physical methods available today possess the sensitivity and specificity required to test this last point. There is only one approach that is capable of detecting minute amounts of material of unknown chemical composition in complex mixtures and it is the biological assay method. It has proved its

value in many instances: it laid down the foundation of the
hormonal concept, it helped to establish the existence of
vitamins and it played a decisive role at the origin of the
neurohumoral theory.

The real significance of the experiments of chemical
transfer of learned information, therefore, is that they
represent an actual bioassay for the demonstration of a
molecular code. If material extracted from the brain of
trained donor animals can induce in naive recipients the
behavior for which the donors were trained, it is highly
probable that the information imparted by the training was
stored in the donor brain in some chemical form. It is evident, however, that before reaching this important conclusion, a number of other possibilities have to be eliminated. This can be done by the choice of an adequate experimental design and the use of appropriate controls. One can
thus eliminate the possibilities of a general facilitation
of learning, the effect of stress or of a sensory or motor
impairment. One can also narrow down the specificity of the
assay by means of "cross-transfer" experiments between closely related behaviors in which either the stimulus or the
response are different. These points have been repeatedly
discussed (DYAL, 1971; FJERDINGSTAD, 1971; UNGAR, 1971a;
UNGAR and CHAPOUTHIER, 1971) and I wish to present here only
our most recent contribution to the problem.

Mice were trained in my laboratory by G. RADCLIFFE to
run two mazes (Fig. 1). One group of mice was trained in
maze 1, another group in maze 2. When the animals reached

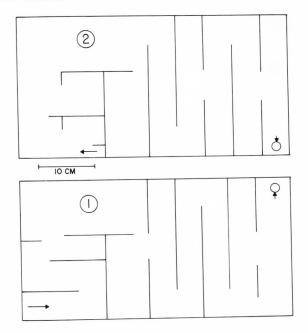

Fig. 1: <u>Diagram of two mazes used for training and testing</u>
<u>of mice</u>
The animals were deprived of water and were allowed
to drink when they reached the target.

the criterion of completing the run in 30 sec or less, their
brains were extracted and the extract injected into recipient
mice. Each pool of extract was injected into two groups of
mice, one group to be tested in the same maze in which its
donors were trained and a second group in the other maze.
Two control groups were also injected with untrained mouse
brain and tested in the two mazes. Fig. 2 shows that the
recipients tested in the maze in which their donors were
trained learned the task significantly faster than those
tested in the other maze. The latter were not different from
the controls.

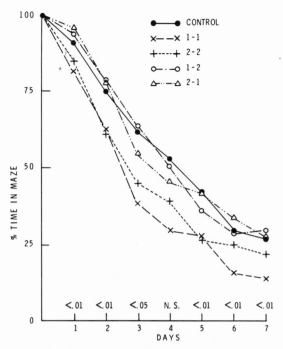

Fig. 2: Maze performances of mice injected with brain

extracts
Abscissa, days of testing; ordinate, time to reach
target in per cent of preinjection performance.
Control mice were injected with extract of brain
from untrained donors; (1-1), mice injected with
brain from mice trained in maze 1 and tested in
maze 1; (2-2), donors trained in maze 2, recipients tested in maze 2; (1-2), donors trained
in maze 1, recipients tested in maze 2; (2-1),
donors trained in maze 2, recipients tested in
maze 1. Probabilities calculated between groups
(1-1) + (2-2) and groups (1-2) + (2-1) by t-test.
N = 10 in each group.

The results suggest a much higher degree of specificity than hitherto suspected. The active substance seems to
be a fairly large molecule, close to the limit of dialyzabil-
ity. More work will still be needed to explore the limit of

specificity of the assays. We shall probably reach a point where cross-reactions appear as it is known to occur with even the most exquisitely specific immune or enzyme reactions.

SCOTOPHOBIN

In my laboratory the ultimate objective of the transfer experiments has always been to use them as bioassays for the isolation of the active substances. This has now been accomplished for the dark avoidance experiments. We have isolated, identified and synthesized a pentadecapeptide, called scotophobin (Fig. 3) which reproduces the behavior of the donor animals from whose brain it was originally extracted (UNGAR, 1972c; UNGAR et al., 1968, 1972b).

```
 1   2   3   4   5   6   7   8   9  10  11  12  13  14
Ser-Asp-Asn-Asn-Gln-Gln-Gly-Lys-Ser-Ala-Gln-Gln-Gly-Gly-
 15
Tyr-NH₂
```

Fig. 3: <u>Amino acid sequence in scotophobin</u>

The behavioral effect of synthetic scotophobin has been confirmed in mice (MALIN and GUTTMAN, 1972) and in fish (GUTTMAN et al., 1972; BRYANT et al., 1972). See also papers at this Symposium by THINES and DOMAGK and by MCCONNELL and MALIN.

I should like to limit this discussion to recent developments in our work. To study the specificity of the behavioral effect, we submitted scotophobin-treated animals
to some tests used in our laboratory (startle response to
sound stimulus, avoidance of step down from a platform,
maze learning) and found no significant behavioral changes
in comparison with uninjected controls. Scotophobin was
found to be without effect on the circadian activity cycle
of rats (C. RICHTER, personal communication) and on GABA
activity when administered iontophoretically (D. CURTIS,
personal communication). We also tested scotophobin on isolated smooth muscle preparations without seeing any effect.

The most important advances were made by means of an
ultramicroanalytical method which allows the detection of
20 ng of scotophobin in partially purified brain extracts.
This method consists in converting amino acids and peptides
into fluorescent derivatives by the dansyl reagent and sep-
arating them by two-dimensional thin-layer chromatography
(NEUHOFF et al., 1969). We found a fluorescent spot in the
extracts of brain taken from dark avoidance trained rats
(Rf 0.16 - 0.18 in both dimensions) that was absent from
untrained brain (UNGAR et al., 1972a; UNGAR and BURZYNSKI,
1972). The spot has the same Rf value as dansylated synthetic scotophobin. Identity of the two substances was ascertained by co-chromatography.

By densitometry of the dansyl spot we were able to obtain quantitative data on the formation of the peptide in
the brain during training and on its regional distribution.

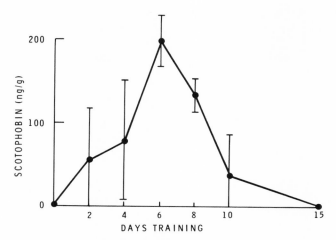

Fig. 4: <u>Quantitative estimation of scotophobin in rat brain</u>
<u>at successive stages of training</u>
Each point is a mean value obtained with 3-6 brains
$\pm$ S.D.

Fig. 4 shows that scotophobin increases progressively in the
brain during the first six days of training and decreases
afterwards, in spite of the persistence of the learned behavior. This observation confirms the bioassay data (UNGAR
et al., 1968) and raises a problem that will be discussed
below.

The regional distribution (Tab. I) shows a predominant
localization in the cortex. A more detailed study will be
attempted both by chemical methods and by autoradiographic
detection of ^{14}C-labeled scotophobin. This is now being
synthesized by Dr. W. PARR and the autoradiographic study
will be done by Professor G. WERNER of the Max-Planck-Ins-

Tab. I: <u>Regional distribution of scotophobin in rat brain</u>
<u>after 6 days training for dark avoidance</u>

	ng/g	%	ng/brain	%
Fronto-parietal cortex	222	35.7	129	40.2
Temporo-occipital cortex	184	29.6	102	31.8
Subcortical region	28	4.5	6	1.9
Brain stem and cerebellum	188	30.2	84	26.1

titute for Brain Research in Frankfurt.

Preliminary results suggest that, although synthetic
scotophobin induces dark avoidance in the goldfish, this
animal, when trained for dark avoidance, produces a different substance. After dansylation, "fish scotophobin"
has a different Rf value from that of "rat scotophobin".
It is hoped to isolate a sufficient amount of the former
to determine the differences between the two substances.

ATTEMPTS AT ISOLATING OTHER BEHAVIOR-INDUCING PEPTIDES

It is obvious that identification of one behavior-inducing substance does not prove by itself the existence of a
molecular code in the nervous system. The most urgent task,
therefore, is to isolate more of these substances. A number
of them have been studied in my laboratory since 1964. They
were all dialyzable and inactivated by incubation with pro-
teases. Tab. II lists them and shows the enzymes that abolish

Tab. II: <u>Enzymic inactivation of peptides extracted from</u>
<u>brain</u>

	Trypsin	Chymotrypsin
Morphine tolerance (UNGAR and COHEN, 1966)	0	+
Sound habituation (UNGAR and OCEGUERA-NAVARRO, 1965)	0	+
Dark avoidance (UNGAR et al., 1968)	+	0
Step down avoidance (UNGAR, 1971a)	+	+
Blue avoidance (UNGAR et al., 1972b)	+	0
Green avoidance (UNGAR et al., 1972b)	+	+
Motor adaptation (HELTZEL et al., 1972)	+	0

+ = inactivation; 0 = no inactivation

their behavioral activity. We selected from this list four
substances that could be obtained in comparatively large
quantities by mass training of the donors.

One of these is the substance inducing <u>habituation to</u>
<u>a sound stimulus</u>. We have trained almost 6,000 rats up to
date and Dr. T.L. INNERARITY was able to isolate enough material to determine the probable amino acid composition (Ala,
Glx, Gly, Lys, Pro, Tyr) but the amount is not sufficient for
a quantitative amino acid analysis or a sequence study.
Additional material is being purified.

ZIPPEL and DOMAGK (1969) have shown that behavior of goldfish based on color discrimination can be chemically transferred. We are training donor goldfish to avoid either the blue compartment or the green compartment of a tank. Extracts of brain taken from these donors reproduce in naive and unreinforced recipients the color-avoidance behavior of the corresponding donors (Fig. 5). The two substances, inducing blue avoidance and green avoidance, have different enzyme specificities. Gel filtration data suggest that both are comparatively small peptides of approximately the same molecular size. We estimate that between 10,000 and 20,000 brains will be necessary for the identification of each of the two substances.

A fourth substance under study is extracted from the brain of goldfish trained to adapt their swimming behavior to a float attached to them. The procedure was first described by SHASHOUA (1968) as a learned motor adaptation that induces changes in precursor incorporation into RNA. Extracts of brain taken from fish wearing the float for four days induce a significantly more rapid adaptation of the recipients than untrained fish brain extracts (Tab. III). The active material is dialyzable but elution data from gel filtration columns suggest a somewhat larger peptide than the others.

It is hoped that other laboratories will undertake similar investigations so that within the next few years a dozen or so of the behavior-inducing molecules will be identified.

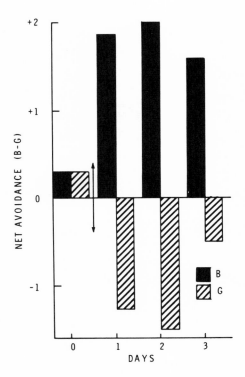

Fig. 5: <u>Blue avoidance and green avoidance in goldfish</u>
<u>injected with brain extract from blue-avoiding</u>
<u>and green-avoiding donors</u>

Abscissa, days after injection of extract (indicated
by arrow); ordinate, mean net avoidance (blue avoidance minus green avoidance) out of five tests.
Solid columns, injected with blue-avoidance extract;
striped columns, injected with green avoidance extract. N = 48 in each group.

Tab. III: Swimming adaptation in recipients of trained brain and in controls

	TC$^+$	$\pm$ S.D.	N	p$^{(*)}$
Experimental				
Crude extract	138	38	36	< .001
Dialyzate	136	32	28	< .001
Control				
Crude extract	200	33	35	

*Compared with controls (t-test)

$^+$TC = time to reach criterion of normal swimming behavior (min.)

VIEWS ON THE OPERATION OF A MOLECULAR CODE

Speculations on the operation of the molecular code may seem premature at this stage. Continuation of research, however, requires working hypotheses which, like the scaffold of a construction, hold temporarily together the experimental data. Tentative interpretations are also necessary to show that the molecular concept is neither "impossible" nor "unnecessary", as maintained by some of its opponents. Three categories of hypotheses have been formulated:

1. The so-called "non-neurological" interpretations, derived from the "field theory", tend to replace neural nets

with "molecular nets" (SCHMITT, 1962). Experience would be recorded in "tape recorder" molecules and memory traces retrieved from them by a mechanism independent of specific neural pathways and connections (MCCONNELL, 1965; LANDAUER, 1964; ROBINSON, 1966). These interpretations ignore the highly differentiated organization of the nervous system and view the brain "as having all the finer structure of a bowlful of porridge" (HEBB, 1949).

2. A somewhat more widespread view is that the molecular code results from the transduction of electrical wave patterns into molecular structures (HALSTEAD and RUCKER, 1970; HYDEN, 1959). It makes the tacit assumption that these patterns encode all the relevant information. There is, however, no evidence at present that they have any meaning when they are taken out of the context of the pathways from which they were recorded and that they contain anything but some quantitative information. Furthermore, we know of no mechanism by which the transduction could take place.

3. The most important elements of neural information are probably represented by specific neural circuits. In the hypotheses of SZILARD (1964), ROSENBLATT (1967), BEST (1968), molecular markers operate as "signposts" directing the traffic of nerve impulses, thus creating the circuits representing the newly acquired information.

In the speculations I published during the past few years (1968, 1970a,b, 1972a,b,c), the molecular code is based on the system of recognition molecules. Postulated

by the concept of "chemospecificity of pathways", this system was first outlined by SPERRY (1963) and was supported
by the findings of GAZE (1970) and of JACOBSON (1971). It
postulates that the development of neural pathways during
embryonic life is controlled by chemical labels that allow
neurons belonging to the same pathway to recognize each
other and establish synaptic contacts. This labeling may
derive from the basic biological process by which homotypic
cells recognize each other (MOSCONA and MOSCONA, 1963) and
which may have reached a higher degree of differentiation
in the nervous system than in other tissues (GARBER and
MOSCONA, 1972).

The genetically determined recognition system can account for the organization of the innate pathways that control reflex and instinctive behavior. It is conceivable that
the same system could be adapted to the processing of acquired information by the addition of a non-genetic mechanism by which the labels of different pathways could be
combined. My assumption is that when, in the course of the
learning process, new connections are created between the
existing pathways, these are encoded as the combination of
the labels that mark the pathways involved. For example, a
new junction between two pathways, labeled $\underline{a}$ and $\underline{b}$, could
result in the formation of the complex $\underline{ab}$. If the labels
are peptides, their combination could be catalyzed by enzymes known as transpeptidases or peptide synthetases. We
observed the presence of this type of enzyme in the brain
and preliminary results suggest the increase of at least
one of them during the early phase of dark avoidance train-

Tab. IV: <u>Glutamyl transpeptidase (α- and γ-GT) activity in</u>
<u>normal and dark avoidance trained rat brain</u>

	α-GT	±	S.D.	N	γ-GT	±	S.D.	N
Normal	11.6		1.9	6	13.6		2.9	6
2 day training	18.45*		1.9	4	14.65		3.9	4
6 day training	13.3		3.3	4	15.3		0.6	4

*Significantly different from normals (p < .01 by <u>t</u>-test)

Results expressed as µmoles of ß-naphthylamide released
from a synthetic substrate per g of brain per hour
(ORLOWSKI and SZEWCZUK, 1972).

ing (Tab. IV). Systematic investigation of this enzyme is
currently being done in our laboratory.

We can assume that any stimulation of a given pathway
results in increased synthesis of its specific label,
whether learning takes place or not. During learning two
or more pathways are stimulated at the same time so as to
create new synaptic junctions between them. If the syn-
thetase is induced at the same time, it may catalyze the
merging of the peptide fragments into a new molecular
species. This may take place in certain types of stra-
tegically located cells: "memory cells" of SZILARD (1964),
"modifiable neurons" of BRINDLEY (1967) or perhaps in
glial cells.

Once these molecules are synthesized they mark the new
connection, just as the original labels mark the innate junc-

tions, and direct the impulses into new channels. The para-
dox of the gradual disappearance of the substance after
the first week of training, in spite of persistence of
the learned behavior (Fig. 4), can have two possible explanations: once the junction is well consolidated, the
coded molecule may no longer be necessary or, more probably, it is still there but in amounts undetectable by
the methods used. Autoradiography may give an answer to
this question.

CONCLUSIONS

Evidence has been accumulating during the last seven
years for the presence in the brain of substances encoding
acquired information. This has been demonstrated by using
recipient animals as decoding devices. Once the substance
has been identified by a behavioral bioassay, its presence
can be detected and measured by chemical methods.

It is hoped that when this is accomplished for several
different substances, each encoding a different item of information, all reasonable objections to the molecular hypothesis will be overcome. There are, however, other objections based on the feeling that the chemical transfer of
information is "impossible", no matter what the experimental results might be. If this feeling is shared by influ-
ential scientists, it may be an obstacle to further work
by denying financial support and outlets for publication.
It is therefore important to fit the experimental data

into a conceptual framework that can be tested experimentally and that shows the molecular hypothesis to be in agreement with many well-established facts.

ACKNOWLEDGEMENT

The experimental work reported in this paper is supported by a grant from the U.S. Office of Education.

LITERATURE

BEST, R.M.: Encoding of memory in the neuron.
Psychol. Rep. 22, 107-155 (1968).

BOOTH, D.A.: Neurochemical changes correlated with learning and memory retention.
In: Molecular Mechanisms in Memory and Learning (G. UNGAR, ed.), pp. 1-57, Plenum Press, New York (1970).

BRINDLEY, G.S.: The classification of modifiable synapses and their use in models for conditioning.
Proc. Roy. Soc. B. 168, 361-376 (1967).

BRYANT, R.C., N.N. SANTOS, and W.L. BYRNE: Synthetic scotophobin in goldfish: specificity and effect on learning.
Science 177, 635-636 (1972).

COHEN, H.D.: Learning, memory, and metabolic inhibitors.
In: Molecular Mechanisms in Memory and Learning (G. UNGAR, ed.), pp. 59-70, Plenum Press, New York (1970).

DYAL, J.A.: Transfer of behavioural bias and learning en-
    hancement: A critique of specificity experiments.
    In: Biology of Memory (G. ADAM, ed.), pp. 145-161,
    Akadēmiai Kiadō, Budapest (1971).

FJERDINGSTAD, E.J., ed. Chemical Transfer of Learned Infor-
    mation, North-Holland Publishing Company, Amsterdam-
    London (1971).

FOERSTER, H.: Das Gedächtnis; Eine quantenmechanische Un-
    tersuchung. F. Deuticke, Vienna (1948).

GARBER, B.B., and A.A. MOSCONA: Reconstruction of brain tis-
    sue from cell suspensions I and II.
    Develop. Biol. 27, 217-234 & 235-243 (1972).

GAZE, R.M.: The Formation of Nerve Connections.
    Academic Press, London-New York (1970).

GERARD, R.W., and J.W. DUYFF, ed. Information Processing in
    the Nervous System.
    Excerpta Medica, Amsterdam (1962).

GUTTMAN, H.N., G. MATWYSHYN, and G.H. WARRINER: III, Synthet-
    ic scotophobin-mediated passive transfer of dark avoid-
    ance.
    Nature New Biol. 235, 26-27 (1972).

HALSTEAD, W.C., and W.B. RUCKER: The molecular basis of mem-
    ory.
    In: Molecular Approaches to Learning and Memory (W.L.
    BYRNE, ed.), pp. 1-14, Academic Press, New York (1970).

HEBB, D.O.: The Organization of Behavior.
    Wiley, New York (1949).

HELTZEL, J.A., R.A. KING, and G. UNGAR: Possible molecular
    coding for a learned motor adaptation in the goldfish.
    Abstract of presentation to Society for Neuroscience
    2nd Annual Meeting (1972).

HYDEN, H.: Biochemical changes in glial cells and nerve
    cells at varying activity.
    In: Biochemistry of the Central Nervous System (Fourth
    International Congress of Biochemistry), pp. 64-89,
    Pergamon Press, New York-London (1959).

JACOBSON, M.: Developmental Neurobiology, Holt, Rinehart
    and Winston, New York (1970).

KATZ, J.J., and W.C. HALSTEAD: Protein organization and
    mental function.
    Comp. Psychol. Monog. 20, 1-38 (1950).

LANDAUER, T.K.: Two hypotheses concerning the biochemical
    basis of memory.
    Psychol. Rev. 71, 167-179 (1964).

MCCONNELL, J.V.: A tape recorder theory of memory.
    Worm Runner's Digest 7 (2), 3-10 (1965).

MCCONNELL, J.V., and D. MALIN (in this volume).

MALIN, D., and H.N. GUTTMAN: Action of synthetic scotophobin
    in mice. Science (in press).

MONNE, L.: Functioning of the cytoplasm.
    Advan. Enzymol. 8, 1-69 (1948).

MOSCONA, M.H., and A.A. MOSCONA: Inhibition of adhesiveness
    and aggregation of dissociated cells by inhibitors of
    protein and RNA synthesis.
    Science 142, 1070-1073 (1963).

NEUHOFF, V., F. von der HAAR, E. SCHLIMME, and M. WEISE:
Zweidimensionale Chromatographie von Dansyl-Aminosäu-
ren in pico-Mol-Bereich, angewandt zur direkten Charakterisierung von Transfer-Ribonucleinsäuren.
Hoppe-Seyler Z. Physiol. Chem. 350, 121-128 (1969).

ORLOWSY, M., and A. SZEWCZUK: Determination of γ-glutamyl
transpeptidase in human serum and urine.
Clin. Chim. Acta 7, 755-760 (1962).

PERKEL, D.H., and T.H. BULLOCK: Neural coding.
Neurosci. Res. Prog. Bull. 6, 221-343 (1968).

ROBINSON, C.E.: A chemical model of long-term memory and
recall.
In: Molecular Basis of Some Aspects of Mental Activity
(O. WALAAS, ed.), Vol. 1, pp. 29-35, Academic Press,
New York (1966).

ROSENBLATT, F.: Recent work on theoretical models of biological memory.
In: Computer and Information Sciences (J. TOU, ed.),
Vol. 2, pp. 33-56, Spartan Books, Washington, D.C.
(1967).

SCHMITT, F.O.: Macromolecular Specificity and Biological
Memory.
MIT Press, Cambridge, Massachusetts (1962).

SHASHOUA, V.E.: RNA changes in goldfish brain during learning.
Nature 217, 238-240 (1968).

SPERRY, R.W.: Chemoaffinity in the orderly growth of nerve
fiber patterns and connections.
Proc. Nat. Acad. Sci. U.S.A. 50, 703-710 (1963).

SZILARD, L.: On memory and recall.
Proc. Nat. Acad. Sci. U.S.A. 51, 1092-1099 (1964).

THINES, G., and G.F. DOMAGK (in this volume).

UNGAR, G.: Molecular mechanisms in learning.
Perspectives Biol. Med. 11, 217-232 (1968).

UNGAR, G.: Role of proteins and peptides in learning and
memory.
In: Molecular Mechanisms in Memory and Learning (G.
UNGAR, ed.), pp. 149-175, Plenum Press, New York (1970a).

UNGAR, G.: Molecular mechanisms in information processing.
Int. Rev. Neurobiol. 13, 223-253 (1970b).

UNGAR, G.: Chemical transfer of acquired information.
In: Methods in Pharmacology (A. SCHWARTZ, ed.), Vol. 1,
pp. 479-513, Appleton-Century Crofts, New York (1971a).

UNGAR, G.: Bioassays for the chemical correlates of acquired information.
In: Chemical Transfer of Learned Information (E.J.
FJERDINGSTAD, ed.), pp. 31-49, North-Holland Publishing Company, Amsterdam-London (1971b).

UNGAR, G.: Le code moléculaire de la mémoire.
La Recherche 3, 19-27 (1972a).

UNGAR, G.: Molecular organization of neural information
processing.
In: The Structure and Function of Nervous Tissue (G.
H. BOURNE, ed.), Vol. 4, p. 215-247, Academic Press,
New York (1972b).

UNGAR, G.: Molecular coding of information in the nervous
    system.
    Naturwissenschaften 59, 85-91 (1972c).

UNGAR, G., and S.R. BURZYNSKI: Detection of a behavior-in-
    ducing peptide (scotophobin) in brain by ultramicro-
    analytical method.
    Fed. Proc. 31, 398Abs (1972).

UNGAR, G., and G. CHAPOUTHIER: Mécanismes moléculaires de
    l'utilisation de l'information par le cerveau.
    L'Année Psychologique 71, 153-183 (1971).

UNGAR, G., and M. COHEN: Induction of morphine tolerance
    by material extracted from brain of tolerant animals.
    Int. J. Neuropharmacol. 5, 183-192 (1966).

UNGAR, G., and C. OCEGUERRA-NAVARRO: Transfer of habituation
    by material extracted from brain.
    Nature 207, 301-302 (1965).

UNGAR, G., L. GALVAN, and R.H. CLARK: Chemical transfer of
    learned fear.
    Nature 217, 1259-1261 (1968).

UNGAR, G., D.M. DESIDERIO, and W. PARR: Isolation, identi-
    fication and synthesis of a specific-behavior-inducing
    brain peptide.
    Nature 238, 198-202 (1972a).

UNGAR, G., L. GALVAN, and G. CHAPOUTHIER: Possible chemical
    coding of color discrimination in goldfish brain
    Experientia in press (1972b).

ZIPPEL, H.P., G.F. DOMAGK: Versuche zur chemischen Gedächt-
    nisübertragung von farbdressierten Goldfischen auf un-
    dressierte Tiere.
    Experientia 25, 938-940 (1969).

RECENT EXPERIMENTS IN MEMORY TRANSFER

James V. McConnell and David H. Malin

Mental Health Research Institute
The University of Michigan
Ann Arbor, Michigan, USA

ABSTRACT

UNGAR and his colleagues have recently identified and
synthesized a polypeptide that appears to cause avoidance
of a dark chamber when injected into mice in doses of 0.5
micrograms. UNGAR has called this substance "scotophobin",
and has provided samples for testing in our laboratory.
Our initial attempts to replicate UNGAR's work yielded
success, but only when we injected our mice with 3 to 6
times as much scotophobin as UNGAR found most effective.
When we controlled for possible degradation of the material,
however, we obtained results similar to those from UNGAR's
laboratory. Additional research suggests that the effect
is highly specific to the experimental situation in which
scotophobin was originally obtained and that scotophobin
does not merely cause an increase in general emotionality
or a generalized "fear of the dark". New behavioral measures of the strength of the effect have proved at least as

343

sensitive as the older way of merely counting the number of
seconds that injected animals spent in dark or white chambers.

In 1967, GAY and RAPHAELSON reported that they had
been able to transfer chemically a tendency to avoid the
dark from donor to recipient rats. GAY and RAPHAELSON used
a simple piece of apparatus that contained three connected
chambers. Two of these chambers were white, while the third
was black. When untrained donor rats were put into the middle white chamber, for the most part they showed a strong
preference for the dark - that is, the untrained animals
entered the dark chamber frequently and spent more time
there than they did in the white chambers. The donor animals were then given a set of training trials in which they
were punished for entering the black compartment. The rats
soon learned to avoid the dark chamber and spent most if
not all of their time in the white compartments. Then they
were sacrificed, their brains removed, and a crude RNA extract was made from the donor brains. This RNA extract -
which surely contained considerable protein as well as
nucleic acid - was then injected into untrained rats. The
recipient animals were subsequently put into the apparatus
for testing. GAY and RAPHAELSON reported that, after the
injection, the animals appeared to avoid the dark chamber
and spend most of their time in the white compartments.
Control animals injected with the same sort of extract from
untrained donors still showed a strong preference for the
dark.

Georges UNGAR and his colleagues were among the first

groups to replicate the GAY and RAPHAELSON experiment successfully. There followed from UNGAR's laboratory an exciting series of studies in which the protein responsible for
this "memory transfer" was first isolated, then characterized as a pentadecapeptide, then synthesized. The synthetic
polypeptide - which UNGAR has called "scotophobin" - appeared to have the same biological activity as did the nat-
urally-occurring substance. Surely these reports by UNGAR,
PARR and their co-workers must rank as very exciting contributions to our knowledge about the biochemistry of memory.

Shortly after scotophobin was synthesized, UNGAR was
kind enough to bring a sample to us at Michigan. We are
pleased to state that, on the whole, our attempts to repeat
UNGAR's initial findings have been successful. Our first
batch of scotophobin was hand-delivered to us by UNGAR; it
came in a 1.0 mg/ml solution in methanol which had been
transported at room temperature for several days. Subsequently the scotophobin was refrigerated for several weeks
during the course of our experiments.

In our first study, we used mice as recipients. The
animals were first screened for dark preference by being
put into the apparatus and allowed to wander about freely
for three minutes. Only those animals that spent at least
half the time in the dark box were used as test animals.
The mice were then divided into control and experiment
groups. The experimental mice each received one intraperitoneal injection of scotophobin dissolved in 0.25 ml

of distilled water; the control animals were injected with
a similar amount of distilled water. The mice were tested
several times after the injection by being placed in the
experimental apparatus and allowed to wander about freely
for three minutes. We recorded the number of seconds that
each mouse spent in the dark compartment.

UNGAR and his colleagues got their best results when
they injected their experimental mice with 0.5 micrograms
of scotophobin. Pilot work in our laboratory gave us little
or no positive results when we used such a low dose. We then
increased the dose level to 1.5 micrograms and 3.0 micrograms. The results for 1.5 micrograms are shown in Fig. 1.

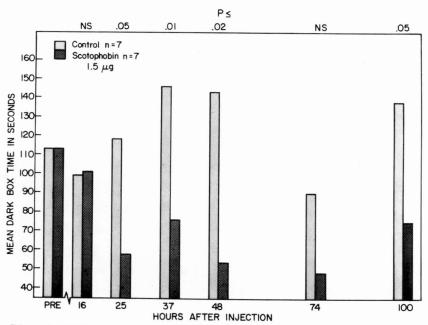

Fig. 1: <u>Effect of 1.5 µg scotophobin on time spent in dark
box</u>

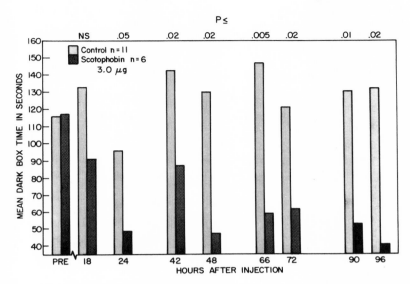

Fig. 2: Effect of 3.0 µg scotophobin on time spent in dark box

As the data indicate, the mice injected with scotophobin spent significantly less time in the black compartment than did the control animals injected with distilled water.

The results for the 3.0 microgram dose of scotophobin are shown in Fig. 2. Again the data suggest that the mice injected with scotophobin showed a significant avoidance of the dark chamber that the control animals did not show.

As pleased as we were with these results, it bothered us that we had to use from 3 to 6 times as great a dose of scotophobin as did UNGAR and his colleagues. It occurred to us that some degradation of the scotophobin might have taken place. In a subsequent series of studies, we used both a new batch of the material (again kindly provided by Professor

UNGAR) and new methods of handling the chemical as well.
This second batch of scotophobin arrived as a powder under
vacuum. It was dissolved in de-ionized distilled water and
the solution was divided into aliquot parts sufficient for
individual experiments. Each aliquot was placed in a Ver-Tis
vac-seal vial, lyophylized, and then was sealed under the
lyophylizer vacuum. The vials were then placed in a refrigerated vacuum desiccator. They were removed from refrigera-
tion and rehydrated one by one as needed. We screened seven
new groups of control and experimental mice and then injected them either with scotophobin or with distilled water.
Seven different dose levels of scotophobin were used, ranging from 0.25 micrograms to 3.0 micrograms. The results of
these experiments with our second batch of scotophobin are
shown in Fig. 3.

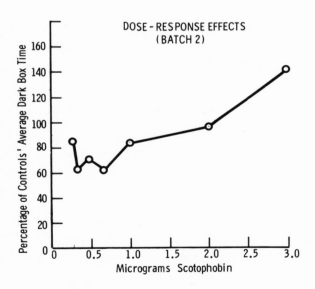

Fig. 3: <u>Dose-response effects of scotophobin</u>

The data in Fig. 3 are recorded in terms of the percentage of the control animals' average time spent in the
dark chamber. A score of less than 100 % suggests that the
controls were spending <u>more</u> time in the dark than were mice
injected with scotophobin, while any point on the graph
greater than 100 % indicates that the controls were spending <u>less</u> time in the dark compartment than were the experimental animals. As Fig. 3 shows, using the new batch and
the better chemical controls, we found that doses around
0.5 micrograms gave us statistically significant results,
while the higher doses did not.

It is apparent from this dose-response curve that mice
injected with 0.5 micrograms of scotophobin tend to spend
less time in the dark chamber after injection than they do
before being injected. So we asked ourselves, how could this
occur? Does scotophobin actually induce a fear of the dark
compartment in the recipient animals? Or is the effect more
general? Perhaps all the scotophobin does is to make the
mice more emotional and hence afraid to explore any new
environment. In short, we wondered if there were any stimulus specificity to this "transfer" effect induced by the
scotophobin injections. In our next series of experiments,
then, we set out to measure the emotionality of mice injected with scotophobin when they were placed in one of
several different test situations.

There are many different measures of emotionality in
animals, of which the simplest and perhaps most reliable is
counting the number of fecal bolluses that the animal drops.

In our studies, we first let our mice become habituated to
the UNGAR apparatus by giving them three sessions of free
exploration without punishing them in any way. Then the mice
were injected with 0.5 micrograms of scotophobin or with
distilled water. Then, 48 hours later, they were forced
through the doorway of one of the white chambers and were
locked into the compartment. Three minutes later the animals were removed and the number of fecal droppings they
had left behind was recorded. Twelve hours later, the same
procedure was followed using the black chamber. Twelve hours
after that, the mice were placed into an entirely different
situation, namely, a small transparent box with a grid floor.
Again, the number of fecal bolluses left by each mouse was
recorded.

The results of the first half of this study appear in
Fig. 4. The control animals showed greater emotionality in
the white chamber and less emotionality in the black chamber
than did the animals injected with scotophobin. Since neither
group had experienced the transparent box before, the scores
for this chamber are higher than for either the white or
black compartment, but again the control animals had higher
scores than did the mice injected with scotophobin. Taken
altogether, these scores show a statistically significant
interaction.

Since, in this study, the order of presentation of the
white and black chambers was confounded, we ran the study
again this time putting both groups of mice into the black
chamber first and then into the white chamber. The results
of the second half of this experiment appear in Fig. 5. As

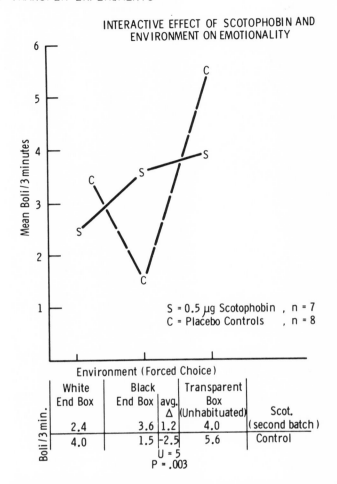

Fig. 4: <u>Interactive effect of scotophobin and environment on emotionality</u>

can be seen, again the experimental animals dropped more fecal bolluses in the black chamber and fewer bolluses in the white chamber than did the control animals. And again these results are statistically significant. It would appear that scotophobin injections do cause an increase in emotionality that is somehow related to the dark chamber.

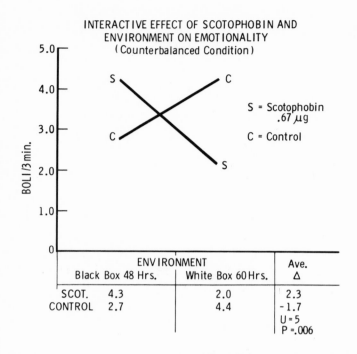

Fig. 5: <u>Interactive effect of scotophobin and environment</u>
<u>on emotionality (counterbalanced condition)</u>

But is it the dark compartment - with all its myriad
cues and stimuli - or merely darkness itself that causes the
emotional response? In a pilot study, we gave mice injected
with 0.5 micrograms of scotophobin a chance either to turn
on or turn off a light, thinking that the experimental animals might turn on the light if it was darkness that they
feared. However, this did not seem to be the case. So, next,
we constructed a new series of test situations. One was the
dark chamber as it actually was used in the UNGAR apparatus.
Another test situation utilized the same chamber, but with

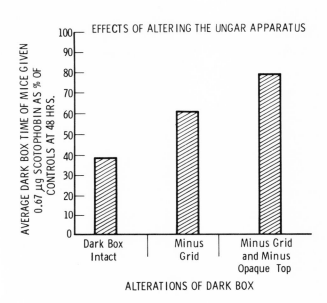

Fig. 6: <u>Average dark box time of mice given 0.67 μg scoto-
        phobin as % of controls at 48 hrs.</u>

the standard grid floor removed. The third situation utilized the same chamber but with both the grid floor and the
standard opaque cover to the box removed. Mice injected with
0.67 micrograms of scotophobin were tested in all three situations. As the data in Fig. 6 indicate, the more like the
original apparatus the test situation was, the more the injected animals tended to avoid the dark chamber.

We interpret these results as follows: scotophobin does
not cause a generalized increase in the recipient's emotionality - instead, the fear response is specific to the black

chamber. Furthermore, the response mediated by the scotophobin is not merely that of "avoiding the dark". Rather,
scotophobin appears to mediate the avoidance of a specific
experimental chamber only one of whose relevant stimulus
dimensions is darkness.

These experiments on emotionality led us to a further
set of studies that, in a way, are the most interesting of
all.

In our first "memory transfer" experiments, we used
planarian flatworms as subjects rather than mice or rats.
In this early work, the donor worms were given classical or
PAVLOVIAN conditioning trials. The conditioning stimulus was
the onset of a bright light over the worm's conditioning
trough; the unconditioned stimulus was an electric current
passed through the water in the trough. At first, the donor
worms did not respond very frequently to the light. After
100 to 200 trials, however, they responded at least 90 % of
the time. We then cut the donors in bits and fed them to
hungry, untrained cannibal planarians. A day or so later,
the cannibals were given the same sort of training that the
donors had been subjected to. We found that, right from the
first few trials, cannibals that had eaten conditioned donors
did significantly better than cannibals that had ingested
untrained donors.

Critics of our early work suggested that the effect we
found was not "memory transfer" at all, but rather was the
transfer of some sensitizing chemical created in the bodies

Fig. 7: <u>T-maze training apparatus used in planarian</u>
<u>experiments</u>

of the donors by the light and shock. Subsequent studies in
several laboratories suggested this criticism was unwarrant-
ed. However, in an attempt to get away from using electric
shock, we soon switched to training our donor animals in a
small T-maze, shown in Fig. 7.

One arm of the maze was painted light gray, the other

dark gray. We adjusted the brightness of the arms so that
the planarians, in general, showed neither a preference for
the light arm nor for the dark arm. Half the donor animals
were trained to go to the light arm, the other half to the
dark arm. In half the trials, the dark arm was on the right,
while in the other half of the trials, the dark arm was on
the left. If the worm went to the correct arm, it was re-
turned to its home bowl to rest for a period of at least
five minutes. If the animal made an incorrect choice, it
was immediately picked up and started over again and was
then forced to go into the correct arm. The animals were
given 10 trials a day until they reached the criterion of
9 correct choices out of 10 for two days in a row. Under
these conditions, it typically takes a planarian 200 or
more trials to learn the maze.

As soon as the donor animals reached criterion, they
were cut in small pieces and were fed to one of three groups
of untrained cannibals. The cannibals were all given two
feedings of donor tissue, 3 days apart. Group I consisted
of 8 cannibals, each of which was trained to go to the same
color arm as had been the donor it ate (we called this the
"positive" transfer or "+" group). The eight cannibals in
Group II were trained to go to the opposite color arm as
had been the donors they ate (we called this the "negative"
transfer or "-" group). The eight cannibals in Group III
were given two feedings of both light-trained and dark-
trained donors - that is, the Group III animals were fed
"conflicting instructions". Half the Group III animals were
trained to go to the light arm, half to the dark arm. We

called this group the "mixed" transfer or "$\pm$" group. The
eight animals in Group IV were fed untrained donors (this
was our "zero" transfer group). Twenty-four hours after
their second feeding, all the cannibals were assigned code
numbers and were trained to criterion by an observer who
did not know which cannibal had eaten which donor.

Tab. I: Mean trials to criterion in a T-maze for four groups
         of planarians fed different types of donor animals

	Group I (+ Group)	Group II (- Group)	Group III ($\pm$ Group)	Group IV (0 Group)
Mean trials to criterion	113.8	166.3	263.8	228.8

The results of this experiment are shown in Tab. I.
The Group I animals reached criterion significantly sooner
($p < 0.05$ or better) than all other groups. The Group II
animals reached criterion significantly sooner than did
Group III and IV ($p < 0.05$ or better). The Group III and
Group IV animals were not significantly different, although
the difference did approach significance (p about 0.10).

It seems to us that this study offers rather good data
for drawing several interesting conclusions about the
"memory transfer" effect. To begin with, the fact that the
Group I animals were significantly superior to all other
groups - but particularly to the Group II animals - suggests
that the transfer effect is stimulus-specific. Second, the

fact that the Group II animals were significantly superior
to those in Group IV suggests that nonspecific as well as
specific factors can be transferred, but that with the proper
experimental design we can differentiate between these two
types of factors. Third, the fact that the Group III animals were significantly inferior to the animals in Groups I
and II suggests that the chemical mediating the tendency to
approach the light arm is distinctly different from the
chemical mediating the tendency to approach the dark arm,
for the two chemicals appear to be antagonistic to each
other.

There is another bit of data that sheds further light
on this "chemical antagonism". During the training of the
cannibals, we kept close track of the behavior that all the
worms showed at the choice point in the maze. We found that
the Group III animals showed much more emotionality - that
is, much more vacillation at the choice point, more turning
around, more refusals to run the maze at all - than did the
animals in any other group.

Surprisingly enough, GAY and RAPHAELSON reported much
the same sort of thing in their first study of dark-avoidance in rats. In describing their work informally, GAY and
RAPHAELSON noted that some of their experimental animals
would actually enter the dark chamber, but would do so re-
luctantly and hesitantly. Sometimes they would squeak while
doing so. Other rats would come close to the dark chamber,
act as if they wanted to enter but instead would lie down
with just their noses in the dark. Although GAY and

RAPHAELSON did not quantify their observations, they were of the opinion that their experimental animals produced more fecal bolluses near the dark chamber than anywhere else in the apparatus.

Keeping all these findings in mind, we have recently begun to record a great deal more of the behavior our animals show in the UNGAR apparatus than merely the amount of time the animals spend in the dark chamber We have categorized three additional response patterns that our animals show. The first is what we call "turning away". The animal approaches within a body length of the black doorway with its head pointed at the door and then turns and walks in the other direction. The second response pattern is that of "hesitation in the doorway". The animal sits stationary on the edge of the door into the black compartment with its body at an oblique angle to the threshold and alternately looks into the black box and then into the white chamber. This is a particularly unusual and striking behavior. The third response is what we call "sitting in the far white corner". The animal settles down in one of the corners furthest away from the entrance to the black chamber for at least five seconds and usually grooms itself fairly vigorously.

All of these behavior patterns suggest some elevated level of emotionality that is directed towards the black chamber. If scotophobin does indeed increase the recipient's "fear of the dark", then we would expect that mice injected with synthetic scotophobin would show more of these behaviors than would animals injected with distilled water. The

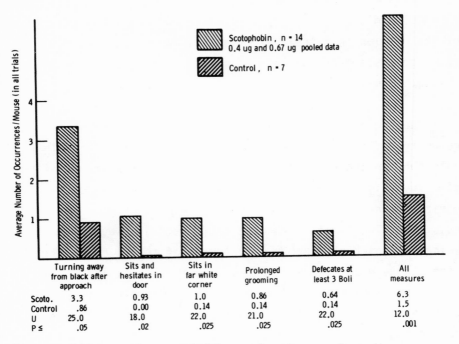

Fig. 8: Effects of scotophobin on frequency of various behaviors in the black-white box

results of our most recent observations are shown in Fig. 8.

As the data indicate, the experimental animals show significantly more of these emotional behaviors than do the controls. Since we often detect these behavior patterns even in experimental animals that eventually spend a fair amount of time in the dark chamber, it is our thought that these behavior patterns might be at least as sensitive a measure of the transfer effect as is the gross amount of time the recipient animal spends in the dark compartment.

In his 1914 text on comparative psychology, the great

American behaviorist John B. WATSON advised his readers
that the first thing they should do when beginning an experiment was to watch their animals closely in the experimental situation for a long period of time. Only then,
WATSON concluded, would we know what kind of an experiment
to run and what part of the animal's behavior we ought to
be measuring. It seems to us that WATSON's advice is as
pertinent to the "memory transfer" experiments in 1972 as
it was to the learning experiments WATSON began more than
60 years ago.

LITERATURE

GAY, R., and A. RAPHELSON: "Transfer of learning" by in-
    jection of brain RNA: A replication.
    Psychonomic Science 8 (9), 369-370 (1967).

MCCONNELL, J.V.: Memory transfer through cannibalism in
    planarians.
    Journal of Neuropsychiatry 3 (1), 542-548 (1962).

MCCONNELL, J.V., and J.M. SHELBY: Memory transfer experi-
    ments in invertebrates.
    In: UNGAR, G. (Ed.) Molecular Mechanisms in Memory
    and Learning, Plenum Press: New York 71-101 (1970).

UNGAR, G.: Transfer of learned behavior by brain extracts.
    Journal of Biological Psychology 9 (1), 12-27 (1967).

UNGAR, G.: Molecular neurobiology: Reflections on the
    first years of a new science.
    Journal of Biological Psychology 11 (2), 6-9 (1969).

UNGAR, G., D.M. DESIDERIO, and W. PARR: Isolation,
    identification, and synthesis of a specific-behavior
    induced brain peptide.
    Nature (London) <u>238</u>, 198-202 (1972).

THE EFFECT OF SYNTHETIC SCOTOPHOBIN ON THE LIGHT TOLERANCE OF TELEOSTS (CARASSIUS AURATUS AND TINCA TINCA)

G. Thines, G.F. Domagk and E. Schonne

Department of Comparative Physiology
Université de Louvain, Pellenberg, Belgium

ABSTRACT

Groups of goldfishes and tenches were kept in a light-dark choice aquarium. After intracranial injections of synthetic scotophobin the light tolerance of both species was found to be increased, whereas injections of Ringer solution into control animals did not change their behavior.

Attempts to induce behavioral changes in animals by injecting them with brain extracts prepared from donors previously trained in controlled learning tasks have given rise to controversial views. Experimental results have often been contradictory and unreproducible (DYAL, 1971; GOLDSTEIN et al., 1971). Fishes were used for this kind of study only in a limited number of cases.

ZIPPEL and DOMAGK (1969) were able to establish definite transfer effects in goldfish trained for color dis-

crimination; their results have been confirmed by UNGAR.
Also learned taste discrimination has been transferred successfully in goldfish; in these experiments evidence was
obtained for peptides of low molecular weight being the
transferring factors (ZIPPEL and DOMAGK, 1971a, 1971b).
- Positive transfer experiments in goldfish have also been
published by FJERDINGSTAD (1969) and BRAUD (1970).

Mass training of thousands of rats in a dark avoidance procedure and subsequent extraction of the pooled
brains allowed UNGAR et al.,(in press) to isolate a uniform
and highly active pentadecapeptide. The structure of this
peptide was shown to be as follows:

H_2N-ser-asp-asn-asn-gln-gln-gly-lys-ser-ala-gln-gln-gly-
gly-tyr-$CONH_2$.

PARR and HOLZER (1971), using the solid phase method, were
able to synthesize milligram amounts of this compound, which
UNGAR named "scotophobin". Intraperitoneal injections of
nanogram doses of scotophobin induce a transient dark fear
in rats and mice.

UNGAR has encouraged us to test the behavioral effect
of scotophobin in fish and insects. In the meantime similar
experiments have been published by GUTTMAN and coworkers
(1972).

In our first experiments 3 groups of 5 goldfishes
(average weight 120 g) each were kept in tanks as shown in
Fig. 1; the water temperature was 20^0 C. The tanks were
kept in a quiet, constant temperature room under 12 hours

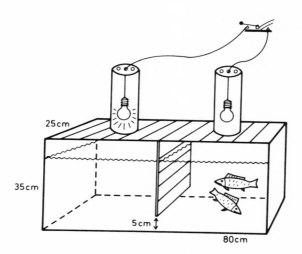

Fig. 1: <u>Tanks used for testing the light-tolerance of gold-fish</u>

The numbers in the figure represent lengths in cm.
All walls of the tank with exception of the front
panel were covered with black paper. Bright sand
was evenly distributed over the floor. Charcoal
filters of equal size were placed in both halves
of the tank.

of light, 12 hours of darkness rhythm. Every day at noon-
time a 30 min test period was given, in which alternatively
one of the two 40 W bulbs was illuminated. Every 30 seconds
the number of fishes being in the light was registered. The
position of the light was changed in 3 min periods. The
total sum of fishes found in the light during 60 counts is
presented as one value in Fig. 2. After an adaptation period
of 3 days group "A", which served as a control, was injected
intracranially with 50 µl of goldfish Ringer solution. The
groups "B" and "C" received 50 µl Ringer containing 500 and
1000 ng of scotophobin respectively. All the fishes survived
the injections and did not show any evident disturbances.

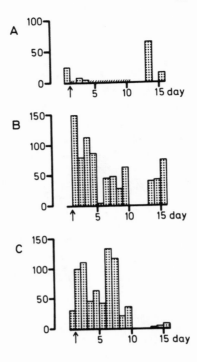

Fig. 2: <u>Effect of intracranial injections of scotophobin</u>
<u>into goldfish</u>

Number of animals found in the light before and
after the injection of Ringer solution (control
group A) and scotophobin (B and C); maximum value =
300. The day of injection is indicated by the arrows. No observations were made on the 11th, 12th
and 13th day.

The tests described above continued for a period of 15 days.

After the injections the fishes showing dark avoidance
increased markedly in groups "B" and "C", whereas the control animals preferred the darkness continuously. Whilst
group "C", injected with the high dose of scotophobin, re-
turned to the innate dark preference after 10 days, the ani-

mals of group "B" retained their light tolerance over more
than 16 days. It should be kept in mind that groups of
Carassius are seldom uniform in their behavior towards
light; nevertheless the 3 animal groups used in our experiments showed comparable levels of dark preference during the pre-injection period.

For our series of experiments we have chosen a species
even more light-avoiding than the goldfish. As was shown in
previous studies (THINES, 1970), the tench, Tinca tinca,
tolerates only a very low level of illumination (10 to 15
lux), beyond which the fish will react always in a negative
manner.

The technique used was the same as described for the
goldfish experiments except that the bulbs of the apparatus
shown in Fig. 1 were replaced by 5W ones, and that 2 observation periods were given per day. 3 groups of 4 tenches
were observed in separate tanks. 2 individuals died in the
course of the 4 months' experiment.

The results of the tench experiments are presented in
Fig. 3. The strong dark preference observed at the end of
the 4 weeks of adaptation was not altered by an intracranial injection of 50 µl of goldfish Ringer. 2 weeks
later 2 fish groups were injected with 50 and 200 ng scotophobin respectively. Ringer solution was injected into the
control animals. Scotophobin induced a marked dark avoidance
lasting for about 4 days. - A second injection of the same
doses of scotophobin, for which experimental and control
animals were exchanged, gave about the same results.

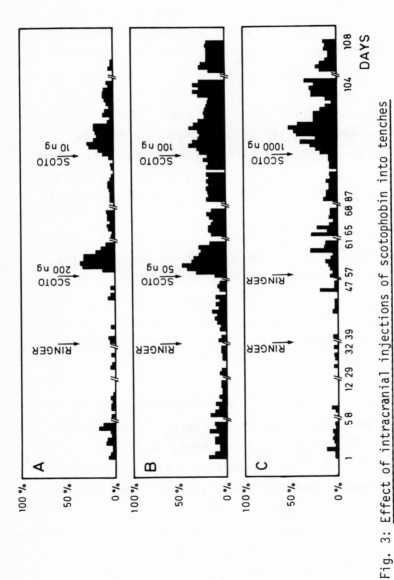

Fig. 3: Effect of intracranial injections of scotophobin into tenches

The bars represent the % of the animals found in the light after various injections of Ringer solution or scotophobin. The day of injection is indicated by the
arrows.

In order to gain information about the dose response we have injected 10, 100 and 1000 ng scotophobin, respectively, into the animals of our 3 groups. All animals reacted positively with an effect lasting for 5 days. However, even with the highest dose, the time spent in the light never exceeded 50 per cent of the test session.

ACKNOWLEDGEMENTS

The skilful assistance of Mademoiselle Genevieve CLAEYS as well as financial support given to us by the Fonds de Developpement Scientifique (F.D.S.) de l'Université de Louvain and of the Deutsche Forschungsgemeinschaft (SFB 33) are gratefully acknowledged. Thanks are also due to Prof. G. UNGAR and W. PARR for repeated generous gifts of synthetic scotophobin.

LITERATURE

BRAUD, W.G.: Extinction in goldfish: facilitation by intra-cranial injection of RNA from brains of extinguished donors.
Science 168, 1234-1236 (1970).

DYAL, J.A.: Transfer of behavioral bias: reality and specificity.
In: Chemical Transfer of Learned Information (E.J. FJERDIGSTAD, ed.), pp. 219-263, North Holland Publ., Amsterdam (1971).

FJERDINGSTAD, E.J.: Memory transfer in goldfish.
    J. biol. Psychol. 9, 20-25 (1969).

GOLDSTEIN, A., P. SHEEHAN, and J. GOLDSTEIN: Unsuccessful
    attempts to transfer morphine tolerance and passive
    avoidance by brain extracts.
    Nature 233, 126-129 (1971).

GUTTMAN, H.N., G. MATWYSHYN, and G.H. WARRINER: Synthetic
    scotophobin-mediated passive transfer of dark avoid-
    ance.
    Nature New Biology 235, 26-27 (1972).

PARR, W., and G. HOLZER: Synthesen von Scotophobin-Analoga.
    Z. physiol. Chem. 352, 1043-1047 (1971).

THINES, G.: Periodic variation in the photic behavior of the
    tench (Tinca tinca L.) in self-regulating conditions.
    I. Interdiscipl. Cycle Res. 1, 4, 367-377 (1970).

UNGAR, G., D.M. DESIDERIO, and W. PARR:
    Nature, in press.

ZIPPEL, H.P., and G.F. DOMAGK: Versuche zur chemischen Ge-
    dächtnisübertragung von farbdressierten Goldfischen
    auf undressierte Tiere.
    Experientia 25, 938-940 (1969).

ZIPPEL, H.P., and G.F. DOMAGK: Transfer of taste preference
    from trained goldfish (Carassius auratus) into un-
    trained recipients.
    Pflügers Arch. 323, 258-264 (1971a).

ZIPPEL, H.P., and G.F. DOMAGK: Experiments concerning the
    transfer specificity of brain extracts in the taste

discrimination of goldfish.
Pflügers Arch. <u>323</u>, 265-272 (1971b)

PEPTIDES AND BEHAVIOR

D. de Wied

Rudolf Magnus Institute for Pharmacology
Medical Faculty, University of Utrecht
Vondellaan 6, Utrecht, The Netherlands

ABSTRACT

Evidence has accumulated that peptides influence
behavior by acting on various structures in the brain. Peptides derived from the anterior pituitary (ACTH, growth
hormone), the intermediate lobe (α-MSH), and the posterior
lobe (vasopressin) of the pituitary have been shown to mod-
ulate conditioned behavior (DE WIED, 1969). Angiotensin II,
a peptide derived from α-2 globulin, is involved in the
control of drinking behavior (FITZSIMONS, 1970) and scotophobin, a peptide extracted from the brain, has been claimed to carry specific information (UNGAR et al., 1968).

ACTH, MSH AND ACTH-ANALOGUES

Pituitary peptides affect acquisition and maintenance
of conditioned behavior. Adenohypophysectomy (DE WIED, 1964)

or ablation of the whole pituitary (APPLEZWEIG and BAUDRY, 1955; DE WIED, 1968) markedly reduces the ability of the rat to acquire a shuttle-box avoidance response. Treatment of hypophysectomized rats with adrenal maintenance doses of ACTH restores the rate of acquisition of a shuttle-box avoidance response toward nearly normal levels (APPLEZWEIG and MOELLER, 1959; DE WIED, 1968). The effect of ACTH, however, is not mediated by the adrenal cortex. Administration of α-MSH which contains the sequence 1-13 of ACTH and even smaller analogues like $ACTH_{1-10}$ or $ACTH_{4-10}$ also restore the rate of acquisition of a shuttle-box avoidance response of the hypophysectomized rat. These peptides are devoid of corticotrophic activities. Thus, the influence of ACTH on behavior is an extra target effect, presumably directed to central nervous structures which are involved in the formation of conditioned and other adaptive responses.

ACTH and related peptides also affect conditioned behavior in intact rats. ACTH administered during shuttle-box training results in an increased resistance to extinction (MURPHY and MILLER, 1955). An even more pronounced effect is found when ACTH is given during extinction (DE WIED, 1967). Although the long term administration of ACTH is associated with hypercorticism, the influence of ACTH on behavior in intact rats is also independent of its action on the adrenal cortex. Synthetic ACTH analogues like the decapeptide $ACTH_{1-10}$ or the heptapeptide $ACTH_{4-10}$ increase resistance to extinction of a shuttle-box as well as a pole-jumping avoidance response. Interestingly, the sequence $ACTH_{11-24}$ is ineffective in this respect (DE WIED et al., 1968).

ACTH also affects passive avoidance behavior of intact rats (LISSAK et al., 1957; LEVINE and JONES, 1965).
Using a simple step-through passive avoidance procedure
(ADER et al., 1972) we found that ACTH administered 1
hour prior to the first retention trial, markedly increased
avoidance latency. The same occurred when $ACTH_{1-10}$ was giv-
en in contrast to $ACTH_{11-24}$ or $ACTH_{25-39}$ which only in
higher doses somewhat increased avoidance latency. Thus,
the active core of ACTH in both "fear" motivated tests
appears to reside in the amino end of the molecule.

ACTH inhibits extinction of an appetitive response too
(SANDMAN et al., 1969, 1971; LEONARD, 1969, GRAY, 1971;
GUTH et al., 1971) and the same holds for $ACTH_{4-10}$, as we
found recently (GARRUD and DE WIED, 1972, non published
observations). Other effects of ACTH are the occurrence
of a stretching crisis in dogs which can also be evoked
with the heptapeptide $ACTH_{4-10}$ (FERRARI et al., 1963) and
a suppressive effect on aggressive behavior in intact and
adrenalectomized mice (BRAIN, 1971; PASLEY and CHRISTIAN,
1972).

One of the first analogues of ACTH which was synthetized was the decapeptide $ACTH_{1-10}$ in which the phenylalanine residue in position 7 was replaced by the D-isomer
($ACTH_{1-10}$ (7-D-phe)). This peptide, as well as $ACTH_{4-10}$
(7-D-phe), was found to facilitate extinction of a shuttle-
box and a pole-jumping avoidance response (BOHUS and DE

WIED, 1966; DE WIED and GREVEN, 1969; VAN WIMERSMA GREIDA-
NUS and DE WIED, 1971). This suggested that 7-D-form peptides could antagonize the action of normally occurring ACTH
and related peptides. However, $ACTH_{1-10}$ (7-D-phe) also facilitated extinction in hypophysectomized rats. Thus, L-
and D-form ACTH analogues act in a non-direct antagonistic
manner. The fact that the introduction of a D-form amino
acid reversed the behavioral effect of the all-L-peptides
supports the specificity of the influence of ACTH analogues
on conditioned avoidance behavior. Further support was obtained by successive replacement of each of the amino acid
residues by a D-isomer substitute in the hexapeptide 8-lys-
$ACTH_{4-9}$ (GREVEN and DE WIED, 1972). Only replacement in
position 7 appeared to reverse the behavioral effect of
this hexapeptide. None of the replacements in other positions facilitated extinction of a pole-jumping avoidance
response. In contrast, these peptides invariably delayed
extinction of the avoidance response like the all-L ACTH
analogues. In most cases replacement by a D-antipode po-
tentiated the inhibitory effect of all-L analogues on extinction (GREVEN and DE WIED, 1972).

VASOPRESSIN AND RELATED PEPTIDES

Removal of the posterior and intermediate lobe of the
pituitary in rats does not affect acquisition of a shuttle-
box avoidance response. It, however, facilitates extinction
(DE WIED, 1965). Treatment of posterior lobectomized rats
with pitressin, a relatively crude extract of posterior

lobe origin, administered as a long acting preparation, normalizes extinction behavior of posterior lobectomized rats (DE WIED, 1965). Pitressin also affects extinction of a shuttle-box avoidance response in intact rats; it increases resistance to extinction (DE WIED and BOHUS, 1966). Pitressin appeared to exhibit a long term effect on the maintenance of an avoidance response which was still demonstrable several weeks after discontinuation of the treatment. This is in contrast to the influence of ACTH and analogues which can be observed only for a relatively short period of time.

The principle present in the posterior lobe of the pituitary which is responsible for the long term effect on extinction, was determined in a pole-jumping avoidance test (DE WIED, 1971). It was found that a single injection of 60 mU (1 μg) of a synthetic lysine vasopressin preparation increased resistance to extinction for a considerable time. Other structurally and physiologically related peptides like oxytocin, angiotensin II, insulin or growth hormone were without effect in this respect.

Lysine vasopressin not only affected extinction of active avoidance behavior, it also affected a passive avoidance response. Injected 1 hr prior to the first retention trial it markedly increased avoidance latency. This effect was maintained for several days in contrast to that of ACTH and ACTH analogues which generally has disappeared 24 hr after injection.

PITUITARY NEUROGENIC PEPTIDES

The studies on avoidance acquisition in hypophysectomized rats demonstrated that a small entity of ACTH, i.e.,
$ACTH_{4-10}$, was able to restore a behavioral deficiency of
these animals. In view of this and of the marked behavioral effects of ACTH analogues and of vasopressin on extinction of conditioned behavior, we postulated that the
pituitary gland might contain small peptides which possibly operate in the formation of adaptive responses (DE WIED,
1968). In order to test this hypothesis, a study was undertaken with the aim of isolating relatively small peptides
related to ACTH, MSH or vasopressin from pituitary tissue
(DE WIED et al., 1970). Starting from a fraction with high
MSH activity from hog pituitary material (SCHALLY et al.,
1962) peptides were purified by ion exchange chromatography
on CMC, gelfiltration on Sephadex G25, paperchromatography
and high voltage electrophoresis (LANDE et al., 1971).
Various active fractions were found. The first fraction
which was isolated in pure form contained all amino acids
of lysine vasopressin except glycine. Structure analysis
studies confirmed that we had isolated desglycinamide
lysine vasopressin. This peptide was highly active in stimulating the rate of acquisition of a shuttle-box avoidance
response in hypophysectomized rats. It also increased resistance to extinction of active and passive avoidance behavior in intact rats and it had, like lysine vasopressin,
a long term effect. Interestingly, desglycinamide lysine
vasopressin is almost devoid of the classical endocrine
activities of lysine vasopressin (DE WIED et al., 1972a).

Tab. I: <u>Amino acid sequence of various behaviorally active peptides</u>

ACTH$_{1-10}$ H-ser-tyr-ser-met-glu-his-phe-arg-try-gly-OH

ACTH$_{11-24}$ H-lys-pro-val-gly-lys-lys-arg-arg-pro-val-lys-val-tyr-pro-OH

Lysine vaso-
pressin   H-cys-tyr-phe-gln-asn-cys-pro-lys-gly-NH$_2$

Desglycinamide
lysine vaso-
pressin   H-cys-tyr-phe-gln-asn-cys-pro-lys-OH

Scotophobin   H-ser-asp-asn-asn-gln-gln-gly-lys-ser-ala-gln-gln-gly-gly-tyr-NH$_2$

DS$_{1-15}$ H-ser————————glu

ZS$_{8-15}$ Z-lys-ser-ala-glu-gln-gly-gly-tyr-NH$_2$

HS$_{8-15}$ H-lys————————

More peptides have been isolated. Some of these are related to ACTH and MSH. For example, one fraction which
was recently isolated contains the amino acid residues
found in $ACTH_{1-8}$. The amino acid composition of several
other isolated peptides, however, does not suggest a
close relation of these peptides with ACTH analogues
or vasopressin analogues (see Tab. I).

SCOTOPHOBIN-LIKE PEPTIDES

The possibility of acquired behavior being conveyed
to naive recipients by chemical principles of brain origin
has been claimed by various investigators (UNGAR et al.,
1968; UNGAR, 1969). One such fraction which could transfer
a conditioned dark-avoidance response, has been isolated
from the brain and characterized as a neutral pentadecapeptide and designated scotophobin (UNGAR, 1970, 1971).
Desacetyl scotophobin (DS_{1-15}) has been synthetized (ALI
et al., 1971, 1972) but it differed from the natural product in terms of both its activity and its R_f values. The
structural assignment, however, shifted to the equivalent
$gln^{5, 11}$ sequence and this is identical with the natural
product (DESIDERIO et al., 1971; PARR and HOLZER, 1971).
DS_{1-15} together with two other peptides whose structures
correspond to residues 8-15 of the parent molecule were
tested in various behavioral tests used to explore the
behavioral activity of ACTH and vasopressin analogues, in
an attempt to determine the active core of DS_{1-15}. Since
the biological activity of DS_{1-15} as compared to the nat-

ural compound was only 2 %, the dose level used was increased by a factor 50. The effect of a single subcutan-
eous injection of 5 µg of the three peptides was tested
on extinction of a pole-jumping avoidance response, on
avoidance latency in a passive avoidance test, in a light-
dark preference and in an open field test (DE WIED et al.,
1972b). The peptides were active only in the active and
passive avoidance test and virtually inactive in the other
tests employed.

DS$_{1-15}$ delayed extinction of the pole-jumping avoidance response. This effect was maintained for at least 5
successive days. The HS$_{8-15}$ and ZS$_{8-15}$ similarly increased
resistance to extinction and the effect of these two
sequences also lasted between 5 and 7 days.

The results obtained in the passive avoidance test
may have a bearing on the mode of action of scotophobin
and are therefore described in more detail. Passive avoidance was tested in a simple step-through avoidance situation (ADER et al., 1972). Animals were placed on a small
illuminated runway leading to a dark box. The rats received 1 trial on day I and 3 trials on day II in which latency
to enter the dark box was measured. After the 3rd trial on
day II animals received electric shocks (0.25 mA; 1 sec)
or sham-shocks in the dark box immediately after entry.
Retention trials were run (1 per day) on subsequent days.
Peptides or placebo (0.5 ml saline) were injected subcutaneously only once, immediately after the shock or the
sham-shock on day II. None of the peptides affected la-

tencies in non-shocked rats on the next day. The latency of
shocked rats treated with saline was slightly increased at the
first retention trial. However, treatment with DS_{1-15} markedly increased the latency to enter the dark box. ZS_{8-15}
similarly increased avoidance latency but HS_{8-15} was much
less active. Latencies remained high during subsequent retention trials.

The influence of DS_{1-15} and analogues resembles that
of ACTH analogues and vasopressin analogues. All these peptides increase resistance to extinction of active and passive avoidance behavior, while gross behavior is not mate-
rially affected. They differ only in their duration of action. DS_{1-15} appeared to be active for 5-7 days, as determined by its effect on the rate of extinction of a pole-
jumping avoidance response. A single injection of an ACTH
analogue is not active for more than 6 hours in this test,
while the effect of lysine vasopressin on extinction may
last for several weeks (DE WIED, 1971).

The influence of DS_{1-15} and related peptides in the
passive avoidance test in which the animals avoided the
dark box, was found only if the rats had been exposed to
mild electric footshock in the box on the day before. This
is also true for ACTH analogues and vasopressin analogues
which do not affect avoidance latency in non-shocked rats.
It might be therefore that the discrepancy in the data reported in the literature on scotophobin-like compounds is
the result of differences in aversive stimulation to which
animals are unintentionally exposed. The fact that the

biological activity of DS_{1-15} resides in the second half
of the molecule makes it difficult to assign a specific
function to this compound. It may be, however, that this
would not have been the case if we had used natural scotophobin or the synthetic $gln^{5, 11}$ pentadecapeptide.

CONCLUSION

During the last decade evidence has accumulated that
peptides affect behavioral performance by an influence on
the central nervous system. ACTH and related peptides (α-
MSH, $ACTH_{1-10}$, $ACTH_{4-10}$) stimulate avoidance learning in
hypophysectomized rats and delay extinction of active and
passive avoidance behavior as well as of appetitive responses in intact rats. Vasopressin, and a vasopressin analogue,
also stimulate acquisition of avoidance behavior in hypophysectomized rats and increase resistance to extinction of
active and passive avoidance behavior in intact rats. As
yet unidentified peptides extracted from the pituitary have
been found to affect acquisition of conditioned behavior
in hypophysectomized rats and extinction in intact rats.

Behaviorally active peptides are also present in the
brain. A pentadecapeptide which has been claimed to contain
the property of stimulating dark-avoidance was found to
affect active and passive avoidance behavior in essentially the same way as ACTH and analogues, and vasopressin and
analogues.

The difference in the behavioral effect of these structurally unrelated peptides is in the duration of their action.

LITERATURE

ADER, R., J.A.W.M. WEIJNEN, and P. MOLEMAN: Retention of a passive avoidance response as a function of the intensity and duration of electric shock.
Psychon. Sci. 26, 125-128 (1972).

ALI, A., J.H.R. FAESEL, D. SARANTAKIS, D. STEVENSON, and B. WEINSTEIN: Synthesis of a structure proposed for scotophobin.
Experientia 27, 1138 (1971).

ALI, A., J.H.R. FAESEL, D. SARANTAKIS, D. STEVENSON, and B. WEINSTEIN: Synthesis of a structure related to scotophobin.
Intern. J. Peptide Protein Res. IV (1972), in press.

APPLEZWEIG, M.H., and F.D. BAUDRY: The pituitary-adrenocortical system in avoidance learning.
Psychol. Rev. 1, 417-420 (1955).

APPLEZWEIG, M.H., and G. MOELLER: Axiety, the pituitary-adrenocortical system and avoidance learning.
Acta Psychol. 15, 602-603 (1959).

BOHUS, B., and D. DE WIED: Inhibitory and facilitatory effect of two related peptides on extinction of avoidance behavior.
Science 153, 318-320 (1966).

BRAIN, P.F.: Possible role of the pituitary-adrenocortical
    axis in aggressive behavior.
    Nature 233, 489 (1971).

DESIDERIO, D.M., G. UNGAR, and P.A. WHITE: The use of mass
    spectrometry in the structural elucidation of scoto-
    phobin - a specific behavior inducing brain peptide.
    Chem. Communs, 432 (1971).

DE WIED, D.: Influence of anterior pituitary on avoidance
    learning and escape behavior.
    Amer. J. Physiol. 207, 255-259 (1964).

DE WIED, D.: The influence of the posterior and intermediate
    lobe of the pituitary and pituitary peptides on the
    maintenance of a conditioned avoidance response in
    rats.
    Int. J. Neuropharmacol. 4, 157-167 (1965).

DE WIED, D., and B. BOHUS: Long term and short term effect
    on retention of a conditioned avoidance response in
    rats by treatment respectively with long acting
    pitressin or α-MSH.
    Nature 212, 1484-1486 (1966).

DE WIED, D.: Opposite effects of ACTH and glucocortico-
    steroids on extinction of conditioned avoidance be-
    havior.
    Exc. Med. Int. Congress Series no. 132, 945-951 (1967).

DE WIED, D.: The anterior pituitary and conditioned avoid-
    ance behavior.
    Exc. Med. Int. Congress Series no. 184, 310-316
    (1968).

DE WIED, D., B. BOHUS, and H.M. GREVEN: Influence of pitui-
    tary and adrenocortical hormones on conditioned avoid-
    ance behavior in rats.
    In: Endocrinology and Human Behavior (R.P. MICHAEL,
    ed.), pp. 188-199, Oxford University Press (1968).

DE WIED, D.: Effects of peptide hormones on behavior.
    In: Frontiers in Neuroendocrinology (W.F. GANONG and
    L. MARTINI, eds.), pp. 97-140, Oxford University
    Press, New York (1969).

DE WIED, D., and H.M. GREVEN: Opposite effect of structural
    analogues of ACTH on extinction of an avoidance re-
    sponse in rats by replacement of an L-amino acid or
    a D-isomer.
    Proc. Int. Union of Physiol. Sciences, vol. VII;
    XXIV Int. Congress of Physiol. Sciences, Washington
    (August 25-31, 1968), (1969).

DE WIED, D., A. WITTER, and S. LANDE: Anterior pituitary
    peptides and avoidance acquisition of hypophysectomized
    rats.
    In: Progress in Brain Research. Pituitary, Adrenal and
    the Brain (D. DE WIED and J.A.W.M. WEIJNEN, eds.),
    Elsevier Publ. Comp. 32, 213-220, Amsterdam (1970).

DE WIED, D.: Long term effect of vasopressin on the main-
    tenance of a conditioned avoidance response in rats.
    Nature 232, 58-60 (1971).

DE WIED, D., H.M. GREVEN, S. LANDE, and A. WITTER: Dis-
    sociation of the behavioural and endocrine effects of
    lysine vasopressin by tryptic digestion.
    Brit. J. Pharmacol. 45, 118-122 (1972a).

DE WIED, D., D. SARANTAKIS, and B. WEINSTEIN: Is scotophobin
    a memory code word?
    (1972b) in prep.

FERRARI, W., G. GESSA, and L. VARGIU: Behavioral effects
    induced by intracisternally injected ACTH and MSH.
    Ann. N.Y. Acad. Sci. 104, 330-343 (1963).

FITZSIMONS, J.T.: The renin-angiotensin system in the con-
    trol of drinking.
    In: The Hypothalamus (L. MARTINI, M. MOTTA and F.
    FRASCHINI, eds.), pp. 195-212, Academic Press, New
    York (1970).

GARRUD, P., and D. DE WIED: unpublished observations (1972).

GRAY, J.A.: Effect of ACTH on extinction of rewarded be-
    havior is blocked by previous administration of ACTH.
    Nature 229, 52-54 (1971).

GREVEN, H.M., and D. DE WIED: The influence of peptides
    derived from ACTH on performance structure activity
    studies.
    Progress in Brain Research, in press.

GUTH, S., S. LEVINE, and J.P. SEWARD: Appetitive acquisi-
    tion and extinction effects with exogenous ACTH.
    Physiol. Behav. 7, 195-200 (1971).

LANDE, S., A. WITTER, and D. DE WIED: Pituitary Peptides:
    An octapeptide that stimulates conditioned avoidance
    acquisition in hypophysectomized rats.
    J. biol. Chem. 246, 2058-2062 (1971).

LEONARD, B.E.: The effect of sodium barbitone, alone and
    together with ACTH and amphetamine, on the behavior of

the rat in the multiple "T" maze.
Int. J. Neuropharmacol. 8, 427-435 (1969).

LEVINE, S., and L.E. JONES: Adrenocorticotropic Hormone
(ACTH) and passive avoidance learning.
J. comp. physiol. Psychol. 59, 357-360 (1965).

LISSAK, K., E. ENDRÖCZI, and P. MEGGYESI: Somatisches Verhalten und Nebennierenrindentätigkeit.
Pflügers Arch. ges. Physiol. 265, 117-124 (1957).

MURPHY, J.V., and R.E. MILLER: The effect of adrenocortico-
trophic hormone (ACTH) on avoidance conditioning in
the rat.
J. comp. physiol. Psychol. 48, 47-49 (1955).

PARR, W., and G. HOLZER: Synthesen von Scotophobin-Analoga. Peptide mit gedächtnisübertragender Wirkung.
Z. Physiol. Chem. 352, 1043 (1971).

PASLEY, J.N., and J.J. CHRISTIAN: The effect of ACTH, group
caging and adrenalectomy in peromysus leucopus with
emphasis on suppression of reproductive function.
Proc. Soc. exp. Biol. Med. 139, 921-926 (1972).

SANDMAN, C.A., A.J. KASTIN, and A.V. SCHALLY: Melanocyte-
stimulating hormone and learned appetitive behavior.
Experientia 25, 1001-1002 (1969).

SANDMAN, C.A., A.J. KASTIN, and A.V. SCHALLY: Behavioral
inhibition as modified by melanocyte-stimulating
hormone (MSH) and light-dark conditions.
Physiol. Behav. 6, 45-48 (1971).

SCHALLY, A.V., H.S. LIPSCOMB, and R. GUILLEMIN: Isolation
    and amino acid sequence of α_2-corticotrophin releas-
    ing factor (α_2-CRF) from hog pituitary glands.
    Endocrinology 71, 164-173 (1962).

UNGAR, G., L. GALVAN, and R.H. CLARK: Chemical transfer of
    learned fear.
    Nature 217, 1259-1261 (1968).

UNGAR, G.: Chemical transfer of passive avoidance.
    Fed. Proc. 28, 647 (1969).

UNGAR, G.: Chemical transfer of information.
    In: Handbook of Neurochemistry (A. LAJTHA, ed.),
    vol. 6, pp. 251-253, Plenum Press, New York (1970).

UNGAR, G.: Chemical transfer of acquired information.
    In: Methods in Pharmacology (A. SCHWARZ, ed.), vol. 1,
    pp. 479-513, Appleton-Century-Crofts, New York (1971).

WIMERSMA GREIDANUS, Tj. B. VAN, and D. DE WIED: Effects of
    systemic and intracerebral administration of two op-
    posite acting ACTH-related peptides on extinction of
    conditioned avoidance behavior.
    Neuroendocrinology 7, 291-301 (1971).

STUDIES WITH DARK AVOIDANCE AND SCOTOPHOBIN

Helene N. Guttman, G. Matwyshyn, and M. Weiler

Department of Biological Sciences
University of Illinois at Chicago Circle
Chicago, Illinois 60680

ABSTRACT

Stability of scotophobin, 5,11-deamidoscotophobin and N-acetyl 5,11-deamidoscotophobin is compared using micro-dansylation followed by chromatographic analysis.

Methods for the routine testing and analysis of re-sults of dark avoidance experiments are considered. The methods allow analysis of results with animals which vary with respect to reaction initiation time, and duration of reaction.

Using the paradigm of dark avoidance activity analysis followed by dark-avoidance training, it is shown that there is low-grade, variable and transient dark avoidance activity for the following compounds: 5,11-deamidoscotophobin, N-acetyl 5,11-deamidoscotophobin, 8-15 amino acid fragment of deamidoscotophobin, and degraded scotophobin.

Proof is presented that scotophobin passes the blood-brain barrier.

Scotophobin does not exert its action by causing a general reduction of activity of goldfish.

Participation of various amino acid residues of scotophobin in binding and activity is discussed. During manifestation of dark avoidance activity, scotophobin is bound to a large molecular weight cellular site. Synaptic membranes or S-100 proteins are candidates for binding materials.

Theoretical considerations on relationship of induction of behavioral alteration and of other inducible, biological processes are briefly considered.

INTRODUCTION

The suggestion made in 1950 (KATZ and HALSTEAD) that learned behavior is encoded and stored in the brain in the form of specific chemical compounds went counter to the then-favored idea of storage in specific electrical impulses. The next development separated learning into subclasses, i.e., that which is only transiently stored and that which is stored for a long time. Implicit as a concomitant for either type of storage is an operational recall system, since proof of storage usually is by a challenge which demands recall. The transient type of storage, currently undisputed as being electrical in nature, was assumed to precede perma-

nent storage until the recent work of DANIELS (1971) and
KESNER and CONNER (1972). These workers' results suggest
that the events leading to short- and long-term storage
may be separate, parallel processes rather than one continuing sequence.

In the decade 1950-1960, tremendous strides took place
in the area of molecular genetics and the general coding
capabilities of chromosomal DNA. In the following decade,
the steps required to translate the encoded information
from DNA to proteins were to a large extent clarified.
Moreover, mechanisms were suggested to explain how changes
in the subcellular environment of cells ranging from pro-
karyotes through higher plants and animals can initiate or
stop the readout of certain genetic information: one example is that which normally occurs during development.
Proteins that are always synthesized by the organism without special commands are called constitutive, whereas those
which are only synthesized at particular times are called
inducible, repressible, or depressible (dependent upon the
control mechanism). It has been suggested that constitutive
syntheses are those for which inducers, or derepressers, are
continuously present.

In one specialized type of "learning", acquisition of
immunity, the signal for the new synthesis to commence is
the presence of an immunogen. Initial acquisition of immunity, through a multiple step process involving several
cell types,takes some time, but after immunity has been
acquired, immunological response to the same immunogen is

almost immediate. This rapid recall is termed <u>immunological</u>
<u>memory</u>. Unexposed (naive) subjects can be immunized <u>pas-</u>
<u>sively</u> by injection with a specific antibody, the proteina-
ceous product of the originally encoded information. However, these passively-immunized animals lose their immunity
once the supply of exogenously-supplied antibody is ex-
hausted.

Using the pinciple of parsimony of nature with respect
to the mechanisms used to control metabolic processes, we
agree with the SZILARD (1954) that the multiple-step process by which a new specific behavioral task can be learned,
stored as a memory, and recalled upon command has much in
common with the immune response (when one discounts the obvious difference that the immune response uses the lymphatic
system and learning-memory-recall uses the nervous system).
Even the passive aspects of both processes resemble each
other. Passive transfer of specific behavioral responses
<u>via</u> brain extracts, or synthetic peptide replicates of those
found in brains of specifically-trained animals, is now
used as an assay for specific behavior-altering molecules
(reviewed by RUSSELL et al., 1972). Perhaps more important
than its assay utility is the further parallel that this
passive reaction has with the immune response and its use-
fulness in elucidating mechanisms.

Rather than dwell at any greater length on general
theoretical considerations associated with the overall neurological meaning of specific behavioral transfer, we will
present instead the results of a series of experiments

designed to lay the foundation for clarification of the
role played by small peptides in the elicitation of passive transfer of specific behavioral response. Only experiments associated with dark avoidance are considered at
this time since resolution of molecular mechanisms associated with the paradigm is most advanced.

METHODS

Common goldfish, 7-7.5 cm long (Auburndale Goldfish
Co., Chicago) were used. Smaller fish yield erratic results. Fish from the only other major supplier in the
U.S., Ozark fisheries, are not as strikingly dark pref-
erring. Fish were maintained in well-aerated, uncrowded
tanks ($\leq$ 5 fish/11.4 liter tank) in rooms held at 20-23°C.

Male albino rats were obtained from Holtzman Co.,
Madison, Wisconsin and male albino (ICR) mice were obtained from Scientific Small Animals, Arlington Heights,
Illinois. Rats were housed 1-2/cage and mice 5/cage in a
23-25°C animal room regulated for 12 hour light:dark
cycles.

Preuse handling procedures for rodents, screening
procedures for rodents and goldfish and descriptions of
training and testing apparatus have been published (GUTT-
MAN and GRONKE, 1971; GUTTMAN et al., 1972).

Microdansylations were done according to the methods

of GRAY and HARTLEY (1963) and the results visualized by
two-dimensional chromatography on polyamide plates (NEUHOFF
et al., 1969).

Synthetic scotophobin, synthesized by PARR and HOLZER
(1971) was generously supplied by Dr. G. UNGAR. N-Acetyl
5,11-deamidoscotophobin, 5,11-deamidoscotophobin, and the
octapeptide comprising the 8-15 amino acid residues of
deamidoscotophobin were synthesized and generously supplied
by Dr. B. WEINSTEIN.

RESULTS AND DISCUSSION

1. Is the Passive-Transfer Reaction Simply a Manifestation
   of a General Alteration in Activity?

To answer this question we injected goldfish with synthetic scotophobin and tested for differences in dark avoidance responses of animals inserted into the test tank on
the dark or lighted portion. If dark avoidance activity was
merely due to reduction in swimming activity, one would expect to obtain good dark avoidance results only with fish
which enter the shuttle box via the lighted portion whereas
fish which enter via the dark portion should show little or
no dark avoidance since, in this case, it would take more
rather than less activity to avoid the dark. As can be seen
(Tab. I), the results are independent of shuttle-box entry
position, thus ruling out imposed sluggishness as the mode
of action of scotophobin.

Tab. I: <u>Dark avoidance by goldfish as a function of test</u>
<u>initiation from the dark or light side of shuttle-</u>
<u>box</u>

	Test initiated side of shuttle-box	
Nature of material injected IC	Dark	Light
Placebo	175*	174
Synthetic scotophobin ∿ 100 ng	86	82

*Mean dark box time/180 second trial. N = 10

A similar but more indirect interpretation can be made
from the results of UNGAR (1970) in which injection of natural scotophobin did not inhibit the normal step-down response of rats. A general and non-specific activity alteration effect by scotophobin would have required such an inhibition.

2. Individuality of Initiation Response of Mice and Goldfish
   to Scotophobin and Implication to Means of Data Analysis.

We use the screening criterion of > 160 seconds spent
in the dark per 180 second trial for the selection of rodents and goldfish to be used in dark avoidance experiments.
Thus, for subsequent activity analyses, it is clear that any
animal whose dark box time (DBT) is reduced to 100 seconds
is reacting. Consequently, we consider the first time
period in which DBT is reduced to 100 seconds as the clear
initiation point of dark avoidance in standard assays.

Tab. II: Heterogeneity of dark avoidance (DA) responses of
mice to natural rat scotophobin

Initiation of DA (hours after injection)	DBT at strongest reaction	Time after injection at which strongest reaction shown
3	0	3
3	0	6-12
3	0	24
6	30	6
6	0	36
6	35	24
6	30	12
6	18	6
6	15	6
6	42	6

| Controls (N=10) | 145[*] | |

[*] Mean. Range for all raw scores at 10 test periods = 118-180.

Mice which received natural rat scotophobin purified
to the stage of passage through one Sephadex G-25 column,
i.e., RNA-free (GUTTMAN and GRONKE, 1971), were tested for
dark avoidance initiation time, and finally time period
and magnitude of the best dark avoidance reaction (Tab. II).
It can be seen that there is heterogeneity with respect to
all three parameters; initiation time, DBT during the

strongest reaction, and the absolute time during which the
strongest reaction is manifested. We therefore consider it
important to correct for the heterogeneous reaction kinetics
from animal to animal when assessing the quality of natural
and synthetic compounds for activity as specific behavior
alterants. To do this we do not calculate the mean DBT of
experimental and control animals at fixed intervals. Rather,
we measure DBT at several intervals and then for each individual, calculate the DBT at the strongest reaction time.
These data are then used for subsequent statistical analyses. Measurement and analysis of this sort eliminate
problems resulting from kinetic differences among populations of animals.

We have done extensive studies on the heterogeneity
of responses of goldfish to synthetic scotophobin (GUTTMAN
et al., 1972) and have found much more striking differences
in reaction initiation time than just seen with natural
scotophobin in mice (Fig. 1). The method just outlined was
used (GUTTMAN et al., 1972) to show that synthetic scotophobin causes a highly significant dark avoidance reaction
in goldfish (p = 0.0001).

3. Microdansylation as a Standard Means of Assessing Chemical Quality of Potential Behavior Altering Peptides
   prior to their Use in Biological Assays.

Although we routinely store synthetic scotophobin as
a dry lyophilizate, in vacuo, in siliconized ampules at
6^0 we have found that the preparation gradually loses

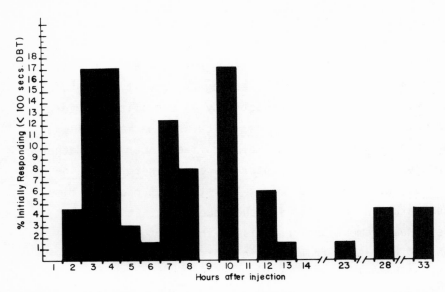

Fig. 1: Heterogeneity of dark avoidance reaction initiation
in goldfish which received 100 ng synthetic rat
scotophobin. N=75. (From GUTTMAN et al., 1972).

biological activity. We therefore assess quality of all
peptides, at the time of their use in biological experiments, by microdansylation followed by two-dimensional
chromatographic analysis (GRAY and HARTLEY, 1963; NEUHOFF
et al., 1969). For our analyses, 10 μg of a particular peptide (or the same volume of diluent for control samples) is
dansylated and 0.25 μg (0.5 μl) plated on several replicate
polyamide plates.

In our hands there is a good, but not absolute, degree
of repeatability of results with plates made from reagent
control dansylations (Tab. III) as well as with synthetic

Tab. III: Chromatographic patterns of microdansylated
          reagent controls

| Spot No. | R_f Ranges* | | % times found[+] |
	Solvent 1	Solvent 2	
1	0.000 - 0.032	0.000 - 0.032	90
2	0.061 - 0.114	0.000 - 0.039	90
3	0.146 - 0.366	0.012 - 0.038	90
4	0.423 - 0.587	0.000 - 0.064	90
5	0.732 - 0.835	0.023 - 0.114	90
6	0.599 - 0.674	0.133 - 0.195	90
7	0.595 - 0.702	0.209 - 0.304	90
8	0.241 - 0.370	0.195 - 0.259	90
9	0.132 - 0.241	0.250 - 0.323	50
10	0.000 - 0.050	0.480 - 0.542	50
11	0.000 - 0.058	0.811 - 0.870	20
12	0.000 - 0.043	0.938 - 0.968	20
13	0.586 - 0.688	0.731 - 0.815	20
14	0.604 - 0.680	0.953 - 1.000	20

*Results of 50 determinations: 90 = present in > 90 % of
the plates; 50 = present in 40-60 % of the plates; 20 =
present in 10-30 % of the plates.

[+]Reagent control plates are always run along with samples
and spots from control plates subtracted from the total
found on experimentals (see Fig. 2).

peptides (Tab. IV). None of the preparations of synthetic
scotophobin (Tab. IV; Fig. 2) we received were completely

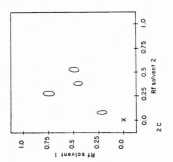

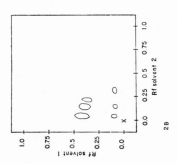

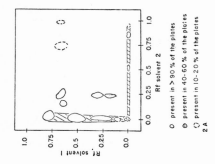

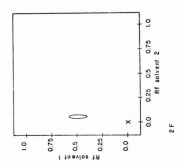

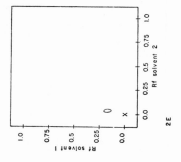

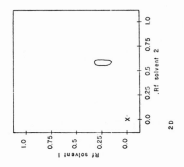

Fig. 2: <u>Microdansylation of potential dark avoidance</u>
<u>peptides</u>

In each case results of resolution with solvent
system 1 (water:90 % formic acid, 100:1.5 v/v)
are shown along the ordinate and results with
solvent system 2 (benzine:acetic acid, 9:1 v/v)
along the abscissa. A. Control plate. Sample =
only the reagents used for dansylation. In all
further figures the spots which appear here have
been deducted from the final result. B. Synthetic
scotophobin as received. Fully active in inducing
dark avoidance. C. Synthetic scotophobin after
storage for 2 months. Biological activity poor.
D. 5,11-Deamidoscotophobin. E. N-Acetyl 5,11-
deamidoscotophobin. F. 8-15 residues of deamidoscotophobin.

homogeneous (at least a few minor impurities were always
noted).

The evanescent nature of synthetic scotophobin sharply
contrasts with the stability of 5,11-deamidoscotophobin,
N-acetyl 5,11-deamidoscotophobin and an octapeptide composed
of residues 8-15 of deamidoscotophobin (Fig. 2 and 3).
Scotophobin was synthesized by the Merrifield method (PARR
and HOLZER, 1971), purified, lyophilized, and sent to us <u>in
vacuo</u>, while the other three peptides were synthesized by
WEINSTEIN using classical synthetic methods and stored for
about 12 months at room temperature in screw-capped tubes.
It is clear that only scotophobin is very labile. Further
experiments are required to assess the role of the structural differences (Fig. 3) between scotophobin (5,11 gln)
and deamidoscotophobin (5,11 glu) and trace contaminants
from the Merrifield synthesis or subsequent purification
steps in labilizing scotophobin.

Tab. IV: Chromatographic characteristics of scotophobin and related peptides

Preps (see Fig. 2)	R_f Ranges (solvent 1; solvent 2)*	
	Main spot	Impurities
Scotophobin (B)+	0.045 -0.102; 0.091 -0.153	0.340 -3.497; 0.033 -0.105
		0.327 -0.448; 0.114 -0.168
		0.290 -0.396; 0.190 -0.243
		0.026 -0.060; 0.162 -0.210
		0.030 -0.074; 0.296 -0.355
Scotophobin (C)	NOT EVIDENT	0.150 -0.264; 0.078 -0.156
		0.468 -0.536; 0.500 -0.566
		0.405 -0.505; 0.343 -0.425
		0.670 -0.799; 0.266 -0.387
5,11-Deamidoscotophobin (D)	0.448 -0.299; 0.600 -0.657	0.024 -0.081; 0.564 -0.614
N-Acetyl 5,11-deamido-scotophobin (E)	0.098 -0.162; 0.057 -0.096	
8-15 Residues of deamido-scotophobin (F)	0.246 -0.364; 0.052 -0.109	

*Spots due to reagents (Tab. III) have been deducted. +See also Fig. 2.

SCOTOPHOBIN: Ser Asp Asn Asn Gln Gly Lys Ser Ala Gln Gln Gly Gly TyrNH$_2$ (A)

5,11-DEAMIDOSCOTO- Ser Asp Asn Asn Glu Gln Gly Lys Ser Ala Glu Gln Gly Gly TyrNH$_2$ (B)
PHOBIN:

PERAMIDOSCOTO-   Ser Asn Asn Asn Gln Gln Gly Lys Ser Ala Gln Gln Gly Gly TyrNH$_2$ (C)
PHOBIN:

Fig. 3: Structure of the dark avoidance peptide and two analogues; A. Scotophobin, B. 5,11 -
Deamidoscotophobin, C. Peramidoscotophobin
Compound B was incorrectly considered to be scotophobin (ALI et al., 1971). It is
named here in a manner consistent with the name and structure which has priority
(compound A).

4. Cellular Mode of Action of Exogenously-Supplied Scotophobin.

   This is obviously only a progress report. The thread
which connects these studies is the diversity of experimental approaches which will be required to solve this problem. Clearly, the insights gained through the scotophobin
studies will make subsequent study of other behavior altering peptides much simpler.

   Early and even a few recent transfer experiments using
"purified" fractions from donor brains suggested that the
active behavior alterant was RNA (JACOBSON et al., 1965;
GOLUB and MCCONNELL, 1971). In all of these rodent-to-rodent
transfer experiments, injection route was IP, making positive
results due to RNA incompatible with the finding of LUTTGERS
et al.,(1966) that RNA does not pass the blood-brain barrier.
In addition, several groups obtained positive transfer results with RNA-free natural preparations (e.g.,UNGAR and
FJERDINGSTAD, 1971; GUTTMAN and GRONKE, 1971) injected IP
into rodents. The finding that synthetic scotophobin transfers dark avoidance to mice (DESIDERIO et al., 1971; MALIN
and GUTTMAN, submitted) and to goldfish (GUTTMAN et al.,
1972) suggested that it should be possible to retrieve IP-
injected scotophobin from brains of recipients and, once
and for all, close the peptide-RNA argument. We therefore
combined this experimental goal with first studies on localization of exogenously-supplied scotophobin in goldfish.

   For these studies synthetic scotophobin was labeled
with ^{125}I and then retrieved from brains of animals which

were harvested at the time they were actively showing dark
avoidance. Thus the picture is now complete with respect to
proof of passage through the blood-brain barrier of one
specific behavioral alterant, scotophobin.

Taking advantage of the different response-initiation
times (Fig. 1), we studied the mode of action of exogenously-supplied scotophobin in recipient goldfish. That is, we
compared localization of ^{125}I-labeled scotophobin of responders (DBT less than 100S) with those which had not yet
initiated their responses (i.e.,preresponders, DBT 170-180S)
(GUTTMAN et al., 1972). Since all fish respond eventually,
there is some experimental smear caused by the mere presence
of ^{125}I-scotophobin.

The results (Tab. V) show greater differential between
preresponders and responders after tissue was extracted with
TCA. The actual amount of ^{125}I-scotophobin (specific activity) in the TCA precipitable tissue is reduced from that in
unextracted tissue. Therefore only a proportion of ^{125}I-
scotophobin, initially present in TCA soluble form, is bound.
The absolute amounts bound and the differential between preresponders and responders are indicative of target tissues
involved in scotophobin utilization. The failure of cerebellum to show ratio differential for the TCA precipitable
fractions presaged the recent report (UNGAR and BURZYNSKI,
1972) that little natural scotophobin is found in cerebellum of rats trained to avoid dark. The large quantities of
^{125}I-scotophobin fixed into the pituitary and spinal cord
(in addition to that in the vagal lobes which have the

Tab. V: <u>Incorporation of ^{125}I-Scotophobin in various tissues</u>
<u>(expressed as CPM/mg tissue or CPM/mg TCA ppt of</u>
<u>tissue)</u>

Fractions are: $\dfrac{\text{responders} \quad \text{(R)}}{\text{prerespondres (P)}}$

Tissue	Sp. Act. tissue	Ratio $\frac{R}{P}$	Sp. Act. TCA ppt fraction	Ratio $\frac{R}{P}$
<u>Spinal cord</u>	$\frac{251}{60}$	4.8	$\frac{795}{138}$	5.76
<u>Brain</u>				
Cerebellum	$\frac{144}{153}$	0.94	$\frac{408}{356}$	1.14
Pituitary	$\frac{3426}{1618}$	2.12	$\frac{967}{186}$	5.19
IIMSE*	$\frac{566}{220}$	2.57	$\frac{526}{77}$	6.83
Olfactory lobes	$\frac{108}{138}$	0.78	---	---
Optic tectum	$\frac{382}{300}$	1.27	---	---
Vagal lobes	$\frac{409}{36}$	11.36	$\frac{983}{79}$	12.44
<u>Other tissue</u>				
Gall Bladder	$\frac{15}{14}$	1.07	---	---
Intestine	$\frac{4}{6}$	0.67	---	---
Kidney	$\frac{170}{19}$	8.95	---	---
Liver	$\frac{7}{6}$	1.17	---	---

*Infundibulum, inferior lobes, medulla oblongata, saccus
vaculosus, eminentia granularis.

highest ratio of activity, responder:preresponder) suggest participation of these tissues in mediation of the passive reaction.

The large portion of ^{125}I-scotophobin which has been converted from TCA-soluble to TCA-insoluble suggests that scotophobin acts by complexing with some particulate cellular component of fairly high molecular weight. Both synaptic junction membranes and one or another of the S-100 proteins are good binding candidates. The identity of the binding material awaits the results of our experiments with ^{14}C-labeled natural goldfish scotophobin.

We used a form of inhibition analysis to gain some insight into the relationship of peptide structure to activity in eliciting a dark avoidance response. The experiment was divided into two parts:(A) analysis of dark avoidance activity of the compounds whose purity was described in Figs. 2C-F, and (B) influence of residues of these compounds on the pattern of retraining which was initiated hours after all direct effects of these compounds on dark avoidance had disappeared, i.e., do these animals show savings when they are trained?

It is worth reiterating that in part (A) <u>undegraded</u> synthetic scotophobin injections act on 100 % and the placebo injections act on 0 % of recipient goldfish since one characteristic of deamidoscotophobin and its derivatives is the variability of results. In all cases, two 180 second trials were given at each testing (part A) or training (part B) session.

Tab. VI: <u>Ability of degraded scotophobin and analogues to</u>
<u>induce dark avoidance in goldfish</u>*

Material Injected	N	Mean DBT
Placebo	10	138
Degraded Scotophobin (See Fig. 2C)		
Reactors	4	21
Non-reactors	3	107
5,11-Deamidoscotophobin (See Fig. 2D)		
Reactors	3	75
Non-reactors	6	141
N-Acetyl 5,11-deamidoscotophobin (See Fig. 2E)		
Reactors	3	45
Non-reactors	4	129
8-15 Aminoacid fragment of deamido-scotophobin (See Fig. 2F)		
Reactors	8	23
Non-reactors	0	

*
Structures are given in Fig. 3. For data analysis method
see Results, section 2.

As can be seen from Tab. VI, animals tested for elicitation of dark avoidance activity by the various peptides are
divided into two groups, reactors and non-reactors. In contrast, animals injected with <u>undegraded</u> scotophobin and the
8-15 fragment of deamidoscotophobin are <u>all</u> reactors. At
this time it is hard to decide upon the amount of weight

to give to failure of a given compound to elicit a result in <u>all</u> animals. Therefore, we just present our findings in the hope that they will provide useful guidelines for others.

Another factor which is difficult to evaluate at this time is the net number of hours during which a potential behavior-altering peptide exerts its activity. Activity of undegraded, synthetic scotophobin lasts for 24-72 hours whereas activity of degraded scotophobin and all the analogues tested is rather transient.

The low-grade activity of all the compounds tested (Tab. VI) suggests that there is at least some binding, but of different strength or duration, of each to the cell site which makes fully-active scotophobin TCA-precipitable.

To determine whether any of the peptides could remain bound to this normal attachment site for scotophobin, we initiated a savings experiment 144 hours after termination of peptide activity tests. Results indistinguishable from those of placebo animals would indicate no binding. Results more unfavorable than those of placebo animals would indicate binding of an immutable analogue (<u>i.e.</u>, can not be converted to scotophobin) to an important attachment site. Results more favorable than those of placebo animals (savings) would indicate binding of the analogue and probably conversion to scotophobin by brain enzymes. Ability to bind also implies that the structural changes between scotophobin and analogue are not at the residues which are involved in binding.

Tab.VII: Effect of prior injection of goldfish with degraded scotophobin or analogues on savings and
number of reinforcements used to train for dark
avoidance

Material originally injected (Tab. VI)	Savings in	
	% time till trained	% reduction of reinforcements*
Degraded scotophobin		
Reactors	56	75
Non-reactors	51	62
5,11-Deamidoscotophobin		
Reactors	33.5	53.5
Non-reactors	36.5	71
N-Acetyl 5,11-deamidoscoto-phobin		
Reactors	31.8	64
Non-reactors	-17.6	64
8-15 Amino acid fragment of deamidoscotophobin		
Reactors	-6.5	50
Non-reactors	-	-

*Between the first and second training trial.

Our results (Tab. VI and Tab. VII) show that degraded
scotophobin has transient initial dark avoidance activity
which is better than that of the analogue peptides and better savings activity of any of the compounds tested. Chro-

matographic analysis of dansylated degraded scotophobin
shows that it was not degraded to 5,11-deamidoscotophobin
(Figs. 2C and D). Deamidoscotophobin and N-acetyl deamidoscotophobin have lowgrade transient dark avoidance activity
(in that it only appears in some fish) and a small but significant amount of savings activity. Since these compounds
are relatively homogeneous (Figs. 2D and E), it is clear
that the results are due to the indicated compounds. The
N-acetyl substituent in effect modifies the first residue,
serine (Fig. 3),but dark avoidance activity (in terms of
(a) distribution of reactors and non-reactors, (b) net activity of reactors) is about the same as for the unsubsti-
tuted deamido derivative. In addition, results of savings
determinations (Tab. VII) for the compounds are similar.
Thus it is possible to modify the first amino acid residue,
serine, without altering binding or activity. Our earlier
experiments (GUTTMAN et al., 1972) showed that we could
modify the last amino acid residue, tyrosine, and not sac-
rifice dark avoidance activity. Thus residues 1 and 15 are
ruled out as binding sites.

The deamido analogues have fair savings ability but
only transient dark avoidance activity. The peramido analogue was shown by others to have poor dark avoidance activity (DESIDERIO et al., 1971). We can therefore suggest
that the 2-asp 5,11-gln residues are necessary for full
dark avoidance activity. However since the 5,11 glu analogue
(i.e.,deamidoscotophobin) still allows binding and savings,
it must be possible for the molecule to be converted to
scotophobin in situ.

The 8-15 amino acid fragment of deamidoscotophobin has good dark avoidance activity and shows either no savings activity or a trace of inhibition. It is therefore clear that this peptide must be bound for at least a short time but that it can not be converted to scotophobin in situ.

We can therefore conclude that the important binding sites are among those residues which have not been modified, i.e.,2-4, 6-10, 12-14. Our results confirm and enlarge upon those of others (DESIDERIO et al., 1971) in which modification of residues 2, 5, and 11 resulted in loss of dark avoidance activity. We can now see that the binding sites and activity sites for scotophobin are not identical. Further experimentation is required for determination of the specific activity of these sites.

ACKNOWLEDGEMENTS

Supported in part by funds from the Research Board of the University of Illinois at Chicago Circle and Sigma Xi.

LITERATURE

ALI, A., J.H.R. FAESEL, D. SARANTAKIS, D. STEVENSON, and B. WEINSTEIN: Synthesis of a structure proposed for scotophobin.
Experientia 27, 1138-1139 (1971).

DANIELS, P.: Acquisition, storage, and recall of memory
    for brightness discrimination by rats following intra-
    cerebral infusion of acetoxycyclohexamide.
    J. Comp. Physiol. Psych. 76, 110-118 (1971).

DESIDERIO, D.M., G. UNGAR, and P.A. WHITE: The use of mass
    spectrometry in the structural elucidation of scoto-
    phobin, a specific behavior-inducing brain peptide.
    Chem. Commun. 9, 432-433 (1971).

GOLUB, A.M., and J.V. MCCONNELL: Empirical issues in inter-
    animal transfer of information.
    In: Chemical Transfer of Learned Information (E.J.
    FJERDINGSTAD, ed.), pp. 1-29, North-Holland, Amster-
    dam (1971).

GRAY, W.R., and B.S. HARTLEY: A fluorescent end-group rea-
    gent for proteins and peptides.
    Biochem. J. 89, 59 (1963).

GUTTMAN, H.N., and L. GRONKE: Passive transfer of learned
    dark and step-down avoidance.
    Psychon. Sci. 24, 107-109 (1971).

GUTTMAN, H.N., G. MATWYSHYN, and G.H. WARRINER III: Syn-
    thetic scotophobin-mediated passive transfer of dark
    avoidance.
    Nature New Biology 235, 26-27 (1972).

JACOBSON, A.L., F.R. BUBASH, and S. JACOBSON: Differential
    approach tendencies produced by injection of RNA from
    trained rats.
    Science 150, 636-637 (1965).

KATZ, M.S., and W.C. HALSTEAD: Protein organization and
    mental function.
    Comp. Psych. Monogr. 20, 1-38 (1950).

KESNER, R.P., and, H.S. CONNER: Independence of short- and
    long-term memory: a neural system analysis.
    Science 176, 432-4, (1972).

LUTTGERS, M., T. JOHNSON, C. BURK, J. HOLLAND, and J.
    MCGAUGH: An examination of "transfer of learning" by
    nucleic acid.
    Science 151, 834-837 (1966).

NEUHOFF, V., F. von der HAAS, E. SCHLIMME, and M. WEISE:
    Zweidimensionale Chromatographie von Dansyl-Aminsäu-
    ren im pico-Mol-Bereich, angewandt zur direkten Cha-
    rakterisierung von Transfer-Ribonucleinsäuren.
    Hoppe-Seyler's Z. Physiol. Chem. 350, 121-128 (1969).

PARR, W., and G. HOLZER: Synthesen von Scotophobin-analoga.
    Hoppe-Seyler's Z. Physiol. Chem. 352, 1043-1047 (1971).

RUSSELL, R.W., G. UNGAR, and E. USDIN: Seminar on the re-
    quirements for testing of hypotheses about molecular
    coding of experience: transfer studies.
    Psychopharm. Bull. 8, 5-13 (1972).

SZILARD, L.: On memory and recall.
    Proc. Soc. Nat Acad. Sci. Wash. 51, 1092-1099 (1964).

UNGAR, G.: Role of proteins and peptides in learning and
    memory.
    In: Molecular Mechanisms in Memory and Learning (G.
    UNGAR, ed.), pp. 149-175, Plenum, New York (1970).

UNGAR, G., and S.R. BURZYNSKI: Detection of a behavior-
    inducing peptide (scotophobin) in brain by ultramicro-
    analytical method.
    Fed. Proc. <u>31</u>, 398 (1972).

UNGAR, G., and E.J. FJERDINGSTAD: Chemical nature of the
    transfer factors; RNA or protein?
    In: Biology of Memory (G. ADAM, ed), pp. 137-143,
    Plenum, New York (1971).

NEW EXPERIMENTAL APPROACHES TO THE INTER-ANIMAL TRANSFER OF ACQUIRED INFORMATION

G.F. Domagk*, E. Schonne and G. Thines

Unité de Biochimie, Université de Louvain
Belgium

ABSTRACT

Experiments have been performed with chicks, mice and goldfish in order to find a procedure for the preparation of larger amounts of "trained brains", which will allow a chemical transfer of information from donor animals into naive recipients.

About 3 years ago ZIPPEL and DOMAGK started a collaboration on the subject of inter-animal transfer of acquired information. In early goldfish experiments (1969, 1971a, 1971b) these workers had used color and taste discrimination tasks in a shock-free training procedure. The original experiments were performed on groups of 2 animals, and the "transfer of memory" was achieved by the i.p. injection of

* Present address: Physiologisch-Chemisches Institut,
                   Göttingen, Germany

dialyzed brain material. Later the study of the chemical
nature of the active substances required the use of groups
of more animals. However, after UNGAR's successful rat experiments which led to the isolation and identification of
scotophobin (UNGAR et al., 1972) it was calculated that
the brains of more than 10,000 trained fishes would be necessary in order to obtain substantial amounts of "memory
molecules".

Last year some Belgian colleagues at the Université de
Louvain became interested in the problem of memory fixation.
We realized that instead of going into competition with the
fish experiments performed by the rich and successful group
at Houston, Texas, we should rather look for some other kind
of animal experiments in which it might be easier to obtain
larger quantities of trained brains. In the literature a
few experiments with newborn chicks had been reported.
These procedures seemed attractive to us since at least the
male subjects are easily available at low price, and these
animals possess relatively large amounts of brain tissue.
First we have repeated the experiments of SPARBER and ROSEN-
THAL (1968, 1971), in which the animals have to learn a
hunger-motivated detour task in the apparatus shown in Fig.
1. The chicks learned very well during the 4 days of training, but neither the feeding of brain homogenates nor the
injections of brain homogenates of dialyzed brain extracts
showed any influence on the learning curves observed in
naive recipients. - In another experimental series we have
tried to transfer an aversion against anthranilic acid
methyl ester induced by a one-trial pecking test in newborn
chicks. But like CHERKIN (1970), who had tried this before,

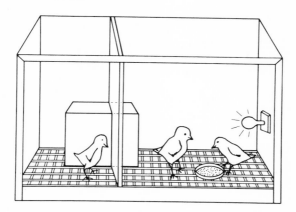

Fig. 1: <u>Training apparatus according to SPARBER and ROSEN-</u>
<u>THAL (1971)</u>

we had only negative results, regardless whether we used
total brain homogenates or fractions thereof.

Another type of experiment we are still engaged with
is based upon a communication by LEVAN and coworkers (1970)
who had shown the chemical transfer of an X-ray conditioned
taste aversion in mice. These experiments are based on an
earlier observation of GARCIA and KIMELSDORF (1960a, 1960b)
who had seen that rats, mice and cats avoid a spontaneously
preferred taste substance after this stimulus had been given
in combination with a whole body irradiation. The mode of
action of this "learning" remained unknown until now. Radio-
logists have tried the irradiation of removed brains and
have discussed a possible formation of H_2O_2 and of free
radicals (LEVAN et al., 1970; LEVAN, 1970).

Since LEVAN's data (Fig. 2) seemed to us to be quite
convincing of a positive transfer effect, we have embarked

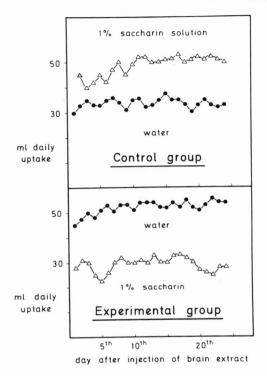

Fig. 2: <u>Inversion of the saccharin drinking preference in</u>
<u>mice by the injection of brain extract prepared</u>
<u>from X-irradiated animals</u>

Data of LEVAN et al., Experientia <u>26</u>, 648 (1970).
The figures represent the amounts drunk by 30 mice
per day.

upon the problem. First we wanted to learn how specific was
the effect of X-rays used in the procedure to induce the
taste aversion. Fig. 3 shows how a group of 10 mice lost
their preference for drinking 10 % ethanol instead of water
after a single dose of 450 roentgens. A 10 minute sham irradiation under the same machine had been given to the same

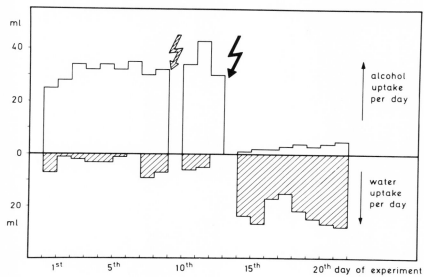

Fig. 3: <u>X-ray induced aversion for drinking alcohol</u>

On days 9 and 13, the 10 mice were not given anything to drink. 30 min before the subsequent irradiation the animals obtained 10 % ethanol, but no
water. ⚡ = sham irradiation; ⚡ = 450 roentgens.

group of animals before and had not effected any change in
their drinking behavior.

In the following experiments, we established the animals' preference for 0.2 % saccharin and 10 % ethanol by
alternate daily presentation of one or other of these solutions over a period of 2 weeks. After a 24 hour period of
thirst one group of the animals received saccharin and 30
minutes later, a 450 r irradiation. The other group of
mice received ethanol after the thirst period, followed by
the irradiation. During the following 3 weeks the animals
showed the behavior expected by our hypothesis: those who

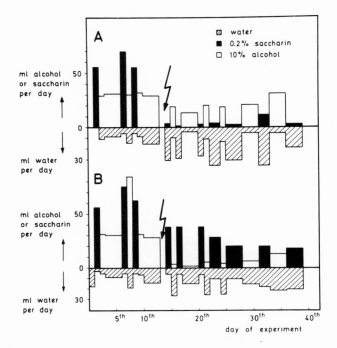

Fig. 4: Specificity of the X-ray induced taste aversion

2 groups of 10 mice received water + saccharin
and water + ethanol alternately, ad libitum. No
liquids were offered on the 13th day. After 24
hours group A received saccharin exclusively,
group B ethanol exclusively, for 30 min, after
which time 450 roentgens were given to both groups.

had drunk the saccharin just before the irradiation preferred water offered simultaneously with saccharin, whereas
the preference for ethanol persisted on the alternate days.
And vice versa, the other group continued drinking saccharin, whereas the preference for ethanol had been converted into an aversion (Fig. 4).

Recent experiments, performed in collaboration with

ZECH at Göttingen, have shown that the X-ray application
for the induction of the taste aversion can well be replaced by other aversive agents, such as an injection of
apomorphine after the thirst period, when they are given
simultaneously with the spontaneously preferred taste
stimulus.

Several attempts of transferring the induced aversion
by an i.p. injection of brain extracts into pre-tested recipient mice have not given very convincing effects so far.
We noted that it might be disadvantageous to do this kind
of experiment with larger groups of animals. ZECH has
developed a technique by which it is possible to register
the drinking behavior of individual animals in large numbers. Only in a very few cases have we observed animals that
after an injection of brain extract drank more saccharin
than water, but according to recent discussions we had with
Dr. FJERDINGSTAD we should perhaps estimate even changes
in the saccharin:water ratio. Another factor, which is not
yet under control, is the ratio of donor brain equivalents
per recipient animal.

Finally, another technique seems to be promising for
the transfer of learned information: it is based on a technique developed by WIJFFELS at Louvain some years ago
(1969). In an apparatus shown in Fig. 5 goldfish learn to
swim through a four-chambered maze in order to find food in
a hidden cup. After a day of exploration the fish is put back
into the compartment "A", which serves as the "living room"
for the time of the experiment. In 2 training sessions and
later 2 test sessions per day the door between "A" and "B"

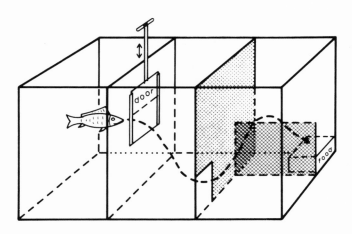

Fig. 5: <u>Four-chambered maze for the training of goldfish in</u>
<u>spatial orientation</u>
(WIJFFELS, 1969).

is opened and at the same time an automatic timer is started
which will stop at the moment the animal reaches compartment
"D". Untrained fish usually do not reach the goal before 15
minutes; well-trained fish may be as fast as 20 seconds. When
brain extracts from trained donors were injected into naive
recipients, the latter gave consistently faster reactions,
which were significant at the 1% level (Sign test).

ACKNOWLEDGEMENTS

The skilful assistance of Mademoiselle Francoise
DELCOURT as well as financial support given by the Fonds
de Developpement Scientifique (F.D.S) de l'Université de
Louvain and of the Deutsche Forschungsgemeinschaft (SFB 33)

are gratefully acknowledged.

LITERATURE

CHERKIN,A.: Failure to transfer memory by feeding trained
    brains to naive chicks.
    J. biol. Psychol. 12, 83-85 (1970).

GARCIA, J., and D.J. KIMELSDORF: Some factors which in-
    fluence radiation-conditioned behavior of rats.
    Radiation Res. 12, 719-727 (1960b).

KIMELSDORF, D.J., J. GARCIA, and D.O. RUBADEAU: Radiation-
    induced conditioned avoidance behavior in rats, mice,
    and cats.
    Radiation Res. 12, 710-718 (1960a).

LEVAN, H., D.L. HEBRON, W.S. MOOS, and H.C. MASON: Induc-
    tion of post-irradiation conditioned avoidance behav-
    ior by intraperiotoneal injection of brain.
    Experientia 26, 648-649 (1970).

LEVAN, H.: Transferability of radiation-induced behavior
    by injection of various organ extracts, paper pre-
    sented at the 139th Meeting of the American Associa-
    tion for the Advancement of Science, Chicago, Dec.
    1970.

ROSENTHAL, E., and S.B. SPARBER: Transfer of a learned
    response by chick brain homogenate fed to naive
    chicks.
    Pharmacologist 10, 168 (1968).

SPARBER, S.B., and E. ROSENTHAL: An apparent transfer ef-
    fect in chickens fed brain homogenates from donors
    trained in a detour task.
    In: Chemical Transfer of Learned Information (E.J.
    FJERDINGSTAD, ed.), pp. 165-180, North Holland Publ.,
    Amsterdam (1971).

UNGAR, G., D.M. DESIDERIO, and W. PARR: Isolation, identifi-
    cation and synthesis of a specific-behavior-inducing
    brain peptide.
    Nature 238, 198-202 (1972).

WIJFFELS, H.J.C.: Group learning versus individual learning
    as observed in fish learning a maze (Barbus conchonius
    and Carrassius auratus").
    Psychol. Belg. 9, 141-165 (1969).

ZIPPEL, H.P., and G.F. DOMAGK: Versuche zur chemischen Ge-
    dächtnisübertragung von farbdressierten Goldfischen
    auf undressierte Tiere.
    Experientia 25, 938-940 (1969).

ZIPPEL, H.P., and G.F. DOMAGK: Transfer of a taste prefer-
    ence from trained goldfish (Carassius auratus) into
    untrained recipients.
    Pflügers Arch. 323, 258-264 (1971a).

ZIPPEL, H.P., and G.F. DOMAGK: Experiments concerning the
    transfer specificity of brain extracts in the taste
    discrimination of goldfish.
    Pflügers. Arch. 323, 265-272 (1971b).

CHEMICAL TRANSFER OF LEARNED INFORMATION IN MAMMALS AND FISH

Ejnar J. Fjerdingstad

Anatomy Department B
University of Aarhus
Denmark

ABSTRACT

The phenomenon of "chemical transfer of learned information" was investigated in rodents and goldfish, using
a variety of behavioral and biochemical approaches. Operant
and classical conditioning, and positive and negative reinforcement were all found to give transfer effects. In discriminative situations indication for behavioral specificity
of the effect was found. Brain homogenates, crude supernatants, and partly purified RNA extracts were equally effective. The site of injection was unimportant except for
influencing the necessary amounts of brain material. This
would seem to indicate that transfer effects are statistically reliable, occur commonly in many different types of
learning, and are specific to the type of training applied
to donors.

INTRODUCTION

The "transfer" field begins with a paper by J.V. MC-
CONNELL (1962) reporting enhancement of learning in planarians receiving material from classically conditioned
donor subjects orally. Subsequently,four groups independently described similar phenomena in rats (BABICH et al.,
1965; FJERDINGSTAD et al., 1965; REINIS, 1965) and mice
(UNGAR and OCEGUERA-NAVARRO, 1965). Although being quite
clearcut as behavioral results go, these reports were received with considerable scepticism and not rarely off-
hand rejection by a majority of researchers. The arguments
often heard were,1) the overall evidence was not statistically convincing, and 2) even granting that there was
some kind of effect, this was probably a completely general, unspecific,stimulative phenomenon.

There are by now more than 100 positive reports of the
transfer effect in vertebrates alone. Although there are
also a considerable number of reports of negative or un-
clear results, it would seem that the overall evidence
heavily favors the existence of the phenomenon (DYAL, 1971).

During several years of work in the field,the present
author has been able to demonstrate transfer effects with
rodents using a number of different behavioral approaches,
i.e.,light-dark discrimination, right-left discrimination,
and alternation in instrumental conditioning situations,
and of simple approach training, radiation induced avoidance, and the phenomenon of "acoustic priming" for audio-

genic seizure. In fish transfer effects were obtained with avoidance and approach procedures. As can be seen from Tab. I, this was combined with a number of types of extract and routes of administration, in some cases within the same behavioral situation.

The wide range of behavioral techniques and types of brain preparation would seem to indicate that transfer is a common phenomenon, neither limited to only one type of conditioning, nor dependent on method of reinforcement. Although at first it seems confusing that many different types of extract are active, this may, as will be discussed later, be due to absorption of the active principle to other components of the extract. The fact that the route of administration seems unimportant (e.g.,FJERDINGSTAD et al., 1970; MOOS et al., 1969) implies that the active principle should be a rather small and diffusible molecule. That this is so has been shown for one behavioral situation by UNGAR and coworkers (UNGAR, 1971).

EXPERIMENTS WITH RODENTS

Indication of specificity of the transfer effect was found in the experiments with right-left discrimination and alternation in the Skinner box. In the first of these two, groups of donors were trained on an F.R. schedule (increased one step every day) for 10 days with only one bar reinforcing, right or left according to the group to yield one group of "right trained" and another of "left trained" donors.

Tab. I: List of paradigms in which transfer effects were demonstrated

Behavioral technique	Subjects	Extract and Injection	Reference
Light-dark discrimination in "two-alley runway"	Rats	"RNA" I.C.	Fjerdingstad et al., 1965; Røigaard-Petersen et al., 1968
Right-left discrimination in Skinner box	Rats	"RNA" I.C. Supernatant I.P.	FJERDINGSTAD et al., 1970
Alternation in Skinner box	Rats to mice	"RNA" I.P. Supernatant I.P.	Fjerdingstad, 1969a
Approach training	Rats to mice	Supernatant I.P.	Fjerdingstad, 1969b
Radiation induced avoidance	Mice	Homogenate in Ringer I.Cer.	Fjerdingstad, in press (replication of Moos et al., 1969)
Acoustic priming for audiogenic seizure	Mice	Homogenate in Ringer I.Cer.	not published before
"Oxygen reinforced approach training"	Goldfish	"RNA" subdurally	not published before
Avoidance in shuttle box	Goldfish	"RNA" subdurally	Fjerdingstad, 1970

I.C. = intracisternally
I.P. = Intraperitoneally
I.Cer. = intercerebrally

These were then sacrificed, the brains removed and
used for preparation of "right trained" and "left trained"
extracts. In a first experiment the extract was a crude
supernatant of a homogenate in Tris-HCl, two brain equivalents per recipient, injected intraperitoneally (FJERDING-
STAD et al., 1970),while in a second experiment the "RNA
extract II" of RØIGAARD-PETERSEN et al.,(1968) was used
for intracisternal injection of 2/3 brain equivalent per
recipient.

The effect on recipients were observed with the Skinner boxes programmed to reinforce continuously on both bars,
i.e.,so as not to influence choice of bar. It was found
that the majority of recipients went progressively more to
the bar on which the corresponding donor group had been
trained; this effect was most pronounced from the first to
the second session. Fig. 1 shows the pooled results from
the two experiments; it is also apparent that "left trained
extracts" were somewhat more potent than "right trained extracts"; this may, as explained below, be due to interference from preinjection prefence.

Even better indication of specificity was found in
further experiments with Skinner boxes, in which one group
of donors was trained to alternate regularly between bars,
while another was reinforced irrespective of choice, thus
being allowed to establish an individual preference of bar
(which turned out to be the left bar for 70 % of the rats,
a likely explanation for the seemingly higher potency of
the "left trained extract" noted above). In five experi-

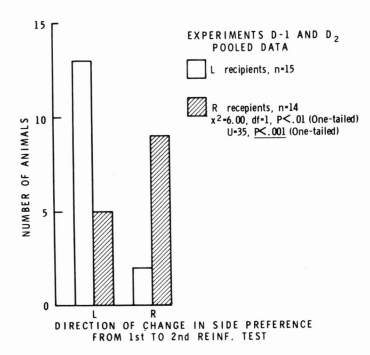

Fig. 1: Results of recipient testing in two experiments on
the transfer of right-left discrimination in the
Skinner box

ments of this type it was consistently found that when recipients were tested with both bars continuously reinforcing, the group receiving "alternation extract" showed a
higher incidence of alternation presses. Fig. 2 shows the
result of recipient testing in one of these experiments
that have been reported in detail elsewhere (FJERDINGSTAD,
1969a; FJERDINGSTAD, 1971). The delay in appearance and
the transient nature of the transfer effect are particularly noticeable here.

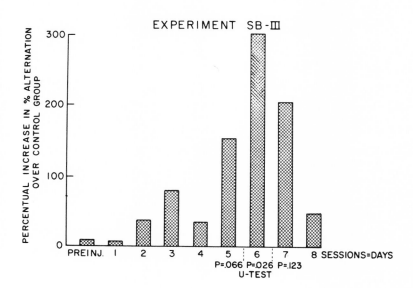

Fig. 2: <u>Behavior of recipients in an experiment on the transfer of alternation behavior in the Skinner box</u>

When considering results like the present, and the many
other examples of specific effects that have been reported
(DOMAGK and ZIPPEL, 1971; DYAL, 1971; GOLUB and MCCONNELL,
1971; ROSENBLATT, 1970; UNGAR, 1971; WEISS, 1971),it seems
that the position should no longer be tenable that transfer
effects may simply consist in a general sensitization or in
enhancement of learning, objections that have not been un-
common.

The question of the chemical nature of the active principle of transfer extracts has been settled, at least for
one particular training situation,by the isolation, sequence
determination, and synthesis of the dark avoidance factor

scotophobin by Dr. UNGAR and his group (UNGAR, this volume;
PARR and HOLZER ibid.). Our group started out, like several
others, impressed with the numerous reports of neuronal RNA
changes caused by experience (see, e.g., HYDEN, 1967;
SHASHOUA, 1968; BATESON et al., 1972), with the assumption
that the active factor would be RNA, which of course is
less likely now in view of the results of UNGAR and his
group and the findings of REINIS (1971) that transfer effects are abolished by inhibition of RNA synthesis with
actinomycin D. However, it is a source of considerable con-
solation that scotophobin was found to be normally quite
firmly bound to the RNA of our extracts, and indeed that
our extraction method proved a most convenient step in preparation of scotophobin, adopted as the standard method of
Dr. UNGAR and his group (UNGAR and FJERDINGSTAD, 1970).

As the last example of our work with rodents I would
like to discuss some attempts to transfer acoustic priming
for audiogenic seizure, which are quite preliminary, but
resulted in very striking effects.

It has been known for some decades that some strains
of mice will go into seizure on exposure to loud noises,
like the sound of an electric bell. These seizures can be
divided into four stages, the final of which is death from
the convulsions (Tab. II). Other strains, however, are resistant and show little or no incidence of seizure on exposure to sound (see FULLER and SJURSEN, 1967). In 1967,
however, HENRY reported that even highly resistant strains,
like the C57BL/6J, could be "primed" for audiogenic seizure

Tab. II: <u>Stages of audiogenic seizures in mice according</u>
<u>to HENRY (1967)</u>

Stage	Response of mouse
I	Wild running
II	Clonic seizures
III	Tonic seizures
IV	Death

by being given one short (30 sec) sound exposure in a brief
sensitive period 16 days after birth, when the animals are
first able to hear. Although the mice did not noticeably
react to this first exposure to the stimulus, they would
on a second exposure show a very high incidence of seizures, the majority of them dying.

Since this phenomenon has some analogies to the well
known imprinting in birds, (i.e., there is a short sensitive
period, the critical period, immediately after the relevant
receptors begin to function) it is possible that it may
represent a type of learning. As such it would be an extreme example of one-trial learning with a very clear-cut
behavioral effect. It would therefore appear very worthwhile to investigate whether transfer could result from
injecting brain extract from primed donors into non-primed
recipients.

In the first of these experiments, donors (these and
the recipients were C57 BL/6J mice) were primed on day 16
after birth by receiving a 30 sec exposure to an Edwards
Electric bell (Cat. No. 340-62) mounted under a wooden lid
closing a cylindrical chromatography jar, 30 cm wide and
45 cm high. Three days later (day 19) they were sacrificed,
the brains homogenized in 3.5 ml cold Ringer per brain and
20 μl of this extract injected intracerebrally (right posterior quadrant) into naive littermates, lightly anaesthe-
tized with ether. Since we were only investigating whether
any effect could be obtained with a "primed extract", and
since it was known that untreated animals should not react
to the sound at all, a control group was not run in this
experiment. On day 21, the recipients were given a 30 sec
exposure to the bell. Three out of 8 seized, going to
stages 1, 3, and 4 respectively (Tab. III). Considering
that none of them should have reacted (as had none of the
donors) this was quite striking.

In a second experiment of this type a control recipient group was included, receiving extract from the brains
of naive littermates to the experimental donors. Again an
effect was found, with 6 animals out of 10 in the experimental group convulsing, while none of the control group
showed any reaction. This was statistically significant
(Tab. III).

We are proceeding with experiments of this type, which
we consider to have many advantages. The rapid test and
training procedures potentially allow the handling of very

Tab. III: <u>Behavior of recipients of extract from acousti-
cally primed donors on exposure to the testing
stimulus</u>

Experiment	Group	Number of Mice Responding	Number of Mice Not Responding
I	Experimentals	3	5
II	Experimentals	6[x]	4
	Controls	0	5

x) P = .05 (Fisher Test)

large numbers of animals. Moreover, it is interesting from
a theoretical standpoint that such a short (30 sec) experience should be transferable, as most other behavioral procedures used in the field require quite extended training
(5 - 10 days of several trials per day) and overtraining
is often specifically recommended (e.g., UNGAR, 1971).

EXPERIMENTS WITH FISH

Our work with fish has mostly been carried out with a
modified model of the shuttle box designed by AGRANOFF (1967).
The box, made from a clear plastic mousecage is 12.5 by 27.5
cm, and 12.5 cm deep, tapering downwards so as to be 4 cm
narrower and shorter here. A black plexiglass wall divides
it into two compartments leaving a space of 2.8 cm through
which the fish may shuttle. Shuttles are recorded by means

of infrared photocells on each side of the barrier. Shock
(pulsed DC shock of 22.5 mA) may be delivered through stainless steel mesh covering the sides of the compartments and
2 W stimulus lights are mounted at the ends of the walls.
The set-up is controlled and responses counted automatically
by electromechanical programming equipment.

During donor training once a minute a light will go on
in the compartment where the fish is present, followed after
an interval (5 sec was found to be preferable to the 20 sec
used by AGRANOFF's group) by shock for 15 sec. Ten or 20
such trials per day for 10 days increases the level of avoiding (swimming away from the light before the shock goes on)
from about 5 per cent to about 70 per cent (group mean).
This is so during the summer too, (Fig. 3) in contrast to
the reports of AGRANOFF (1967).

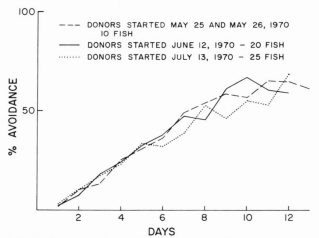

Fig. 3: Learning curves of donor goldfish given 10 trials a
day in the shuttle box with an interval of 5 sec
from the onset of the CS to the onset of UCS

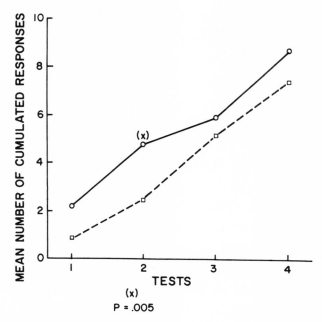

Fig. 4: <u>Avoidance responses to unreinforced light presen-</u>
<u>tations in recipient groups of a shuttle box ex-</u>
<u>periment</u>

Recipients of RNA extracts prepared as described by
RØIGAARD-PETERSEN et al.,(1968) from donors trained in this
way show a significantly higher level of light avoidance
than recipients of extract from naive donor fish (Fig. 4),
when tested with repeated daily presentations of light only,
i.e., unreinforced (see also FJERDINGSTAD, 1970). These results have been replicated successfully by BRAUD (1970).

One drawback of this paradigm is the long periods of
time consumed by training fish one at a time for several
days. In cooperation with Mr. Rodney C. BRYANT at the University of Tennessee an attempt was therefore made to con-

struct a shuttle box allowing the massed training of 12 fish. This was accomplished by increasing the dimensions of the box to 117.5 x 65 x 40 cm, and changing the training schedule by leaving the shock on in one compartment until the onset of the stimulus light in the opposite compartment, in order to keep the fish synchronized. Recording the behavior visually this was found to result in about the same rate of learning as in individually trained fish.

BRYANT (1971) has also used this type of set-up for massed testing of recipients, confirming the results that were obtained with individual testing. The massed training and testing procedure therefore would seem of potential value in research requiring large numbers of donors and recipients, i.e., purification and isolation of further transfer factors.

Another possibility of the shuttle box would be to use several stimulus lights of different color, in training several donor groups, reinforcing only one specific color for a given donor group. The resulting extracts might then be tested by random, unreinforced presentation of all the colors, which should give a very good indication of the specificity of the effect. We are embarking on a project of this type now. Earlier work with transfer of color discrimination in fish has been carried out, but has been concerned with the shift of preference from one color to another as resulting from positive reinforcement (DOMAGK and ZIPPEL, 1971).

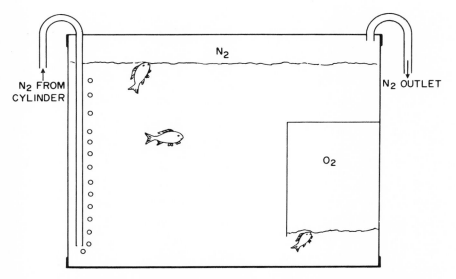

Fig. 5: <u>Apparatus for oxygen-reinforced training of gold-
fish</u>
The water in the tank is washed free of other dissolved gases by a stream of nitrogen that leaves a
nitrogen atmosphere over the surface. After a period
of trying in vain to obtain oxygen from the surface,
the fish will learn to obtain it from the beaker
near the bottom.

Finally, I would like to briefly mention an experiment with fish,of a completely different type. Goldfish
placed in oxygen poor water will normally surface to suck
air, from which they can take up enough oxygen to survive.
In the set-up shown in Fig. 5 we have been able to train
them to go down to take oxygen from an inverted beaker near
the bottom, rather than going to the surface where only
nitrogen is available. This problem is learned very rapidly, probably because of the strongly reinforcing effect of
oxygen in such a situation, as may be seen from Fig. 6. Re-

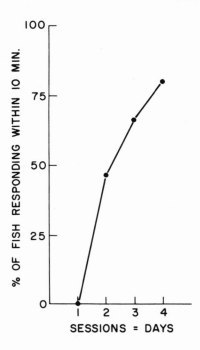

Fig. 6: <u>Learning curve of a group of fish trained in the</u>
<u>set-up shown in Fig. 5</u>

cipients injected with RNA extracts prepared as described
above from donors sacrificed after four days of such training were found to go to the surface significantly less often than recipients of "naive extract" (Fig. 7), even though
testing was carried out without reinforcement, i.e.,with
nitrogen both on the upper surface and in the beaker. Considered together with the experiments of DOMAGK and ZIPPEL
(1971) this shows that also in fish transfer can be obtained with both negative and positive reinforcement procedures
and thus cannot be caused by a simple stress factor resulting from exposure to shock.

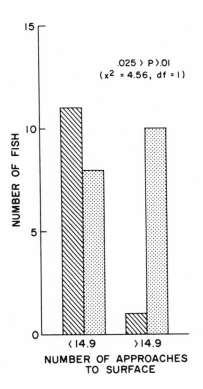

Fig. 7: <u>Result of testing recipients in an experiment with</u>
<u>the set-up shown in Fig. 5</u>

▨ recipients of "trained" extract,

▦ recipients of "naive" extract.

LITERATURE

AGRANOFF, B.W.: Agents that block memory.
    In: The Neurosciences, a Study Program (G.C. QUARTON,
    T. MELNECHUK and F.O. SCHMITT, eds.), pp. 756-764,
    Rockefeller University Press, New York (1967).

BABICH, F.R., A.L. JACOBSON, S. BUBASH, and A. JACOBSON:
    Transfer of a response to naive rats by injection of
    ribonucleic acid extracted from trained rats.
    Science 149, 656-657 (1965).

BATESON, P.P.G., G. HORN, and S.P.R. ROSE: Effects of early
    experience on regional incorporation of precursors in-
    to RNA and protein in the chick brain.
    Brain Res. 39, 449-465 (1972).

BRAUD, W.G.: Extinction in goldfish: facilitation by intra-
    cranial injection of RNA from brains of extinguished
    donors.
    Science 168, 1234-1236 (1970).

BRYANT, R.C.: Transfer of light-avoidance tendency in groups
    of goldfish by intracranial injection of brain extract.
    J. Biol. Psychol. 13, 18-24 (1971).

DOMAGK, G.F., and H.P. ZIPPEL: Chemical transfer of learned
    information in goldfish.
    In: Chemical Transfer of Learned Information (E.J.
    FJERDINGSTAD, ed.), pp. 183-198, North-Holland, Amster-
    dam (1971).

DYAL, J.A.: Transfer of behavioral bias: reality and speci-
    ficity.
    In: Chemical Transfer of Learned Information (E.J.
    FJERDINGSTAD, ed.), pp. 219-263, North-Holland, Amster-
    dam (1971).

FJERDINGSTAD, E.J.: Chemical transfer of alternation train-
    ing in the Skinner box.
    Scand. J. Psychol. 10, 220-224 (1969a).

FJERDINGSTAD, E.J.: Chemical transfer of learned preference.
    Nature 222, 1079-1080 (1969b).

FJERDINGSTAD, E.J.: Memory transfer in goldfish.
    J. Biol. Psychol. 11, 20-25 (1970).

FJERDINGSTAD, E.J.: Chemical transfer of positively rein-
    forced training schedules: evidence that the effect is
    due to learning in donors.
    In: Chemical Transfer of Learned Information (E.J.
    FJERDINGSTAD, ed.), pp. 65-84, North-Holland, Amster-
    dam (1971).

FJERDINGSTAD, E.J., T. NISSEN, and H.H. RØIGAARD-PETERSEN:
    Effect of ribonucleid acid (RNA) extracted from the
    brain of trained animals on learning in rats.
    Scand. J. Psychol. 6, 1-6 (1965).

FJERDINGSTAD, E.J., W.L. BYRNE, T. NISSEN, and H.H. RØIGAARD-
    PETERSEN: A comparison of transfer results obtained with
    two different types of extraction and injection proce-
    dures, using identical behavioral techniques.
    In: Molecular Approaches to Learning and Memory (W.L.
    BYRNE, ed.), pp. 151-170, Academic Press, New York
    (1970).

FULLER, J.L., and F.H. SJURSEN: Audiogenic seizures in elev-
    en mouse strains.
    J. Hered. 58, 135-140 (1967).

GOLUB, A.M., and J.V. MCCONNELL: Empirical issues in inter-
    animal transfer of information.
    In: Chemical Transfer of Learned Information (E.J.
    FJERDINGSTAD, ed.), pp. 1-29, North-Holland, Amster-
    dam (1971).

HENRY, K.R.: Audiogenic seizure susceptibility induced in
    C57BL/6J mice by prior auditory exposure.
    Science 158, 938-940 (1967).

HYDEN, H.: Biochemical changes accompanying learning.
    In: The Neurosciences, a Study Programm (G.C. QUARTON,
    T. MELNECHUK and F.O. SCHMITT, eds.), pp. 765-771,
    Rockefeller University Press, New York (1967).

MCCONNELL, J.V.: Memory transfer trough cannibalism in
    planarians.
    J. Neuropsychiat. 3 (suppl. 1), 42-48 (1962).

MOOS, W.S., H. LEVAN, B.T. MASON, H.C. MASON, and D.L.
    HEBRON: Radiation induced avoidance transfer by brain
    extracts of mice.
    Experientia 25, 1215-1219 (1969).

REINIS, S.: The formation of conditioned reflexes in rats
    after the parenteral administration of brain homogenate.
    Activ. Nerv. Sup. 7, 167-168 (1965).

REINIS, S.: A derepressor hypothesis of memory transfer.
    In: Chemical Transfer of Learned Information (E.J.
    FJERDINGSTAD, ed.), pp. 109-142, North-Holland, Amster-
    dam (1971).

ROSENBLATT, F.: Induction of specific hehavior by mammalian
    brain extracts.
    In: Molecular Mechanisms in Memory and Learning (G.
    UNGAR, ed.), pp. 103-147, Plenum Press, New York (1970).

RØIGAARD-PETERSEN, H.H., E.J. FJERDINGSTAD, and T. NISSEN:
    Effect of ribonucleic acid (RNA) extracted from the
    brain of trained animals on learning in rats.

III. Results obtained with an improved procedure.
Scand. J. Psychol. $\underline{9}$, 1-16 (1968).

SHASHOUA, V.E.: The relation of RNA metabolism in the brain
to learning in the goldfish.
In: The Central Nervous System and Fish Behavior (D.
INGLE, ed.), pp. 203-213, University of Chicago Press,
Chicago (1968).

UNGAR, G.: Bioassays for the chemical correlates of acuired
information.
In: Chemical Transfer of Learned Information (E.J.
FJERDINGSTAD, ed.), pp. 31-49, North-Holland, Amsterdam (1971).

UNGAR, G., and E.J. FJERDINGSTAD: RNA-bound peptides in
chemical transfer of learned behavior.
Mol. Neurobiol. Bull. $\underline{2}$, 9 (1969).

UNGAR, G., and C. OCEGUERA-NAVARRO: Transfer of habituation
by material extracted from brain.
Nautre $\underline{207}$, 301-302 (1965).

WEISS, K.P.: Information specificity in memory transfer.
In: Chemical Transfer of Learned Information. (E.J.
FJERDINGSTAD, ed.), pp. 85-95, North-Holland, Amster-
dam (1971).

CHEMICAL TRANSFER OF A DUMMY REACTION, RELEASED IN YOUNG
MOUTHBREEDING FISH (TILAPIA NILOTICA) DURING THE "CRITICAL
PERIOD", FROM IMPRINTED DONORS INTO UNIMPRINTED RECIPIENTS
AFTER THE "CRITICAL PERIOD"[*]

Hans Peter Zippel and Cornelis Langescheid

Physiologisches Institut
Lehrstuhl II der Universität
34 Göttingen, Humboldtallee 7, Germany

ABSTRACT

The "linking behavior" to the dummy, shown by young
fish during the "critical period", appears to be transferable to unimprinted animals after the "critical period" by
injection of brain extracts from imprinted donors. On the
other hand, control recipients injected with extracts from
unimprinted donors show no such orientation to the model.
Following injection the behavior of the recipients was comparable with that of imprinted animals at an equivalent
age. However, the high level of "linking behavior" manifested by the donor animals at the time of decapitation
could not be transferred.

[*]This material was first presented in a dissertation by
C. LANGESCHEID (Diss. Göttingen, 1972).

INTRODUCTION

During the past few years many experiments have been
performed on the interanimal transfer of learned information. Although the evidence has met with some controversy,
recent findings would appear to have established the fact
that the chemical transfer of acquired information is a
real phenomenon (e.g., UNGAR 1972, for a review see DYAL,
1971).

The present investigations were initiated to ascertain
whether it was also possible to transfer a genetically controlled behavior pattern using injections of "antagonistic"
brain material. Up until the present time, only three other
experiments in this direction have been reported, all of
which have yielded negative results: DYAL and GOLUB (1971)
unsuccessfully attempted to transfer an inborn alcohol-pre-
ference from one strain of mice to another strain which had
a spontaneous dislike for this substance. Similarly, the experiments of REINIS and MOBBS (1970), in which they tried to
transfer the innate aggressive tendency of killer-rats to
non-killer-rats, met with no success. Moreover, the reverse
procedure also proved negative. In like manner, LAGERSPETZ
(1971) was not able to transfer the "pacifying" factor from
comparatively inactive mice to animals with a very high activity.

The present experiments were conduced with young mouthbreeding cichlids (Tilapia nilotica). In order to obtain a
complete understanding of the results it is first necessary

to give a brief summary of the animals' normal behavior.
Shortly after spawning the female takes both eggs and sperm
into her mouth where fertilisation subsequently occurs. The
eggs are retained within the mouth for some sixteen days,
during which time they are constantly moved around by the
special "chewing" movements of the female. The young fish
are hatched after four to five days and following the absorption of the yolk-sac and the filling of the swim-blad-
der, they are released from the mouth for the first time.
At night and during periods of danger, the young return to
the mother's mouth. With increasing age the bond between
mother and offspring gradually lessens and after 3-4 weeks
the young begin to live independently (BRESTOWSKY, 1968).

Prior to this period the young fish will react readily
to a dummy (e.g., a spherical model) presented in place of
the mother. The young often react to the dummy for many
minutes, remaining in contact with the surface and trying
to penetrate it (PETERS and BRESTOWSKY, 1961). It was established, first by BREDER (1934) and then by PETERS (1937),
that, in cichlids, the "bond" between parents and offspring
is innate, and BAUER (1968) has shown that visual stimuli
are of paramount importance in the direction of this behavior. Moreover, BRESTOWSKY (1968) has described two main
components, in mouthbreeding cichlids, which influence the
"linking behavior" to the mother: firstly, an appetitive
orientation component which motivates the animals to respond
to a well defined visual stimulus and secondly, a tendency
to make contact with this stimulus.

This "linking behavior" is normally manifested about seven days after fertilisation, when the young are first able to swim. However, this behavior can also be elicited following presentation of a dummy, between the seventh and twelfth days after spawning, in animals which have been bred in isolation and which have thus been deprived of any experience of the mother (BRESTOWSKY, 1968). Once this "critical period" has been passed, the animal can no longer be imprinted to an artificial parental substitute. And furthermore, after this "critical period", unimprinted animals show an active avoidance to presented models.

TESTING CHAMBERS

The testing chambers were constructed, with some modifications, on the design previously used by BRESTOWSKY (1968). The arrangement of these chambers was such that they provided not only the optimum conditions for the experimental animals, but also afforded a completely objective observation and critical analysis of the animals' behavior.

The experimental apparatus (Fig. 1) consists of three identical testing chambers arranged side by side. These chambers are completely independent of each other with the exception of a common water supply, which ensures that the temperature and the quality of the water are the same in all three chambers. Each individual testing tank has a crystal-glass bottom, around the periphery of which is fixed a collar of plexiglass (Ø 50 cm; height 10 cm). Four neon

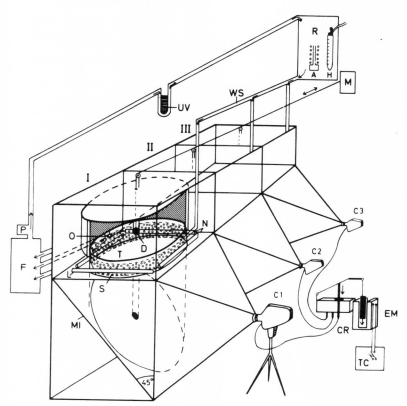

Fig. 1: <u>Testing chambers</u>
A = Aerator, CR = Cable release, D = Dummy, EM = Electromagnet, F = Filter, H = Heater, M = Motor (which moves the dummy), MC = Movie camera, Mi = Mirror, N = Neon light, O = Outlet, P = Pump, R = Reservoir, S = Sieve, T = Testing tank, TC = Time lapse control, UV = UV-light source, WS = Water supply

lights (Osram, 20 Watt/25, white-universal) are arranged in a square around this plexiglass collar. On top of the latter a closed cylindrical drum with a matted black interior finish is affixed. With this "dark-field arrangement" (BRESTOWSKY, 1968) the majority of the light is totally re-

flected from the surface of the water and the crystal-glass
bottom: consequently each fish appears as a light spot on
a dark background. A mirror is arranged below each chamber
at an angle of 45° to the crystal-glass bottom, so that
the reactions to the dummy can be observed from outside
the testing chambers. The position of the animals in the
tank is recorded every five minutes by means of cine cam-
eras (BAUER C1M, Super 8) equipped with a time-lapse facility. The spherical models (optimum for Tilapia nilotica:
black, Ø 40 mm; LANGESCHEID, 1968) are suspended in the
center of each chamber and their movement during testing
is synchronized (amplitude 10 mm; frequency 144 per minute;
BRESTOWSKY, 1968).

With the above experimental set-up a completely objective estimation of the animals' behavior is possible.
External disturbances are almost entirely excluded and the
conditions for each animal under test are identical.

BREEDING, IMPRINTING AND TRANSFER

For the transfer experiments it was necessary to use
completely naive animals: that is, animals which have been
denied experience of the mother, or a similar substitute,
and are thus unimprinted.

The spawn (200-400 eggs) is removed from the mouth of
the female on the first or second day following fertilisation and placed in the breeding apparatus. The latter is

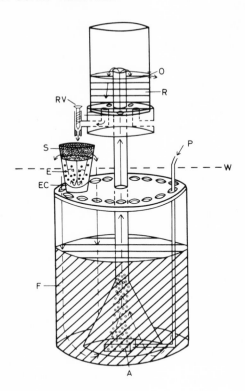

Fig. 2: <u>Breeding apparatus</u>
A = Aerator, E = Eggs, EC = Egg container, F = Fil-
ter, O = Overflow, P = Pump, R = Reservoir, RV =
Regulating valve, S = Sieve, W = Water level

housed in a white chest in order to avoid possible imprinting of the young. It is imperative that the eggs be kept in
constant motion, for unless this is achieved the embryo sinks
into the yolk-sac and dies (PETERS, 1963a). The breeding
apparatus (Fig. 2) is once again a modification of that
previously used by BRESTOWSKY (1968). A water current is
created by a stream of air bubbles from an aerator at the
base of the apparatus. This current flows up into the re-

servoir above and from thence into the egg containers. The
flow of water can be adjusted by means of the regulating
valves so that the optimum movement of the eggs can be obtained. The water then passes out through a sieve around
the top of the containers and after being sucked through
a filter the cycle is repeated.

Four days after fertilisation the young are hatched
and four days later they are able to swim. At this time a
certain number (usually 50 %) of the young fish are removed
from the breeding chest and placed immediately in the testing chambers. As soon as the dummy is presented the animals
show the characteristic "linking behavior" which is manifested during imprinting. These animals are then used as
the imprinted donors. Following a further period of 4 days
(12 days after fertilisation), all except 10 of the imprinted animals are removed from the testing chambers, immediately killed by placing them in an ice-cooled petri dish and
then decapitated. On the same day, the majority of the unimprinted animals are removed from the breeding chest and
are similarly killed and decapitated. These animals then
form the unimprinted donor group. The rest of the unimprinted young remain in the breeding chest and these are
the recipient animals which receive injected extracts on
the 14th day.

The heads of both imprinted and unimprinted donors are
deep-frozen for 20 min at $-40^{\circ}C$, after which they are homogenized (using a Potter-Elvehjem-homogenizer), in an ice-
cooled container. The homogenate is then centrifuged

(15,000 rpm, 20 min at 4°C). The resulting supernatant is
lyophilized for about 20 hours, dissolved in 1 ml aqua dest.
and centrifuged once more (5,000 rpm). This supernatant is
again lyophilized and finally dissolved in 45 µl of a cooled electrolyte solution. Evans blue is added and the solution is sucked up into a capillary (tip diameter 20-30 µ).
On the 14th day the extracts are injected intramuscularly
(approx., 0.3 µl/animal), 10 animals receiving extracts
from unimprinted donors (control recipients) and a further
10 receiving extracts from imprinted donors (recipients).
These animals are then placed in the experimental tanks,
each group occupying a separate testing chamber: group 1,
imprinted animals; group 2, injected recipients; group 3,
injected control recipients. This procedure is summarised
in Fig. 3.

The behavior of the animals to the models is then film-
ed between the 14th and 25th days, and the resulting photo-
graphs are analysed in the following manner: the film is
projected, frame by frame, on to a sheet of white cardboard
on which the area of the testing tank is represented (Fig.
4). This area is divided by means of concentric circles of
increasing diameter around the dummy, into five definite
regions.

Area I, is a region of 5 cm radius immediately surrounding
       the center of the tank, and also that of the dummy.
       It is in this region that the "linking behavior"
       of the animals to the dummy is at its strongest.
Area II, r = 5 cm to r = 10 cm.

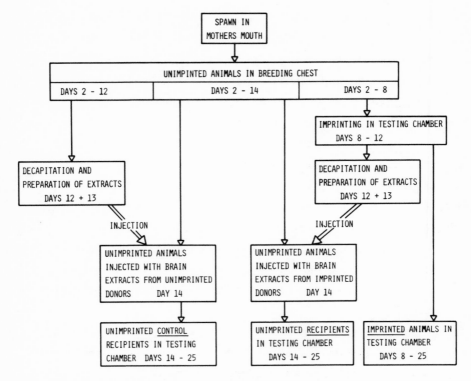

Fig. 3: <u>Summary of the experimental procedure</u>

Area III, r = 10 cm to r = 15 cm.

Area IV, r = 15 cm to r = 20 cm.

Area V,   r = 20 cm to r = 25 cm, extends to the edge of the
        testing chamber and animals found here exhibit
        a very low level of "linking behavior" to the
        model.

The numbers of fish found in each particular region are
counted on each separate photograph, punched on to paper
tape, and then summated by a PDP-12 computer. In this manner a daily estimate of the total number of fish in each

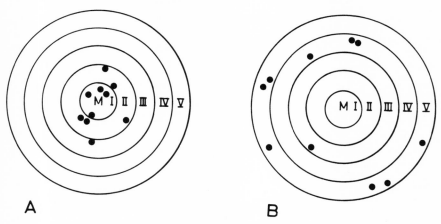

A   B

Fig. 4: <u>Schematic presentation of the typical orientation
to the dummy manifested by imprinted (A) and un-
imprinted (B) animals in the testing chamber</u>
● = Position of a fish at the moment of exposure.
M = Position of the dummy.
See text for explanation of areas I to V.

of the respective regions can be obtained.

RESULTS

Fig. 5 gives a summary of the results obtained from
7 groups of animals, for each of the five regions, over a
period of 12 days following injection. In each case the
number of unimprinted control recipients is taken as zero,
and the numbers of both the unimprinted recipients and the
imprinted controls are represented as the difference, more
(positive values) or less (negative values) as compared
with the actual number of unimprinted control recipients.
On the first day after injection, the behavior of the un-

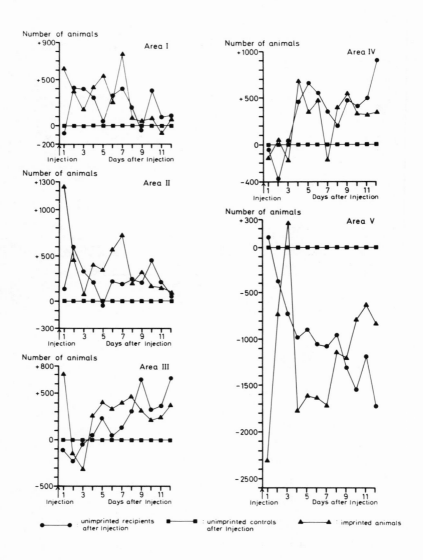

Fig. 5: <u>A summary of the relative "linking behavior" of imprinted and injected animals as compared with the
controls</u>
For further details see text.

imprinted recipients is directly comparable with that of
the unimprinted controls. However, over the remainder of
the testing period, the behavior of the unimprinted recipients is not essentially different from that shown by
the imprinted animals. During this period, the number of
animals in both of these groups is greater than the number
of unimprinted controls in areas I, II, III and IV. In area
V, however, which is furthest from the dummy, this relationship is reversed. Furthermore, in the first 4 days following injection, more of the recipient and imprinted
animals are found in areas I and II, but later in the experimental period areas III and IV are preferred.

From these results it is evident that both the recipient and imprinted groups have a greater affinity for the
dummy than the controls.

In Fig. 6 a comparison is made between the injected
recipients and the injected controls, in terms of the actual
number of animals present in the respective regions, for
each of the twelve days of the experiments. This comparison
includes a statistical analysis which confirms the points
made above.

DISCUSSION

The behavioral findings of the present investigations
differ in some very significant respects from those reported previously by BRESTOWSKY (1968) and BAUER (1968). The

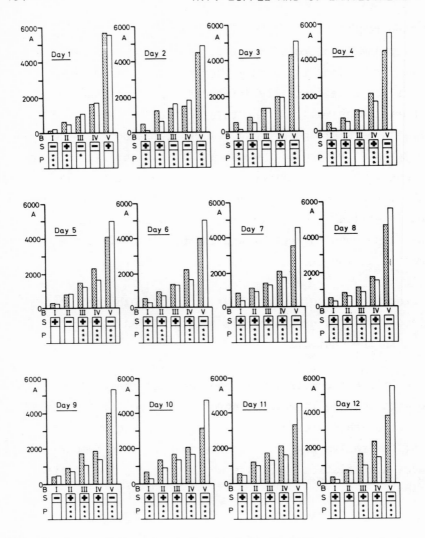

Fig. 6: A summary and statistical evaluation of the "linking behavior": a comparison between injected recipients and injected control recipients

A = Number of animals (stippled histograms = injected recipients; open histograms = injected control recipients).
B = Areas I-V

S = The negative and positive signs indicate whether
the number of recipients is more or less than that
of the controls.
P = Significance level; X^2-test (without Yates-cor-
rection) *** = $p < .001$; ** = $p < .01$; * = $p < .05$.

principal difference is that, after the 12th day, the "linking behavior" of the imprinted animals to the dummy is at a
much lower level than one might expect from the work of the
above authors. These results are made more unusual by the
fact that prior to the 12th day, the results of both investigations are practically identical. The only major modification in the experimental set-up was the use of a less
sensitive photographic film which required a much higher
light intensity. However, because of the similarity in the
results over the initial experimental period, the negative
phototropic behavior of the animals, as described by BREDER
(1934), would not appear to be a sufficient reason to explain the later differences. Moreover, these ambiguities
cannot be explained on the basis of any genetical differences, since the animals used in the present investigations
were in fact obtained from the Tübingen group (PETERS,
BRESTOWSKY). In the light of these considerations it would
seem necessary to investigate this problem further.

The transfer effects reported above are also not so
readily explained. The "linking behavior" of the donors at
the time of decapitation (12th day) is at a much higher
level than that shown by the recipients immediately following injection. However, the behavior of the latter group is
directly comparable with that of the donors at an equivalent
age (15th to 25th day). That is to say, the "linking be-

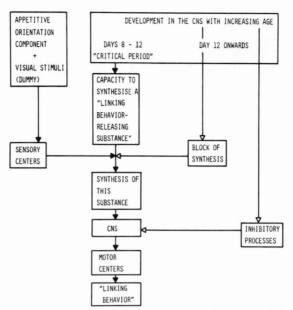

Fig. 7: <u>Hypothetical model for the age-dependency of the</u>
<u>"linking behavior"</u>

havior" of the recipients is primarily governed by the
actual age of the animal, rather than by any "information
quality" of the injected extracts. So that, as in these experiments, even if the extracts are taken from donor animals in which the "linking behavior" is at its highest level, the behavior of the recipients will be equivalent to
that of imprinted animals of the same age.

However, it should be pointed out that these are the
results of preliminary investigations and any conclusions
which are drawn must thus be treated with some circumspec-
tion. Accordingly the following hypothesis (Fig. 7) is proposed as a tentative explanation of the above findings.
During the "critical period", the young fish has the capac-

ity to produce a substance which releases the "linking behavior" whenever the requisite visual stimulus, such as the
mother or an adequate dummy, is presented. From the 12th
day after fertilisation onwards the capacity to synthesise
this substance is presumably lost, or blocked, because if
the animal is presented with the model at this time avoidance behavior is shown. With increasing age, there is a
weakening in the "linking behavior" to the dummy and it is
possible that these changes are brought about through inhibitory processes in the developing nervous system which
directly influence the animal's behavior. All, or some, of
the processes outlined above might well explain the apparent
age-dependency of the observed transfer effects.

In conclusion it would seem worthwhile to speculate why
the results of the present investigations showed positive
transfer effects, while previous work on the same topic by
other workers (see introduction) has met with no success.
The principal difference of the present experiments is that
no attempt has been made to change an animal's normal behavior, but rather to restore the faculty to behave in the
normal manner. That is, animals which have themselves been
denied experience of the mother, or model substitute, during
the "critical period" when they would normally show the
"linking behavior" characteristic of imprinting, are given
this facility by means of brain extracts from imprinted
donors.

ACKNOWLEDGEMENTS

The research was supported by a grant from Deutsche Forschungsgemeinschaft (Sonderforschungsbereich 33)

This manuscript has been carefully revised by A. BURT, DAAD fellow.

LITERATURE

BAUER, J.: Vergleichende Untersuchungen zum Kontaktverhalten verschiedener Arten der Gattung Tilapia (Cichlidae, Pisces).
Z. Tierpsychol. 25, 22-70 (1968).

BREDER, C.M.: An experimental study of the reproductive habits and life history of the cichlid fish Aequidens latifrons (Steindachner).
Zoologica, N.Y. 18, 1-42 (1934).

BRESTOWSKY, M.: Vergleichende Untersuchungen zur Elternbindung von Tilapia-Jungfischen (Cichlidae, Pisces).
Z. Tierpsychol. 25, 761-828 (1968).

DYAL, J.A.: Transfer of behavioral bias: reality and specifity.
In: Chemical Transfer of Learned Information (E.J. FJERDINGSTAD, ed.), pp. 219-263, North-Holland Publ. Comp., Amsterdam (1971).

DYAL, J.A., and A.M. GOLUB: Attempt to modify a genetically
controlled response via injections of brain RNA.
In: Chemical Transfer of Learned Information (E.J.
FJERDINGSTAD, ed.), pp. 257, North-Holland Publ. Comp.,
Amsterdam (1971).

LAGERSPETZ, K.Y.H.: Transfer of the conditioned darklight
preference and activity level by brain extracts.
In: Chemical Transfer of Learned Information (E.J.
FJERDINGSTAD, ed.), pp. 257, North-Holland Publ. Comp.,
Amsterdam (1971).

LANGESCHEID, C.B.: Vergleichende Untersuchungen über die
angeborene Größenunterscheidung bei Tilapia nilotica
und Hemihaplochromis multicolor (Pisces, Cichlidae).
Experientia 24, 963-964 (1968).

PETERS, H.M.: Experimentelle Untersuchungen über die Brut-
pflege von Haplochromis multicolor, einem maulbrüten-
den Knochenfisch.
Z. Tierpsychol. 1, 201-218 (1937).

PETERS, H.M.: Über einige Verhaltensweisen bei Cichliden
und deren Grundlage.
Bull. Inst. océanogr. Monako N° spécial 1 D, 15-33
(1963a).

PETERS, H.M., and M. BRESTOWSKY: Artbastarde in der Gattung
Tilapia (Cichlidae, Teleostei) und ihr Verhalten.
Experientia 17, 261-265 (1961).

REINIS, S., and D.R. MOBBS: Some applications of "memory
transfer" in the study of learning.
In: Molecular Approaches to Learning and Memory (W.L.

BYRNE, ed.), pp. 189-193, <u>29</u>, Academic Press, New York (1970).

UNGAR, R.: Molecular coding of information in the nervous system.
Die Naturwissenschaften, Heft <u>3</u>, 85-91 (1972).

NEUROCHEMISTRY

NEUROCHEMICAL MICROMETHODS

Volker Neuhoff

Max-Planck-Institut für experimentelle Medizin
Arbeitsgruppe Neurochemie
Göttingen, Germany

ABSTRACT

This paper presents some details of refined micromethods employed in the neurochemical analysis of single
neurons and sub-cellular structures and gives some examples of results obtained with these methods.

Micromethods are extremely useful for neurochemical
analysis of tissue samples from defined regions of the
brain, requiring material from only a small number of animals. In addition, micromethods are often the only method
available for biochemical analysis where the quantity of
material is limited. This article will outline the micromethods currently in progress in our laboratory and also
give some examples of results obtained using these procedures.

471

Micro Disc Electrophoresis of Proteins

The theory of disc electrophoresis in polyacrylamide
gels, developed by ORNSTEIN (1964) and DAVIES (1964), remains
valid for micro procedures. This widely used technique
(MAURER, 1971) was first applied to micro scale by PUN and
LOMBROZO (1964) and later modified by GROSSBACH (1965),
HYDEN, BJURSTAM and MC EWEN (1966), MC EWEN and HYDEN (1966)
and NEUHOFF (1968). Micro disc electrophoresis is generally
performed in 5 μl Drummond Microcaps (length 32 mm, inner
diameter 0.45 mm) but 1, 2 or 10 μl Microcaps are also suitable. In order to get the same high resolution of protein
patterns in micro disc electrophoresis as obtained with macro
disc electrophoresis on polyacrylamide gels, higher concentrations of the monomer acrylamide and lower concentrations
of the bis-acrylamide are necessary, depending on the diameter of the capillaries used. 0.1 to 0.5 g protein can be
subjected to electrophoresis in 5 μl capillaries; that means
5,000-10,000 runs can be performed with 1 ml containing 1 mg
protein. The lowest limit for a single band in a gel with
450 μ diameter is 10^{-9} g protein, stainable with amido black
10 B. The resolution power and sharpness of the bands is
normally better in micro than in macro scale. Furthermore,
the micromethod has the advantage of being less time consuming,
the period of electrophoresis is shorter and staining and
destaining require only 5 min and 30 min respectively. Moreover, an electrophoretic destaining is not necessary. Gels
prepared (for technical details see NEUHOFF, 1970) with stock
solutions according to NEUHOFF (1968) are optimal for the
fractionation of water soluble brain proteins. However, the
optimal pH and gel concentration should be determined for

each protein mixture (NEUHOFF and LEZIUS, 1967, 1968; NEU-
HOFF, SCHILL and STERNBACH 1968, 1969a, b, c, 1970; NEUHOFF
et al., 1970).

Micro gels to be dissolved in ammonia (SEAR and ROIZMAN,
1968) for measurements of radioactivity incorporated into
proteins (CASOLA et al., 1969), are prepared with modified
stock solution containing ethylene diacrylate as the crosslinking agent instead of bisacrylamide (CHOULES and ZIMM,
1965). Such gels have the same separation power as normal
ones. If very dilute protein solutions are to be fractionated,
the capillary volume can be increased by connecting a second,
or even a third, 5 µl capillary to the gel capillary by
means of polyethylene tubing. The constant voltage (NEUHOFF
et al., 1967) chosen depends on the concentration of the
lower gel, and the capillaries do not need to be cooled. The
migration of the proteins can easily be followed using an
electrophoretic buffer containing fluorescein and slightly
coloured with bromphenol blue. After electrophoresis, the
gel is pushed out of the capillary by water pressure into
a solution of amido black in acetic acid. The protein bands
are recorded with a Joyce-Loebl Double Beam Microdensitometer, the gels being kept in a suitable chamber (NEUHOFF,
1968, 1970).

Individual proteins separated during electrophoresis
can be re-isolated (for details see NEUHOFF, 1970; NEUHOFF
and SCHILL, 1968). The gels containing the bands to be eluted
are stained for 1 - 5 min at neutral pH with 0.1 % bromphenol
blue and then cut into discs containing single bands with
a razor blade under a stereomicroscope. For the elution of

a protein from the gel disc, the lower third of a 5 μl capillary is filled with a 20 % gel to close the lower end. A
short piece of a capillary, with a larger diameter, is attached to the top and the whole system filled with buffer.
One or more gel discs are put into the solution and subjected to electrophoresis which is terminated as soon as the
bromphenol blue stained protein appears in the buffer phase
of the 5 μl capillary. The protein is removed and can now
be used for enzyme activity tests (NEUHOFF and LEZIUS, 1967,
1968; NEUHOFF et al., 1968, 1969a, b, c, 1970; NEUHOFF et
al., 1970), microdialysis (NEUHOFF and KIEHL, 1969), micro
analysis of amino acid composition (NEUHOFF et al., 1970) or
micro immunoprecipitation (NEUHOFF and SCHILL, 1968). This
electrophoretic elution method can be applied to all small
proteins, like albumin (NEUHOFF and SCHILL, 1968). To elute
large protein particles, like RNA-polymerase, the isolated
gel discs must be frozen at -170° C in water-free glycerol,
slowly thawed and eluted as described elsewhere (NEUHOFF
et al., 1969c).

In order to demonstrate differences between the protein
patterns of various brain regions, one has to apply a special
extraction procedure (ANSORG et al., 1971) because in primarily water soluble proteins only quantitative, rather than
qualitative, differences exist. The prepared brain region
therefore is firstly homogenized with a micro-homogenizer
(NEUHOFF, 1968, 1970) in Tris HCl buffer (pH 7.4) and following centrifugation the sediment is again homogenized 4
or 5 times in buffer to completely extract the water soluble
proteins. The sediment is then homogenized once more in a
mixture of n-butanol/buffer or iso-pentanol/buffer and the

water phase taken for micro disc electrophoresis. This procedure liberates membrane associated water soluble proteins
as can be seen in the protein patterns after electrophoresis.
In order to demonstrate qualitative differences in the protein patterns of the brain regions one has firstly to completely extract the primarily water soluble proteins and
afterwards to homogenize the sediment with 1.5 % Triton X
100 in Tris/HCl buffer. Finally one has to shake this extract with iso-pentanol and to use the water phase for the
electrophoresis (ANSORG et al., 1971). The treatment of the
extract with iso-pentanol prevents the production of artifacts by agglomerated proteins during the electrophoresis.

The nervous specific protein S-100 first isolated by
MOORE (1965), gives a single peak when chromatographed on
DEAE-cellulose, DEAE-Sephadex, G-100 Sephadex and hydroxy-
lapatide, and when centrifuged through a sucrose density
gradient. A single band is obtained following electrophoresis
on starch gel, and when submitted to double diffusion in
agarose a single precipitation band is formed (MOORE, 1965;
VINCENDON et al., 1967; MOORE and PEREZ, 1968; DANNIES and
LEVINE, 1969; CALISSANO et al., 1969; CICERO and MOORE,
1970). When electrophoresed on continuous or discontinuous
polyacrylamide gel systems the protein gives, however, 1 -
7 bands (GOMBOS et al., 1966, 1971a, b, c; MC EWEN and
HYDEN, 1966; UYEMURA et al., 1967, 1971; TARDY et al., 1967;
FILIPOWICZ et al., 1968; MOORE and PEREZ, 1968; CALISSANO
et al., 1969). It is suggested that S-100 protein may have
a number of isometric forms (MC EWEN, 1966; GOMBOS et al.,
1971a, b, c), in the same manner as the lactate dehydrogenase
iso-enzymes revealed by the superior resolving power of poly-

acrylamide gel electrophoresis (UYEMURA et al., 1971). On
the other hand, polyacrylamide gels may cause artifacts
during separation of tissue extracts or purified proteins,
which could lead to some misinterpretation (FANTES and
FURMINGER, 1967; BREWER, 1967; KING, 1970; ANSORG et al.,
1971). The multiple forms of S-100 protein, revealed by
polyacrylamide gel micro disc electrophoresis, are caused
by oxidation of this thiol-protein by ammonium persulfate,
potassium ferricyanate and probably other by-products of
polymerization. The protein runs as a single sharp symmetric
fast migrating component (FMC) if no oxidizing or reducing
agent in appropriate concentration is present in the system.
Ca^{++} does not change the single-banded pattern of the oxidizing agent-free system. Depending on the ratio between protein
and oxidizing agents in the gels respectively added to the
system, up to seven bands or a single sharp symmetric slow
migrating component (SMC) are formed (ANSORG and NEUHOFF,
1971). By fractionated extraction combining 0.01 M Tris/HCl-
buffer, pH 7.4, 1.5 % Triton X 100 and iso-pentanol, a membrane associated protein fraction, which has the same electrophoretic mobility as the oxidized form (SMC) of S-100,can be
isolated from primary water insoluble material of the guinea-
pig brain (ANSORG and NEUHOFF, 1971).

Micro disc electrophoresis on 20 % polyacrylamide gels
in 5 µl caps was used to study the water soluble proteins of
various tissues of the snail, Helix pomatia (OSBORNE et al.,
1971). The following eight areas were fractionated: identifiable giant neurons of the metacerebral ganglion, brain,
liver, kidney, penis retractor muscle, optic tentacle, oesophagus and heart. The electrophoretic pattern of all tissues

showed differences in both slow running and fast running proteins. A large amount of total protein from isolated neurons was associated with the fastest running protein and appeared to consist of three components. This fast running band was also present in high amounts in the brain and the optic tentacle and to a lesser extent in the heart, the penis retractor muscle and the oesophagus. Measurements of the band's Rf value, refractionation of isolated bands, and mixing experiments showed that the fast running bands from the various tissues had the same electrophoretic mobility. This protein band was absent in liver and kidney extracts. On the basis of its distribution, it is suggested that the band is a nervous specific protein. Whether the fast moving band of Helix is identical or similar to the S-100 and/or the 13-4-2 brain specific proteins of vertebrates (MOORE, 1965; MOORE and PEREZ, 1968) is of great interest, especially for comparative biology.

The protein pattern of the rabbit egg was analysed before and after impregnation and during cleavage, using micro disc electrophoresis, and compared with the surrounding fluids (PETZOLDT et al., 1972). The developing embryo showed a specific protein pattern differing from that of the follicle fluid, the oviductal secretion, and also the serum. Quantitative changes in the protein pattern could be seen in several developmental stages, especially after impregnation and at the beginning of cleavage.

Enzyme Kinetics of Dehydrogenase in Microgels

Micro disc electrophoresis is carried out in 5 μl caps
on 20 % polyacrylamide gels under standard conditions (NEU-
HOFF, 1968, 1970). 2 μl or even 1 μl Drummond Microcaps can
also be used for fractionations, in which case an increase
of acrylamide concentration to 25 % is advantageous. The resolving gel buffer is tris/sulfate, pH 8.8 (NEUHOFF, 1968).
Other buffer systems (e.g., tris/HCl, tris/phosphate, trie-
thanolamine/phosphate, or triethanolamine/sulfate) may be
used providing the pH is appropriate. Generally 0.3 - 1.5 μl
of the diluted sample containing the dehydrogenase to be
tested is subjected to electrophoresis. The optimal concentration has to be determined separately for each sample.
Tissue samples are homogenized in a suitable ice cold buffer (1:10 w/v) and centrifuged at 0° C for 2 hours at
24,000 g. The clear supernatant is divided into several
portions and stored at -20° C. Since it was observed that
very dilute solutions stored either at 0° C or -20° C recorded a loss of dehydrogenase activity, the final dilution
of the supernatants is made just prior to electrophoresis.
After electrophoresis, the gels are pressed out of the
capillaries on to a watch glass and then transferred into
a tetrazolium incubation mixture containing NAD or NADP, and
the appropriate substrate depending on the dehydrogenase,
the activity of which is to be measured (for details see
CREMER et al., 1972). The concentrations used must be such
that all reagents are available in excess amounts. Incubation
is carried out in the dark at 25 or 37° C and terminated by
transferring the gels into 7.5 % acetic acid. The incubation
mixture must be kept in the dark because the tetrazolium

system is sensitive to light. The blue formazan, which is
precipitated at the enzymatically active side of the gel as
the final product of the coupling reactions (WILKINSON, 1970),
can then be recorded with a Joyce-Loebl double beam microdensitometer and the peak areas measured by planimetry. The
colour of the tetrazolium band could be maintained for several
weeks in the dark, but in the light the colour faded within
a few days. The formation of the blue dye at the enzymatically
active side of the gel can also be recorded densitometrical-
ly in the dark at intervals of one minute. In this case one
can demonstrate that the peak area increases linearly with
incubation time (CREMER et al., 1972). Even when only 0.8 μl
of a glucose 6 phosphate dehydrogenase (G6PD) solution corresponding to 0.64 pg G6PD was fractionated on a 5 μl gel,
a linear correlation (correlation coefficient r = 0.996)
between peak area and time of incubation could be observed
for up to a period of 80 minutes (NEUHOFF, 1971a). If the
reaction rate or the linear increment from experiments with
different amounts of G6PD is plotted against the amount of
enzyme subjected to electrophoresis, one obtains a regression line which can be used as a calibration plot for the
quantitative determination of enzyme concentration. The
method is sensitive enough to allow quantitative determinations with extracts from only a few μg of fresh tissue; a
single mouse ovum, for example, contains sufficient G6PD to
allow several assays of the G6PD activity (DAMES et al.,
1972; CREMER et al., 1972). Furthermore, the method allows
G6PD-variants to be analysed in a single step procedure as
was shown for the separation of G6PD variants of man and
rat, and man and mouse. For a further characterization of
a G6PD-variant it is possible to measure its binding con-

stants after fractionation on microgels (CREMER et al.,
1972). It is therefore possible to compare a microgel to a
"microcuvette" with a volume from 0.003 to 0.015 µl, since
a single protein band in a 5 µl gel extends from approximately 20 to 100 µm.

Micro-Isoelectric Focussing of LDH from Brain Regions

Since the development of ampholytes (KOLIN, 1954;
SVENSSON, 1961) it is possible to fractionate single protein components according to their isoelectric points, even
from mixtures of different proteins. Isoelectric focussing
(IEF) was initially carried out in saccharose gradients
(VESTERBERG and SVENSSON, 1966) and later in polyacrylamide
gels (DALE and LATNER, 1968; WRIGLEY, 1968). The micro
versions of this very useful method described so far (RILEY
and COLEMAN, 1968; CATSIMPOOLAS, 1968), require 10 µg of a
pure protein or 0.2 - 0.4 mg of a protein mixture for a
single fractionation. However, for the analysis of lactate
dehydrogenase (LDH) isoenzymes from either defined brain
regions or from bioptic tissue samples, a more sensitive
method is needed. The micromethod described by QUENTIN and
NEUHOFF (1972) requires less than one µg wet weight of tissue for isoelectric focussing to allow the determination of
LDH isoenzymes from various regions of the rabbit brain.
The activity of the isoenzymes is determined with a tetrazolium staining procedure which has been shown to be sensitive
enough to detect picogram amounts of dehydrogenases (DAMES
et al., 1972; CREMER et al., 1972). 15 samples of rabbit
brain (medulla oblongata, pons, cerebellum, thalamus, hypo-

thalamus, hippocampus, lamina quadrigemina anterior and posterior, hypophysis, frontal lobe, bulbus olfactorius, retina, nervus opticus, chiasma and tractus opticus) were prepared and homogenized in 0.01 M phosphate buffer, pH 7.0 (1:10 w/v). IEF of the clear supernatants is carried out at a constant voltage of 200 V at room temperature in 5 µl caps filled with an accurate gel/ampholine mixture (acrylamide 6.4 %, crosslinking 2.5 %, ampholine 3.8 %, saccharose 12 %). After about 50 min the electrophoresis is terminated and the gels are transferred to the tetrazolium incubation mixture, containing lactate as the substrate, for measuring the enzymatically active proteins. Incubation is carried out in the dark at 37° C for 90 min and terminated by transferring the gels into 7.5 % acetic acid. The blue formazan is then recorded with a Joyce-Loebl double beam microdensitometer and the peak areas measured by planimetry. An incubation time of 90 min is required to detect those LDH isoenzymes which are present in very low concentrations in the tissue extracts. Even then it is necessary to use extracts with higher protein concentration (e.g., 60 µg wet weight tissue for one IEF) in order to detect the LDH_5 component in several brain regions (lamina quadrigemina posterior, hypothalamus, cerebellum, medulla, hypophysis, pons).

Each analysed brain region has a specific composition of LDH isoenzymes. Seven regions (lamina quadrigemina posterior, hypothalamus, cerebellum, medulla oblongata, hypophysis, pons and thalamus) were quite similar in composition, isoenzymes LDH_5 and LDH_4 always being present although in very low amounts. Isoenzyme LDH_1 had the highest activity

followed by LDH_2 and LDH_3 respectively the only exception
occurring in the hypophysis, where LDH_1 and LDH_2 had similar
activities. The pattern of isoenzymes in the frontal lobe
and the bulbus olfactorius was almost identical to that in
the hippocampus, all five isoenzymes being clearly demonstrable, with LDH_1 having the highest peak. The distribution
of the isoenzymes in the retina, lamina quadrigemina anterior,
nervus opticus, chiasma and tractus opticus was also very
similar, the highest peak of enzymatic activity being LDH_5.
The retina contained all five isoenzymes but here LDH_3 had
the highest activity. The lamina quadrigemina anterior was
the only tissue analysed which showed a decreasing activity
from LDH_1 to LDH_4, with a very high activity of LDH_5. In
order to obtain a similar peak area for LDH_1 in both parts
of the lamina quadrigemina, as much as 60 µg of wet weight
from the posterior part, and only 12 µg from the anterior
region were fractionated. Even then the LDH_5 isoenzyme in
the posterior area was negligible compared to its predominance in the anterior part. The significance of the
uneven distribution of LDH isozenzymes in the different
brain regions is not yet fully understood. The differences
in the LDH isoenzyme patterns in brain regions of close
proximity (e.g.,lamina quadrigemina anterior and posterior),
in contrast to the similarity of LDH isoenzymes in brain
regions which are functionally related (e.g.,retina, n.
opticus, chiasma, tractus opticus and lamina quadrigemina
anterior), lead to the conjecture that the varying isoenzyme patterns in the different regions reflect functional,
rather than morphological, differences.

Micro-Dialysis

In order to determine the amino acid composition of a
single protein separated by micro disc electrophoresis (NEU-
HOFF et al., 1970), the protein is firstly eluted from the
isolated gel disc as described above. It is then necessary
to eliminate the glycine, arising from the electrophoretic
buffer, from the eluate by means of microdialysis (NEU-
HOFF and KIEHL, 1969). Volumes of 10, 15, 25, 50, 100, 250
and 500 µl can be dialyzed in Dynal (polyoxymethylene) microdialysis chambers. One side of the chamber is covered with
a suitable wet dialysis membrane and a rubber gasket; the
sample is introduced and the open side is closed in the
same manner. The filled chamber (the diameter of the sample
part is 2 mm for a 10 µl chamber; the total diameter of the
chamber is 2 cm) is attached to a perspex coated stirring
bar magnet with a stainless steel spring, put into a beaker
containing a suitable dialysis bath and rotated with a
magnetic stirrer. After dialysis, the contents of the chamber
are absorbed with a micro pipette, pierced through one membrane, and transferred into a suitable capillary for hydro-
lysis with HCl. After evaporation of the HCl, dansylation
of the amino acids is performed prior to determination.

Fresh Weight Determination in the Lower Milligram Range

For quantitative microanalysis (e.g., amino acids, phospholipids) it is necessary to know the exact fresh weight of
the sample. A number of very sensitive methods (quartz fibre
balance, X-ray absorption, interferometry etc.) exist for

dry weight determination. But for fresh weight determination
it is always difficult to overcome the problem of humidity
losses during the transportation of the minute sample to a
balance. With the following procedure the fresh weight determination of tissue samples in the range between 1 mg and
0.01 mg is very simple (NEUHOFF, 1971). A 5 µl Drummond
Microcap is carefully heat sealed at one end and completely
filled with a suitable buffer solution. A minute amount of
fresh tissue is then transferred to the top of the capillary
and centrifuged (15 min, 15,000 rpm) to the bottom using a
Heraeus-Christ Haematocrit Centrifuge equipped with a special
adaptor for capillaries. For wet mass determination the
capillary is embedded in water, projected at a suitable
magnification through a microscope and that part of the
capillary filled with the sample is traced on paper. The
volume is then calculated using the formula for a paraboloid
of revolution corrected for the magnification. If the paraboloid part of the heat sealed 5 µl capillary is just filled,
the amount of tissue is about 0.01 - 0.03 mg. For the determination of lighter tissue samples it is necessary to
use capillaries with smaller diameters (2 µl or 1 µl Drummond Microcaps). If more than the paraboloid part of the
capillary is filled with the sample, this part must be calculated separately using the formula for a cylinder corrected for the magnification. The total amount of tissue is obtained by summating both results. If the specific weight of
the tissue is known, the wet weight of minute tissue samples
can be calculated. If one wants to determine the free amino
acids of the sample, micro homogenization is performed directly in the same capillary with a dentist's nerve canal
drill (Beutelrockbohrer), and following centrifugation,

aliquots are taken for the reaction with dansylchloride.

Microdetermination of Phospholipids

80 - 100 μg fresh weight of brain tissue are necessary
for the fluorometric microdetermination of phospholipids.
After determination of the wet weight as described above,
the water phase in the capillary is carefully removed and
the capillary filled with chloroform/methanol (2/1, v/v).
Homogenization is then performed with the nerve canal drill
driven at about 24,000 rpm and the homogenate is allowed to
stand at room temperature for 1 h. After centrifugation the
total lipid extract is removed with a very thin capillary.
This extraction procedure is repeated twice. For the fluorometric determination of total phospholipids, the extracts
are combined in a cuvette, dried under a stream of nitrogen
and analysed after addition of cyclohexane, rhodamine and
acetic acid. The reaction mixture has to be composed according to the expected range of phospholipid content (for
details see SCHIEFER and NEUHOFF, 1971). For the determination of single phosphatides the extracts are quantitatively
transferred to microchromatoplates (32 x 24 mm) spread with
a thin slurry of silica gel. The lipid extract is applied
as a small spot (diameter about 1 mm) in one corner and
chromatographed in two dimensions. After drying with a stream
of nitrogen the lipid spots are stained with iodine vapour
and identified by comparison with chromatograms of authentic
lipids. The whole procedure, from application of the lipid
extract to complete staining with iodine, takes about 10 min.

After evaporation of the iodine, the marked lipid spots are
scraped from the plate with microspatulas, transferred to
small centrifuge tubes and eluted five times from the silica
gel with chloroform/methanol. Control experiments showed that
the recovery of the lipids was 90 - 105 % of the applied
amount. The chloroform/methanol extracts are collected in
cuvettes and dried under a stream of nitrogen. The fluores-
cence is measured after the addition of cyclohexane, rhodamine and acetic acid. The quantity of phospholipids is obtained from calibration curves specially made for each authentic phospholipid (SCHIEFER and NEUHOFF, 1971). In 162 µg
wet tissue from the anterior column of the cat's spinal cord,
the following amounts of phospholipids were found (the corresponding values calculated by BRANTE (1949) are given in
brackets): phosphatidylethanolamine 2.0 µg (2.4 µg), phosphat-
idylinositol and phosphatidylserine 0.7 µg (0.3 µg), phosphatidylcholine 1.7 µg (1.7 µg), sphingomyelin 1.1 µg (1.2
µg). The method is sensitive enough to allow the determination of the total lipid content of single isolated neurons.
Depending on their size and the number of dendrites, they
contain between 7.2 and 32.2 ng phospholipid per cell (SCHIE-
FER and NEUHOFF, 1971).

An analysis of the tissues in the snail, Helix pomatia
(OSBORNE et al., 1972), showed that per 100 µg wet weight of
tissue, the brain contained 1.4 µg phospholipid, the heart
0.87 µg, the liver 0.6 µg and the kidney 0.44 µg. Using the
described microprocedure (SCHIEFER and NEUHOFF, 1971), the
phospholipid composition of the brain was as follows:
phosphoinositides and phosphatidylserine 43.4 %, phosphatidylcholine 25.8 %, phosphatidylethanolamine (including plas-

malogen) 19.9 %, sphingomyelin 10.8 %, and traces of cardio-
lipin. The _in vivo_ incorporation of ^{32}P into individual
phospholipids of the brain and identifiable neurons was
studied after perfusing the nervous system of the snail
with $NaH_2{}^{32}PO_4$ for 90 min. The pattern of ^{32}P incorporation
into the tissues was similar: phosphoinositides and phosphatidylserine > phosphatidylethanolamine > phosphatidylcholine > sphingomyelin. However, the radioactivity associated with phosphoinositides and phosphatidylserine, in
comparison with the other phospholipids, was higher in the
neurons than in the brain.

Determination of Amino Acids, GABA and Serotonin in
Picomole Range

   1-dimethylaminonaphthalene-5-sulfonyl-chloride (dansyl-
Cl) reacts with aliphatic amino- or hydroxyl-groups forming
an intensively fluorescing dansyl derivative (GRAY and HARTLEY,
1963). The dansyl derivatives of amino acids are normally
separated by two-dimensional chromatography on silica gel
(SEILER, 1970) or a 15 x 15 cm polyamide layer (WOODS and
WANG, 1967). Using a 3 x 3 cm polyamide layer as little as
10^{-12} mol of each dansyl-amino acid can be detected (NEUHOFF
et al., 1969; NEUHOFF et al., 1971). If the reaction is
performed with ^{14}C-dansyl-Cl, quantitative determinations
can be done in the same range of sensitivity (for details
see NEUHOFF, 1968a, 1969). For quantitative determinations
it is necessary to know the optimal reaction conditions
(pH, concentration of amino acid and dansyl-Cl etc.), as
the formation of the dansyl derivatives is strongly dependent

on the pH and concentration of the reactants (BRIEL and NEU-
HOFF, 1972; BRIEL et al., 1972; NEUHOFF, 1971). Volumes of
0.1 - 0.5 µl of a dansylated sample are applied with an ultra
thin tip of a capillary (NEUHOFF, 1970), under a stereomicroscope, at the corner of a 3 x 3 cm micropolyamide layer
(NEUHOFF et al., 1971). Chromatography is performed in a
50 ml beaker within 5 min for both dimensions. Radioauto-
grams are obtained with a highly sensitive 3 x 4 cm film,
normally used for radiation dosage meter tags, which is laid
on a radioactive fingerprint, sandwiched between two glass
slides and exposed for 4 - 8 days. For quantitative determination the derivatives can be labelled either in the
dansyl- or in the amino-moiety. Commercial ^{14}C-amino acids
often contain labelled impurities as can be demonstrated by
reaction with unlabelled dansyl-Cl, microchromatography and
radioautography (NEUHOFF et al., 1969). The fluorescent
spots are marked with a soft pencil under a UV-lamp, scraped
from the layer with a special razor splinter knife (NEUHOFF,
1970) and transferred directly into counting vials filled
with scintillation liquid. γ-aminobutyric acid (GABA) and
serotonin can be detected in the same range of sensitivity
as other amino acids, since they migrate on the chromatogram
to places free from other dansyl-amino acids (NEUHOFF and
WEISE, 1970; OSBORNE et al., 1971). The sensitivity of this
micromethod allows the determination of free amino acids in
fresh tissue samples of 1 - 0.05 µg wet weight. With this
technique it was demonstrated, for example, that isolated metacerebral neurons of Helix pomatia contained serotonin in
rather high amounts, while the buccal neurons were free of
serotonin. The amount of GABA was very similar in both cell
types (BRIEL et al., 1971; OSBORNE and COTTRELL, 1972).

When 5-hydroxy-^{14}C-tryptophan was perfused through the snail's central ganglia, only the metacerebral neurons (and not the buccal neurons) synthesized 5-hydroxy-^{14}C-tryptamine. It was estimated from a number of experiments that in vivo the metacerebral neurons form 0.8 ng of serotonin/2 h at 20^{0} C (OSBORNE, 1972a).

Low amounts of GABA were shown for the first time to occur in the brain, optic tentacle, pharyngeal retractor muscle and heart of Helix pomatia (OSBORNE et al., 1971). In addition, taurine, thought to be absent from terrestrial gastropods, was shown to be present in the snail's tissues. In vitro incubation of brain and heart with ^{14}C-labelled glucose demonstrated the formation of ^{14}C-glutamic acid, ^{14}C-glutamine, ^{14}C-alanine and ^{14}C-aspartic acid. These amino acids, in addition to ^{14}C-GABA, were also formed when brain and heart were incubated with ^{14}C-labelled glutamic acid. In all other tissues studied the distribution of amino acids was quite similar, with the exception of ornithine which was present in only low amounts in the brain compared with high amounts found in the heart, pharyngeal retractor muscle and optic tentacle. The brain also differed from the other tissues studied in having a higher concentration of glycine, glutamic acid, serotonin and 5-hydroxytryptophan. Furthermore, 5-hydroxyindole was only found in the brain (OSBORNE et al., 1971). Electrical stimulation of the metacerebral neurons of Helix pomatia, during perfusion with ^{14}C-glucose, increased the production of the metabolites and also produced two unidentified substances. Only three radioactive substances (glutamine, alanine and an unidentified compound), all in very low concentrations, were detected in

metacerebral neurons perfused with ^{14}C-glutamate. Electrical
stimulation increased the concentration of all these substances (OSBORNE, 1972b). The levels of various free amino
acids, serotonin and 5-hydroxyindole in the brain of Helix
pomatia were also measured before and following electrical
stimulation (OSBORNE et al., 1972). After stimulation there
was an increase in the levels of alanine, 5-hydroxyindole
and an unidentified substance, whereas there was a decrease
in the concentrations of tyrosine, tryptophan, ornithine,
lysine, phenylalanine, leucine, isoleucine, proline, valine,
glycine, asparagine and serine. At the same time, there
appeared to be a slight decrease in serotonin and glutamic
acid. The level of the amino acids taurine, methionine, GABA,
glutamine, threonine, arginine, ε-lysine, α-amino-histidine,
cystine and aspartic acid did not appear to be influenced
by the applied stimulation.

The advantages of using molluscan giant neurons for
neurochemical microanalysis, namely their large size and
constant identifiable positions within the CNS, are evident.
In addition, unlike vertebrate cell preparations, they retain
functional activity after dissection and survive for several
hours or even days. The functional activity of the neuron
can also be easily tested by means of either intra- or extracellular recording. Recordings can be made simultaneously
in different parts of the cell, and the latter can be physiologically stimulated. Furthermore, the environment of the
cell can be changed by the addition or substitution of ions
or by adding known concentrations of metabolites, inhibitors,
toxins, drugs,etc. These qualities, in combination with more
sensitive and more refined biochemical techniques, would

permit the thorough study of the biochemistry of identifiable giant neurons in relation to their physiology, morphology and function (for review see OSBORNE and NEUHOFF,
1972).

Isolation of Nerve Cells from Histological Sections
for Quantitative Cytophotometry

In order to analyse a cell cytophotometrically it has
to be isolated from the surrounding cells of the slice.
The thickness of the section should be 20μm in the case of
ganglion cells, because the cells to be analysed are then
arranged in well defined discs. The slices are put on a
glass chamber with two open sides and the space is filled
with liquid paraffin. Cells to be analysed with UV-light
can be isolated from slices embedded in paraffin, because
paraffin does not interfere in UV-spectrophotometric analyses. To isolate stained cells, the slices are kept in
xylol until they are put on the glass chamber. The ganglion
cells are isolated with an upward bent round needle, operated by a De Fonbrune micromanipulator, under a phase contrast microscope. After isolation of 20 - 30 cells, which
remain attached to the slide by capillary adhesion, the
needle is replaced by a capillary with a tip diameter
slightly bigger than that of the isolated cells. A constriction introduced into the capillary improves the pressure
regulation. The isolated cells are taken up by the capillary, together with liquid paraffin, under a microscope
and collect in front of the constriction. The solution containing the isolated cells is then blown into a paraffin

drop on a fresh slide covering an empty glass chamber. The
drop is covered with a cover slip and hermetically sealed.
More than 30 ganglion cells per hour can be isolated quite
easily in this way (NEUHOFF et al., 1968). A cytophotometric
analysis, with a universal microscope spectrophotometer
(Carl Zeiss, UMSP I), of 238 nerve cells isolated from
histological sections, which were obtained from the tri-
geminal nucleus of a brain with subacute sclerosing panen-
cephalitis (SSPE), in comparison with 236 control cells of
the same region, revealed in the SSPE a 40 % increase in
protein, a 34 % increase in RNA and a 29 % decrease in DNA.
Computer analysis of the data, following ninhydrin Schiff
staining, revealed an increase in protein not only within
the cytoplasm, but also within the nucleus and nucleolus.
Since an RNA virus resembling measles has been isolated from
SSPE tissue cultures, it is reasonable to conclude that the
increase in intracellular RNA was due to the presence of
viral nucleic acid, and the increase in the protein content
was due to the presence of virus-specific antibodies (TER
MEULEN et al., 1968; TER MEULEN et al., 1970).

During isolation of human nerve cells from the nucleus
terminalis nervus trigeminus by free hand dissection with
a fine steel wire (HYDEN, 1959), we found that 5 neurons
out of 100 had one dendrite circling back to the same cell
body. This finding was checked by means of careful light
microscopy (NEUHOFF, 1967). If these "feedback neurons" really exist in the nervous system it seems possible that they
may play a role in information transfer.

Two-Dimensional Micro-Immunoprecipitation

A sensitive micromethod is needed for the immunological
characterization of proteins in microscale, especially if
the proteins to be characterized have been isolated, following micro disc electrophoresis, by re-electrophoresis from
gel discs (NEUHOFF and SCHILL, 1968). Accordingly, glass slides
(approx. 2.5 x 2.5 cm) are covered with a 1 mm layer of
0.75 - 1.25 % agarose in a suitable buffer. A rosette of
0.5 mm holes is punched into the agarose layer with a bent
10 µl Drummond Microcap connected via an adaptor to a vacuum
pump (NEUHOFF, 1970). The holes are punched and filled under
a stereomicroscope. The central hole is filled with antiserum and the six other holes, arranged in a circle at 1-2 mm
intervals, are filled with the samples (approx. 0.5 - 1 µl)
to be tested. The formation of the immunoprecipitate, which
can be followed under a stereomicroscope, starts after 5 -
20 minutes depending on the concentration of antigen and antibody. The precipitate can be photographed directly in in-
cident light without staining the protein (NEUHOFF and SCHILL,
1968). Precipitates can also be measured with a microdensitometer (NEUHOFF et al., 1970). The optimal conditions (agarose
concentration, thickness of the layer, distance between the
holes, antigen and antibody concentrations, etc.) depend on
the problem being studied. The 0.5 mm holes in the agarose
layer have just the right diameter to accommodate the isolated gel discs from the micro disc electrophoresis in 5 µl
capillaries. The micromethod for immunoprecipitation is not
only much quicker than the macroscale procedure but is also
much more sensitive. It is possible, for example, to detect

10^{-8} g of albumin as an immunoprecipitate (NEUHOFF and SCHILL, 1968). Furthermore, the two-dimensional micro diffusion method is very suitable for the characterization of polymerase template complexes (NEUHOFF et al., 1970).

Micro disc electrophoresis has been combined with antigen-antibody crossed electrophoresis in vertical agarose gel (DAMES et al., 1972). Micro disc electrophoresis is performed in 5 µl capillaries as described above. For the second electrophoresis, glass cells were made from two microscope slides which had an effective mould space of 75 x 18 x 0.75 mm. The cell is filled with antiserum containing 1 % agarose solution. Subsequent to gelling, an electrophoresed polyacrylamide microgel is laid onto the surface of the antibody-agarose gel and covered with 1 % agarose solution. The second electrophoresis, using barbital buffer in the electrode reservoirs, is performed vertically at room temperature. The vertical arrangement of the immuno-electrophoresis in closed cells guarantees an excellent contact between agarose gel, polyacrylamide micro-gel and the electrode buffers, and apparently reduces endosmotic liquid flow. After electrophoresis, the agarose gels are washed with 0.15 M NaCl for 24 - 36 h to remove excess antibody, and stained with 0.2 % Coomassie brilliant blue G 250 in ethanol/acetic acid/water (4:1:5 v/v/v) for 30 min. The shape ("rockets") of the immunoprecipitation peak in the agarose gel varies according to the duration of immuno-electrophoresis and the antiserum concentration. Peaks which have stopped migrating are sharp and can be used for quantitative determinations. In contrast, peaks which are still migrating are blunt, indicating an incomplete immunreaction. Less than 1 µg of antigen protein and

about 20 µl of antiserum are needed for one micro antigen-
antibody crossed electrophoresis.

High Speed Capillary Centrifugation

By means of a special adaptor it is possible to centrifuge capillaries at a much higher speed than in the
haematocrit centrifuge. Drummond Microcaps can be centrifuged
in either fixed angle or swinging bucket rotors up to 70,000
rpm for unlimited periods. 70,000 rpm for a 5 µl capillary
in a swinging bucket rotor (Heraeus-Christ, Osterode, Germany) is equivalent to 449,000 g, and 60,000 rpm in an angle
rotor to 254,000 g (NEUHOFF, 1968a, 1969). The capillaries
do not break during centrifugation providing they are embedded in a medium with the same density as that of their contents. Since capillary forces, at present not accurately
calculated, are important in capillary centrifugation, the
normal rules or experiences from macroscale centrifugation
are not applicable to high speed capillary centrifugation.
For example, in 2 hours of centrifugation at 60,000 rpm in
a swinging bucket rotor (330,000 g), human albumin (MW
64,000) is centrifuged down to the bottom of a 5 µl capillary. This method is therefore also suitable for rapid concentration of small volumes of low molecular weight proteins.

 Neurochemical Microanalysis Following Post-Tetanic Potentiation of Monosynaptic Reflexes in Cat Spinal Cord.

    A process of potentiation follows after high frequency
activation of synapses and often continues for quite a long
time. This phenomenon of post-tetanic potentiation was discovered by LLOYD (1949) and has since been regarded as of-
fering a possible paradigm of the process required for
learning at synapses (ECCLES and MC INTYRE, 1953; CURTIS
and ECCLES, 1960; KANDEL and SPENCER, 1968). In collaboration with the Department of Pharmacology in this institute,
we are engaged in studies on the biochemical changes underlying this process. Accordingly, following post-tetanic
potentiation of monosynaptic reflexes (CLEVELAND et al.,
1972), isolated single nerve cells and small pieces of
punched grey matter from the spinal cord are analysed,
using the described micro methods, to record any qualitative
and quantitative differences in the amounts of ribonucleic
acids, amino acids, proteins and phospholipids (ALTHAUS et
al., 1972).

    Following post-tetanic potentiation, the spinal cord
is dissected, thorougly washed in ice cold 0.9 % NaCl
solution and stored at - 20° C. The frozen spinal cord is
sectioned in 1 mm slices on a precooled object slide in
the region of the treated segment (L7 or S1). Cylinders of
tissues containing α-motoneurons are punched out of both
the ipsilateral side,where the post-tetanic potentiation
was performed, and the contralateral control side,with a
sharpened canula (inner diameter 1 mm). The wet weight of
a tissue cylinder is about 0.9 - 1.2 mg. After microhomo-

genization in Tris/HCl buffer (pH 7.4, 1:50 w/v), 1.5 µl of
the clear supernatant is used for analysis of the water
soluble proteins by micro disc electrophoresis. The plani-
metric evaluation of the pherograms from 30 experiments
demonstrated that of the cats which reacted successfully to
post-tetanic potentiation, 81 % showed a marked decrease in
slow migrating proteins and 95 % an increase in fast migrating proteins which include the S-100 protein. Control experiments from segments higher than the treated L7 or S1
segments showed no differences in the distribution of the
protein patterns.

Microdetermination of the phospholipids from 11 cats
showed that post-tetanic potentiation caused a marked increase in the percentage of sphingomyelin and phosphat-
idylinositol, while, in contrast,the percentage values
of phosphatidylcholine and phosphatidylethanolamine decreased. The total phospholipid content remained more or
less constant. As regards the free amino acids, the percentage of taurine, alanine, proline and glycine increased,
whilst aspartic acid, GABA and leucine decreased. The RNA-
base composition of the isolated α-motoneurons, which was
analysed according to RÜCHEL (1971), remained unchanged
under the conditions used for the post-tetanic potentiation.

The results demonstrate that post-tetanic potentiation
causes changes in all biochemical constituents of the
α-motoneuron region of the cat's spinal cord which have so
far been studied. Further experiments are necessary to
determine what type of change may be involved in the "memory-

phenomenon" of post-tetanic potentiation, or if the observed
biochemical changes only reflect variations in local metabolism caused by the procedure of stimulation.

LITERATURE

ALTHAUS, H.-H., G. BRIEL, W. DAMES, and V. NEUHOFF: Zellu-
    läre und molekulare Grundlagen der nervösen Erregungs-
    speicherung. Neurochemische Mikroanalysen des Rücken-
    marks der Katze nach posttetanischer Potenzierung mono-
    synaptischer Reflexe.
    In: Sonderforschungsbereich 33 Nervensystem und biolo-
    gische Information, Göttingen 1969-1972, 107-121 (1972).

ANSORG, R., W. DAMES, and V. NEUHOFF: Mikro-Disk-Elektropho-
    rese von Hirnproteinen. II. Untersuchung verschiedener
    Extraktionsverfahren.
    Arzneimittelforschung 21, 699-710 (1971).

ANSORG, R., and V. NEUHOFF: Micro-disc electrophoresis of
    brain proteins. III. Heterogenity of the nervous
    specific protein S-100.
    Intern. J. Neuroscience 2, 151-160 (1971).

BRANTE, G.: Studies on lipids in the nervous system.
    Acta physic. scand. 18: Suppl. 63, 1-184 (1949).

BREWER, J.M.: Artifact produced in disc electrophoresis by
    ammonium persulfate.
    Science 166, 256-257 (1967).

BRIEL, G., E. GYLFE, B. HELLMANN, and V. NEUHOFF: Micro-
    determination of free amino acids in pancreatic islets
    isolated from obese-hyperglycemic mice.
    Acta Phys. Scand. 84, 247-253 (1972).

BRIEL, G., and V. NEUHOFF: Microanalysis of amino acids and
    there determination in biological material using
    dansylchloride.
    Hoppe-Seyler's Z. Physiol. Chem. 353, 540-553 (1972).

BRIEL, G., V. NEUHOFF, and N.N. OSBORNE: Determination of
    amino acids in single identifiable nerve cells of Helix
    pomatia.
    Intern. J. Neuroscience 2, 129-136 (1971).

CALISSANO, P., B.W. MOORE, and A. FRIESEN: Effect of calcium
    ion on S-100, a protein of the nervous system.
    Biochemistry 8, 4318-4326 (1969).

CASOLA, L., H. WEISE, and V. NEUHOFF: In vitro protein syn-
    thesis by optic nerves.
    Hoppe-Seyler's Z. Physiol. Chem. 350, 1175 (1969).

CATSIMPOOLAS, N.: Micro isoelectric focussing in polyacryl-
    amide gel columns.
    Anal. Biochem. 26, 480-482 (1968).

CHOULES, G.L., and B.H. ZIMM: An acrylamide gel soluble in
    scintillation fluid. Its application to electrophoresis
    at neutral and low pH.
    Analyt. Biochem. 13, 336-344 (1965).

CICERO, T.J., and B.W. MOORE: Turnover of the brain specific
    protein, S-100.
    Science 169, 1333-1334 (1970).

CLEVELAND, S., J. HAASE, H.-G. ROSS, and B. VOGEL: Zellulä-
    re und molekulare Grundlagen der nervösen Erregungs-
    speicherung, Motoneurone, post-tetanische Potenzierung
    und recurrente Inhibition.
    In: Sonderforschungsbereich 33 Nervensystem und biolo-
    gische Information, Göttingen 1969-1972, 96-106 (1972).

CREMER, Th., W. DAMES, and V. NEUHOFF: Micro disc electro-
    phoresis and quantitative assay of glucose-6-phosphate
    dehydrogenase at the cellular level.
    Hoppe-Seyler's Z. Physiol. Chem. 353, 1317-1329 (1972).

CURTIS, D.R., and J.C. ECCLES: Synaptic action during and
    after repetitive stimulation.
    J. Physiol. 150, 374-398 (1960).

DALE, G., and A.L. LATNER: Isoelectric focussing in poly-
    acrylamide gels.
    Lancet 1, 847-848 (1968).

DAMES, W., H.R. MAURER, and V. NEUHOFF: Micro antigen-anti-
    body crossed electrophoresis in vertical agarose gels
    following micro-disc electrophoresis.
    Hoppe-Seyler's Z. Physiol. Chem. 353, 554-558 (1972).

DAMES, W., V. NEUHOFF, and Th. CREMER: Mikroelektrophore-
    tische Trennung und Bestimmung von Dehydrogenasen ein-
    zelner Zellen.
    Naturwissenschaften 59, 126 (1972).

DANNIES, P.S., and L. LEVINE: Demonstration of subunits in
    beef brain acidic protein (S-100).
    Biochem. biophys. Res. Commun. 37, 587-592 (1969).

DAVIES, B.J.: Disc-electrophoresis II. Method an Application to human serum proteins.
Ann. N.Y. Acad. Sci. 121, 404-427 (1964).

ECCLES, J., and A.K. MC INTYRE: The effect of disuse and of activity on mammalian spinal reflexes.
J. Physiol. 121, 492-516 (1953).

FANTES, K.H., and I.G.S. FURMINGER: Proteins, persulphate and disc electrophoresis.
Nature 215, 750-751 (1967).

FILIPOWICZ, W., G. VINCENDON, P. MANDEL, and G. GOMBOS: Topographical distribution of fast and slow migrating fractions of beef brain S-100.
Life Science 7, 1243-1250 (1968).

GOMBOS, G., W. FILIPOWICZ, and G. VINCENDON: Fast and slow components of S 100 protein fraction: regional distribution in bovine central nervous system.
Brain Research 26, 475-479 (1971a).

GOMBOS, G., G. VINCENDON, J. TARDY, and P. MANDEL: Hétérogénéité électrophorétique et préparation rapide de la fraction protéique S-100.
C.R. Acad. Sc. Paris 263, 1533-1535 (1966).

GOMBOS, G., J.-P. ZANETTA, P. MANDEL, and G. VINCENDON: Studies of F-S100, a neurospecific protein fraction.
I. In vivo presence of fast and slow migrating components of S-100 protein fraction.
Biochimie 53, 635-644 (1971b).

GOMBOS, G., J.-P. ZANETTA, P. MANDEL, and G. VINCENDON:
    Studies on F-S100, a neurospecific protein fraction:
    I. Molecular heterogeneity of S-100 protein fraction.
    Biochimie 53, 645-655 (1971c).

GRAY, W.R., and B.S. HARTLEY: The structure of a chymotryptic
    peptide from pseudomonas cytochrome c-551.
    Biochem. J. 89, 59 P, 379-380 (1963).

GROSSBACH, U.: Acrylamide gel electrophoresis in capillary
    columns.
    Biochem. Biophys. Acta 107, 180-182 (1965).

HYDEN, H.: Quantitative assay of compounds in isolated
    fresh nerve cells and glial cells from control and
    stimulated animals.
    Nature, Lond. 184, 433-435 (1959).

HYDEN, H., K. BJURSTAM, and B. MC EWEN: Protein separation
    at the cellular level by micro disc electrophoresis.
    Annal. Biochem. 17, 1-15 (1966).

KANDEL, E.R., and W.A. SPENCER: Cellular neurophysiological
    approaches in the study of learning.
    Physiol. Rev. 48, 65-134 (1968).

KING, E.E.: Disc electrophoresis: avoiding artifacts caused
    by persulfate.
    J. Chromatg. 53, 559-563 (1970).

KOLIN, A.: Separation and concentration of proteins in a pH
    field combined with an electric field.
    J. Chem. Phys. 22, 1628-1629 (1954).

LLOYD, D.P.C.: Post-tetanic potentiation of respone in mono-
    synaptic reflex pathways of the spinal cord.
    J. gen. Physiol. 33, 147-170 (1949).

MAURER, H.R.: Disc electrophoresis and related techniques
    of polyacrylamide gel electrophoresis.
    In: Working Methods in Modern Sciences (K. FISCHBECK,
    ed.), Walter de Gruyter, Berlin, New York (1971).

MC EWEN; B., and H. HYDEN: Study of specific brain proteins
    on the semi-micro scale.
    J. Neurochem. 13, 823-833 (1966).

MOORE, B.W.: A soluble protein characteristic of the nervous
    system.
    Biochem. biophys. Res. Commun. 19, 739-744 (1965).

MOORE, B.W., and V.J. PEREZ: Specific acidc proteins of the
    nervous system.
    In: Physiological and Biochemical Aspects of Nervous
    Integration (F.D. CARLSON, ed.), pp. 343-359, Prentice
    Hall, (1968).

NEUHOFF, V.: "feedback-Neurone" im Zentralnervensystem des
    Menschen.
    Naturwissenschaften 54, 287-288 (1967).

NEUHOFF, V: Micro-Disc-Electrophorese von Hirnproteinen.
    Arzneimittelforschung 18, 35-39 (1968a).

NEUHOFF, V.: Simplified technique of high-speed capillary
    centrifugation.
    Analytical Biochemistry 23, 359-362 (1968b).

NEUHOFF, V.: Einfaches Verfahren zur hochtourigen Kapilla-
    renzentrifugation.
    GIT 13, 86-87 (1969).

NEUHOFF, V.: Manual, 1st EMBO-Course on Micromethods in
    Molecular Biology, Max-Planck-Gesellschaft, Dokumenta-
    tionsstelle (1970).

NEUHOFF, V.: Micromethods in Molecular Biology, Springer-
    Verlag, Heidelberg, New York (1972), in press

NEUHOFF, V.: Manual, 2nd EMBO-Course on Micromethods in
    Molecular Biology, Max-Planck-Gesellschaft, Dokumen-
    tationsstelle (1971a).

NEUHOFF, V.: Wet weight determination in the lower milligram
    range.
    Analytical Biochemistry 41, 270-271 (1971b).

NEUHOFF, V., G. BRIEL, and A. MAELICKE: Characterization
    and microdetermination of histidine as its dansyl-com-
    pounds.
    Arzneimittelforschung 21, 104-107 (1971).

NEUHOFF, V., F. von der HAAR, E. SCHLIMME, and E. WEISE:
    Zweidimensionale Chromatographie von Dansyl-Aminosäuren
    im pico-Mol-Bereich, angewandt zur direkten Charakte-
    risierung von Transfer-Ribonucleinsäuren.
    Hoppe-Seyler's Z. Physiol. Chem. 350, 121-123 (1969).

NEUHOFF, V., and F. KIEHL: Dialysiergeräte für Volumen zwi-
    schen 10 und 500 µl.
    Arzneimittelforschung 19, 1898-1899 (1969).

NEUHOFF, V., and A. LEZIUS: Nachweis der Substruktur von
    DNA-Polymerasen, der enzymatisch aktiven Proteinkom-
    ponente und ihrer Enzym-Substrat-Komplexe mit der Micro-
    Disc-Electrophorese.
    Hoppe-Seyler's Z. für Physiol. Chem. 348, 1239 (1967).

NEUHOFF, V., and A. LEZIUS: Nachweis und Charakterisierung
    von DNS Polymerasen durch Micro-Disc-Electrophorese.
    Z. Naturforschung 23b, 812-819 (1968).

NEUHOFF, V., B. MÜHLBERG, and J. MEIER: Strom- und spannungs-
    konstantes Netzgerät für die Micro-Disc-Electrophorese.
    Arzneimittelforschung 17, 649-651 (1967).

NEUHOFF, V., D. MÜLLER, and V. ter MEULEN: Präparation von
    Ganglienzellen für cytophotometrische Untersuchungen.
    Z. Wiss. Mikroskopie 69, 2, 65-72 (1968).

NEUHOFF, V., and W.-B. SCHILL: Kombinierte Mikro-Disk-Elek-
    trophorese und Mikro-Immunpräzipitation von Proteinen.
    Hoppe-Seyler's Z. Physiol. Chem. 349, 795-800 (1968).

NEUHOFF, V., W.-B. SCHILL, and D. JACHERTS: Nachweis einer
    RNA-abhängigen RNA-Replicase aus immunologisch kompe-
    tenten Zellen durch Mikro-Disk-Elektrophorese.
    Hoppe-Seyler's Z. Physiol. Chem. 351, 157-162 (1970).

NEUHOFF, V., W.-B. SCHILL, and H. STERNBACH: Microanalysis
    of pure deoxyribonucleic acid-dependent rubonucleic
    acid polymerase from Escherichia coli.
    Biochem. J. 117, 623-631 (1970).

NEUHOFF, V., and M. WEISE: Determination of pico-mole quan-
    tities of γ-amino-butyric-acid (GABA) and serotonin.
    Arzneimittelforschung 20, 368-372 (1970).

NEUHOFF, V., H. WEISE, and H. STERNBACH: Micro-analysis of
     pure deoxyribonucleic acid-dependent ribonucleic acid
     polymerase from Escherichia coli. VI. Determination of
     amino acid composition.
     Hoppe-Seyler's Z. Physiol. Chem. 351, 1395-1401 (1970).

ORNSTEIN, L.: Disc-electrophoresis I. Background and theory.
     Ann. N.Y. Acad. Sci. 121, 321-349 (1964).

OSBORNE, N.N.: Effect of electrical stimulation on the in vivo
     metabolism of glucose and glutamic acid in an identified
     neuron.
     Brain Research 41, 237-241 (1972b).

OSBORNE, N.N.: The in vivo synthesis of serotonin in an
     identiefied serotonin-containing neuron of Helix pomatia.
     Intern. J. Neuroscience 3, 215-228 (1972a).

OSBORNE, N.N., H.-H. ALTHAUS, and V. NEUHOFF: Phospholipids
     in the nervous system of the gastropod mollusc Helix
     pomatia, and the in vivo incorporation of ^{32}P into the
     phospholipids of identified neurons.
     Comp. Biochem. Physiol. (1972), in press.

OSBORNE, N.N., R. ANSORG, and V. NEUHOFF: Micro-disc electro-
     phoretic separation of soluble proteins from nervous
     and other tissues of Helix (Pulmonate mollusca).
     Intern. J. Neuroscience 1, 259-264 (1971).

OSBORNE, N.N., G. BRIEL, and V. NEUHOFF: Distribution of
     GABA and other amino acids in different tissues of the
     gastropod mollusc Helix pomatia, including in vitro ex-
     periments with ^{14}C glucose and ^{14}C glutamic acid.
     Internat. J. Neuroscience 1, 265-272 (1971).

OSBORNE, N.N., and G.A. COTTRELL: Amine and amino acid micro-
    analysis of two identified snail neurons with known
    characteristics.
    Experientia 28, 656-658 (1972).

OSBORNE, N.N., and V. NEUHOFF: Neurochemical studies on
    characterized neurons.
    Naturwissenschaften (1972), in press.

OSBORNE, N.N., B. POWELL, and G.A. COTTRELL: The effect of
    electrical stimulation on the levels of free amino
    acids and related compounds in the snail brain.
    Brain Res. 41, 379-386 (1972).

PETZOLD, U., W. DAMES, G.H.M. GOTTSCHEWSKI, and V. NEUHOFF:
    Das Proteinmuster in frühen Entwicklungsstadien des
    Kaninchens.
    Cytobiologie 5, 272-280 (1972).

PUN, J.Y., and LOMBROZO: Microelectrophoresis of brain and
    pineal protein in polyacrylamide gel.
    Annal. Biochem. 9, 9-20 (1964).

QUENTIN, C.-D., and V. NEUHOFF: Micro-isoelectric focussing
    for the detection of LDH isoenzymes in different brain
    regions of rabbit.
    Intern. J. Neuroscience 4, 17-24 (1972).

RILEY, R.F., and M.K. COLEMAN: Isoelectric fractionation of
    proteins on a microscale in polyacrylamide and agarose
    matrices.
    J. Lab. and Clin. Med. 72, 714-720 (1968).

RÜCHEL, R.: Mikroelektrophoresen von RNS-Basen, Anwendung
    zur Untersuchung bestimmter Hirnregionen und kritische
    Analyse der Methode.
    Inauguraldissertation, Göttingen (1971).

SCHIEFER, H.-G., and V. NEUHOFF: Fluorometric determiantion
    of phospholipids on the cellular level.
    Hoppe-Seyler's Z. Physiol. Chem. $\underline{352}$, 913-926 (1971).

SEAR, P.G., and B. ROIZMAN: An improved procedure for H^3
    and C^{14} counting in acrylamide gels with an nonaqueous
    scintillation system.
    Analyt. Biochem. $\underline{26}$, 197-200 (1968).

SEILER, N.: Methods of biochemical analysis. Use of the
    Dansyl reaction in biochemical analysis.
    In: Interscience Publ. (D. GLICK, ed.), Vol. 18, 259-
    337, John Wiley & Sohn, New York, London, Sydney,
    Toronto (1970).

SVENSSON, H.: Isoelectric fractionation, analysis, and
    characterization of ampholytes in natural pH gradients:
    I. The differential equation of stae of solute con-
    centrations at a steady state and its solution for
    simple cases.
    Acta Chem. Scand. $\underline{15}$, 325-341 (1961).

TARDY, J., K. UYEMURA, G. GOMBOS, and G. VINCENDON:
    Caractérisation dûn groupe de protéins spécifiques du
    système nerveux.
    J. Physiol. $\underline{59}$, 510 (1967).

TER MEULEN, V., D. MÜLLER, G. ENDERS-RÜCKLE, V. NEUHOFF,
    M.Y. KÄCKELL, and G. JOPPICH: Ist die subakute pro-
    gressive Panenzephalitis eine Masernerkrankung?
    Dtsch. Med. Wschr. 93, 1303-1308 (1968).

TER MEULEN, V., D. MÜLLER, V. NEUHOFF, and G. JOPPICH:
    Immunhistological, microscopcial and neurochemical
    studies on encephalitides. V. Subacute sclerosing
    panencephalitis. Cytophotometric studies on isolated
    nerve cells.
    Acta Neuropath. 15, 128-141 (1970).

UYEMURA, K., J. TARDY, G. VINCENDON, P. MANDEL, and G.
    GOMBOS: Mise en évidence de protêines spécifiques du
    cerveau chez les Mammifêeres.
    C.R. Soc. Biol. 161, 1396-1399 (1967).

UYEMURA, K., G. VINCENDON, G. GOMBOS, and P. MANDEL: Puri-
    fication and some properties of S-100 protein fractions
    from sheep and pig brains.
    J. Neurochem. 18, 429-438 (1971).

VESTERBERG, O., and H. SVENSSEN: Isoelectric fractionation,
    analysis, and characterization of ampholytes in natural
    pH gradients: IV. Further studies on the resolving
    power in connection with separation of myoglobins.
    Acta Chem. Scand. 20, 820-834 (1966).

VINCENDON, G., A. WAKSMAN, K. UYEMURA, J. TARDY, and G.
    GOMBOS: Ultracentirfugal behavior of beef brain S-100
    protein fraction.
    Arch. Biochem. Biophys. 120, 233-235 (1967).

WOODS, K.R., and K.T. WANG: Separation of danysl-amino acids
by polyamide layer chromatography.
Biochim. Biophys. Acta 133, 369-370 (1967).

WRIGLEY, C.W.: Analytical fractionation of plant and animal
proteins by gel electrofocussing.
J. Chromatog. 36, 362-365 (1968).

NEURONAL PLASTICITY, PROTEIN CONFORMATION AND BEHAVIOR

Holger Hydén

Institute of Neurobiology
Faculty of Medicine
University of Göteborg
Göteborg, Sweden

ABSTRACT

The influences of changes in behavior on the chemistry of different parts of the brain have been investigated. The results indicate the following working hypothesis: a sensory input causes an electrical field change
around neurons in various regions of the brain. Ca increases and effects the translation of this electrical
field change into conformational changes in brain specific
protein of the cell membrane. The continuous protein network also reacts to Ca, modifying the membrane tension and
facilitating the protein reaction. The pattern of surface
protein and its conformational state could label and process valued information to neurons on electrical events
where the final step involves the synapse.

511

In any field of the life sciences there is no problem
that is so complicated that - by approaching it from a
suitable angle - we cannot make it even more complicated.
The problems of this meeting, especially, present the structural complexity and the multiplicity of interactions between relevant factors which require both empirical en-
quiries and analytical treatment.

Empirical data from different laboratories are difficult to compare. Part of these difficulties could presumably be resolved by agreement on some satisfactory
training tests which met stringent requirements for controls and excluded complicating stress factors.

Some requirements for storage of new information are
met by the connective possibilities in the three dimensional
Gestalt of the neuron-glia structures with synapses as cru-
cial organs. Translational mechanisms of electrical phenomena into molecular are still in the open. Other requirements are met by proposed molecular mechanisms for identification. If this is looked upon as differentiation of
brain cells which progresses through the brain with time
and experience, then an obvious target is neuronal surface
protein for recognition and outgrowth of connections.

For this discussion, I would like to present some recent data and an attempt to study neural surface protein.
The hypothesis presumes that differentiation of protein on
the inner and outer side of the neuron provides a mechanism
of recognition whereby neurons respond to electrical field

changes and molecules in the environment, passing from
area to area. The mechanism involves in addition a contractile protein on the inside of the plasma membrane and
Ca^{++} to translate electrical field changes into protein
response. The glia may act as stabilizers of the micro-
environment, producers of brain specific proteins and
peptides, and regulators of ions.

The contractile protein close to the inner part of the
nerve cell membrane was detected by a chance observation
(HYDEN and METUZALS, 1972). Two to three filaments, which
are 2 nm in diameter and tightly coiled, build up a continuous network which contains an actin-like substance (Fig. 1).

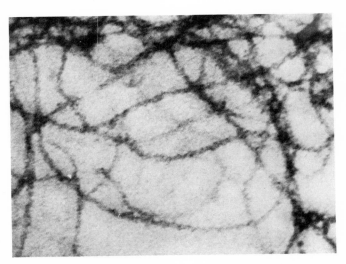

Fig. 1: EM photograph of the continuous network close to
the inner side of nerve cell membrane consisting
of two-three coiled 20 A diameter protein filaments containing an actin-like substance
Uranyl acetate fixed. From HYDEN, METUZALS,
HANSSON and MUSHYNSKI, 1972.

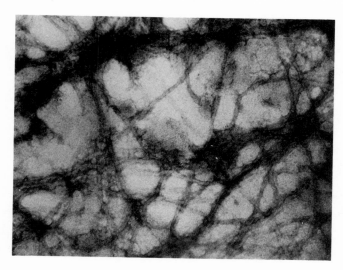

Fig. 2: <u>Addition of 2 mM CaCl$_2$ gives uncoiling of the 20 A</u>
<u>diameter protein filaments in Fig. 1</u>
From HYDEN, METUZALS, HANSSON and MUSHYNSKI, 1972.

The filaments give arrow-heads on treatment with heavy
meromyosin. Furthermore, the filaments react to 2 mM Ca^{++}
with uncoiling (Fig. 2). The localization of this continuous
network at the plasma membrane makes it probable that it
can affect the tension of the membrane. Chemically it may
be related to the "neurin" protein described by PUSZKIN
and BERL (1972).

When an animal is placed in a novel situation which requires a change in behavior, a sequence of protein changes
proceeds throughout the brain beginning with the hippocampal
nerve cells. For eight years we have used reversal and re-
reversal of handedness in rats as behavioral tests (HYDEN

and LANGE, 1970b). This experiment involves few stress factors, and controls in both reversal and re-reversal tests
seem satisfactory.

Two findings seem relevant from the data given in
Tab. I: 1. The incorporation values of the trained animals
are lower than the controls. One exception is the hippocampus. 2. With time and more experience, the responses of
the hippocampus and of the cortex and other areas are inverse.

LEVITAN et al. (1972) at our laboratory prepared synaptosomal fractions such as membranes and mitochondria of
the hippocampus, and found that the trained rats incorporated 60 to 80 % more leucine than did the controls. The
perikaryal microsomes incorporated 30 % more. This agrees
well with previously presented data (HYDEN and LANGE, 1968).
They also confirmed the decreased incorporation of the cortex during initial training when the incorporation in the
hippocampus was increased. This lack of cortical response
does not, however, exclude a response of brain specific
proteins since these constitute only 0.1 - 0.2 % of the
total protein.

We have presented evidence that with time and experience there occurs a differentiation of both neural RNA and
protein (HYDEN and EGYHAZI, 1964; HYDEN and LANGE 1965,
1968, 1969, 1970a). I would like to discuss hippocampal
nerve cells since these neurons are outstanding in their
early response during training,and four observations can

Tab. I: Relative specific activities $\pm$ SEM (Ap/As) for eight different brain areas of controls and experimental rats trained to reverse handedness for 25/min/day twice a day during 4+2+2 days with 14 days intermission between each of the three training periods

Brain area	Training (days)*			Control[+]
	4	6	8	
Cortex	0.40 ± 0.01	0.51 ± 0.02	0.53 ± 0.02	1.40 ± 0.15
Thalamus, nucleus dorsomedialis	0.28 ± 0.01	0.35 ± 0.02	0.39 ± 0.03	0.95 ± 0.15
Entorhinalis	0.25 ± 0.002	0.32 ± 0.01	0.46 ± 0.02	0.84 ± 0.02
Septum	0.19 ± 0.004	0.29 ± 0.01	0.26 ± 0.01	0.78 ± 0.12
Corpus mamillare	0.39 ± 0.02	0.48 ± 0.04	0.43 ± 0.04	0.87 ± 0.09
Nucleus dentatus	0.39 ± 0.02	0.34 ± 0.01	0.41 ± 0.03	0.81 ± 0.03
Hippocampus	0.69 ± 0.04	0.38 ± 0.03	0.47 ± 0.03	0.75 ± 0.06
Formatio reticularis	0.32 ± 0.01	0.36 ± 0.03	0.52 ± 0.03	0.90 ± 0.03

* Four trained rats were used for each training period.

[+] Three control rats were used.

From HYDEN and LANGE, 1972.

be considered in a broader context.

First, we observed that the amount of S100 protein
increased in the hippocampal nerve cells during training
(HYDÉN and LANGE, 1970b).

Second, if anti-S100 protein antiserum is injected
intraventricularly, it binds to the hippocampal S100 and
further performance is blocked in the animals, but there
is no motor impairment (HYDÉN and LANGE, 1970b).

Third, the content of Ca increased in the hippocampus
during reversal of handedness (HALJAMÄE and LANGE, 1972).
This was not a circulatory effect since there was no concomitant rise of Na, K, or water content.

Fourth, the S100 protein of trained rats splits in
two fractions in comparison with control S100 if separation is carried out by electrophoresis under certain standard conditions (HYDÉN and LANGE, 1969). The S100 is a heterogeneous protein made up of 3 fractions at least and
each has a molecular weight of 7000 (GOMBOS et al., 1971).
Recent data indicate different primary composition of
S100 protein of different molecular weight (DANNIES and
LEVINE, 1971; STEWART, 1972).

We have interpreted the increased migration rate of
part of the S100 protein and the Ca increase to mean that
part of the protein has undergone a conformational change
during training. Since anti-S100 antiserum can inhibit
further learning, the S100 in the hippocampal nerve cells
seems to serve a special function for behavior. The S100
protein is only one of several brain specific proteins,

however, and other such proteins could react in a similar
way.

    In the study of neuronal surface protein, we are still
in the beginning. We have used nerve cells isolated by
micro-dissection and 0.2 mm sections of brain tissue and
iodinated the surface protein by the method of MARCHALONIS
et al. (1971), using lactoperoxidase, hydrogen peroxidase
and (^{125} I)iodide. After solubilization, the radioiodinated
proteins were analysed by micro-disc electrophoresis. The
radioactivity per mm gel was determined by spectroscopy.
Electronmicroscopic radioautography is used for localization of the covalently bound iodide. The aim is to study
whether the composition of neuronal surface protein differs
from area to area and whether this surface protein differ-
entiates with time and experience into a pattern, laterally
arranged over the dendrites and cell body surface (HYDEN
et al., to be published).

    A working hypothesis may be discussed on the basis of
these findings. A sensory input sets up an electrical field
change around some millions of neurons in different loci.
Ca increases and serves to translate the electrical field
change into conformational changes of brain specific protein of the cell membrane. The continuous protein network
also reacts to Ca and changes the membrane tension and
facilitates the protein reaction. The pattern of surface
protein and its conformational state could label and process valued information to millions of neurons on electrical
events where the last step is the synaptic event.

LITERATURE

DANNIES, P.S., and L. LEVINE: Structural properties of
      bovine brain S-100 protein.
      J. biol. Chem. 246, 6276-6283 (1971).

GOMBOS, G., J.-P. ZANETTA, P. MANDEL, and G. VINCENDON:
      Etude de la fraction protéique neurospécifique S 100:
      II. Hétérogénéité moléculaire de la fraction pro-
      téique S 100.
      Biochimie 53, 645-655 (1971).

HALJAMÄE, H. and P.W. LANGE: Calcium content and con-
      formational changes of S-100 protein in the hippo-
      campus during training.
      Brain Res. 38, 131-142 (1972).

HYDEN, H., and E. EGYHAZI: Changes in RNA content and base
      composition in cortical neurons of rats in a learn-
      ing experiment involving transfer of handedness.
      Proc. nat. Acad. Sci. 52, 1030-1035 (1964).

HYDEN, H., and P.W. LANGE: A differentiation in RNA re-
      sponse in neurons early and late during learning.
      Proc. nat. Acad. Sci. 53, 946-952 (1965).

HYDEN, H., and P.W. LANGE: Protein synthesis in the hippo-
      campal pyramidal cells of rats during a behavioral
      test.
      Science 159, 1370-1373 (1968).

HYDEN, H., and P.W. LANGE: Synthesis of acidic proteins in
      nerve cells during establishment of new behavior.
      Symp. Soc. Int. Cell Biol. 8, 335-350 (1969).

HYDEN, H., and P.W. LANGE: S100 brain protein: Correlation
     with behavior.
     Proc. nat. Acad. Sci. 67, 1959-1966 (1970a).

HYDEN, H., and P.W. LANGE: Correlation of the S100 brain
     protein with behavior.
     In: Biochemistry of Brain and Behavior (R.E. BOWMAN
     and S.P. DATTA, eds.), pp. 327-346, Plenum Press,
     New York (1970b).

HYDEN, H., and P.W. LANGE: Protein changes in different
     brain areas as a function of intermittent training.
     Proc. nat. Acad. Sci. 69, 1980-1984 (1972).

HYDEN, H., P.W. LANGE, and G. RAMIREZ: To be published.

HYDEN, H., J. METUZALS, H.-A. HANSSON, and W. MUSHYNSKI:
     A contractile net-work of microfilaments at the nerve
     cell membrane.
     To be published.

LEVITAN, I.B., G. RAMIREZ, and W. MUSHYNSKI: Amino acid
     incorporation in the brains of rats trained to use
     the non-preferred paw in retrieving food.
     Brain Res., in press.

MARCHALONIS, J.J., E.R. CONE, and V. SANTER: Enzymic
     iodination.
     Biochem. J. 124, 921-927 (1971).

PUSZKIN, S., and S. BERL: Actomyosin-like protein from
     brain.
     Biochim. biophys. acta 256, 695-709 (1972).

STEWART, J.A.: Tissue specific brain S-100.
     Biochim. biophys. acta 263, 178-192 (1972).

PHOSPHORYLATION OF NON-HISTONE ACID-EXTRACTABLE NUCLEAR PROTEINS (NAEP) FROM BRAIN.

Edward Glassman, Barry Machlus and John Eric Wilson

Division of Chemical Neurobiology, Department of Biochemistry, School of Medicine, University of North Carolina, Chapel Hill, North Carolina 27514, USA

ABSTRACT

The effect of a short training experience on the phosphorylation of acid-extractable proteins from the nuclei of rat brain cells has been studied. It appears that learning an avoidance conditioning task has a permanent biochemical or physiological effect on the rat so that increased incorporation of phosphate into brain NAEP is triggered by subtle reminders which have not yet been determined.

Because of the hypothesized relationship of nuclear proteins to RNA synthesis, the effect of a short training experience on the phosphorylation of acid-extractable proteins from the nuclei of rat brain cells has been examined.

521

The training apparatus was a modification of the electrified
runway described by COLEMAN et al. (1971b). It consisted
of an electrified grid floor and an escape area that was
elevated 8 cm. The rats were given 5 min to explore the
apparatus, and then were lowered onto the start area. Five
seconds later, a foot shock was activated and remained on
for 25 seconds. The trial lasted 30 seconds. At the end of
this time the entire procedure was repeated. Training continued for 5 minutes and consisted of 10 trials. Animals
were judged to have made an avoidance response if they ran
from the grid floor to the platform before the onset of
the shock. The average number of avoidances was 6 out of
the 10 trials. Learning curves and further details are
shown in COLEMAN et al. (1971a), who reported that 15 to
20 minutes of this training procedure produces increased
incorporation of radioactive uridine into polysomes of rat
brain.

A double isotope method was used throughout. Two male
rats were injected under light ether anesthesia through the
optic foramen into the basal forebrain. One rat of the pair
received 0.1 mCi of $H_3^{32}PO_4$ in 100 µl, while the other rat
received 0.1 mCi of $H_3^{33}PO_4$. The rats were coded and return-
ed to their cages. After 25 minutes, one rat was placed in
the training apparatus for 5 minutes of adaptation, after
which it was trained for 5 minutes. The rat was then immediately sacrificed by decapitation and its brain was homogenized gently with the brain of the other rat, which had
been kept quiet. The order in which the trained and untrained rats were sacrificed was randomized.

Fractions containing nuclei were isolated and extracted with 0.2 $\underline{N}$ HCl. The soluble proteins were precipitated with acetone, dried with ethanol and ether, and then dissolved in 1 ml of 7 % guanidinium chloride. About 250 γ of protein were applied to a 0.9 x 15 cm Amberlite IRC-50 column, and eluted first with 100 ml of a linear gradient of guanidinium chloride from 7 % to 14 %, and then with 25 ml of 40 % guanidinium chloride in 0.1 $\underline{M}$ potassium phosphate, pH 6.8. One ml samples were collected. The flow rate was one drop every 30 seconds, and protein recovery was over 95 %. Protein was determined turbidimetrically at 400 mμ after adding TCA, and radioactivity was determined in a scintillation counter.

To correct for experimental errors in the amounts of radioactive phosphate injected and distributed throughout each brain, the ratio of ^{32}P to ^{33}P in AMP isolated from the original homogenate was used as a correction factor. The ratio of ^{32}P to ^{33}P in GMP, CMP and UMP of the trained and untrained rats was similar to that in AMP.

There are 4 main protein peaks following column chromatography of the total acid-soluble nuclear proteins from brain on Amberlite IRC-50 resin. These corresponded to, in order of elution, non-histone acid-soluble proteins (NAEP), which are not retarded by the column under the conditions used, the lysine rich histones, the slightly lysine rich histones, and the arginine rich histones.

Data comparing the amount of radioactive phosphate in NAEP in 32 pairs of trained and untrained rats sacrificed

immediately after the 5 minute training period show no
significant difference in the average amount of radioactivity in the AMP from the brains of trained and untrained
rats, while the trained rats showed an average of 109 $\pm$
18 % (S.D.) more radioactivity in brain NAEP than untrained rats. The increase in radioactivity is not due to an
increase in amount of brain NAEP, since the amount of brain
NAEP does not significantly change as a result of training.
On the other hand, comparison of 5 trained and 5 untrained
rats showed an average of 59 $\pm$ 40 % (S.D.) less radioactivity in brain histones of the trained rats. Additional studies on histones are underway.

Additional fractionation of NAEP by electrophoresis
on polyacrylamide gels demonstrated that NAEP consists of
at least fourteen proteins, but only three bands were radioactive. To show that the phosphate was incorporated into
NAEP, and not into contaminating RNA or phospholipids, the
NAEP containing ^{32}P and ^{33}P from trained and untrained rat
brains was subjected to treatment with various enzymes.
The data showed that RNase, DNase, and phospholipase-C were
ineffective in removing the radioactive phosphate from
NAEP. Only proteolytic enzymes had any effect; as expected
pronase was the most effective of these enzymes. To examine
this further, NAEP prepared separately from trained and untrained rats was hydrolyzed, and amino acid analysis carried
out. The increase of radioactive phosphate in brain NAEP
from the trained rat was accountable for as an increase in
radioactivity in phosphoserine. Further, when trained and
untrained rats were compared separately, the molar ratio

of phosphoserine to serine in NAEP in the trained rats is
more than twice that in the untrained ones. This finding of
an increase in the amount of phosphoserine relative to
serine in the brain NAEP of the trained rat is consistent
with the increase in radioactive phosphate. Since this increase in amount of phosphoserine in brain NAEP of the
trained rat was also observed in the absence of any injection of radioactive phosphate these data eliminate the injection or the use of the AMP correction factor as playing
any role in these phenomena.

This difference between the amounts of phosphoserine
from brain NAEP from trained and untrained rats may be due
to increased phosphorylation or decreased dephosphorylation
of NAEP in the trained animal, or to any one of a number of
other alternatives. It is not possible at present to shed
light on this question because of technical difficulties,
and thus we refer only to the difference in the amount of
phosphoserine, and do not specify a mechanism of how it
comes about.

The exact cells that are responding to this stimulus
are not known. In preliminary experiments the area of the
brain involved in this chemical response was localized by
grossly dissecting the brain into 4 parts. Only the lower
half of the cerebrum, the part containing the amygdala,
entorhinal cortex, hypothalamus, mid-brain tegmentum, post-
eriorventral hippocampus and other structures showed more
radioactive phosphate in brain NAEP of the trained rat. More
work is necessary for more precise localization, possibly
by autoradiography. It is of interest that this is also the

region involved in the increased incorporation of radioactive uridine into RNA of mice undergoing jump box training as shown in previous studies (KAHAN et al., 1970; ZEMP
et al., 1967).

Another important problem is that the behavioral or
environmental agents that caused such chemical changes in
the brain of the trained rat have not been elucidated. It
may be that the learning, per se, or the special stresses
and emotional and motivational effects of learning are
responsible for triggering such chemical responses. We have
performed experiments on rats lacking the pituitary or
adrenal glands and shown that secretory responses by these
glands are not necessary for the increase of radioactive
phosphate in brain NAEP. To determine whether the effect
on NAEP was due to the stimulation the trained animal received from the handling and shocks or to its physical activity, the effects of two additional behaviors on this
chemical response were studied. In these behaviors animals
were denied access to the safe area, but either received a
fixed schedule of random shocks, or else received shocks
paired with the training record of a previously trained rat
(yoked mice). There were no differences between the rats
undergoing these behaviors and quiet rats with respect to
the amount of radioactivity incorporated into brain NAEP.
These results show that the generalized environmental stimulation that the trained rat receives is not the cause of
the increased radioactive phosphate in brain NAEP in the
trained rats.

To test whether the performance or recall of the task

had some relevance, rats were trained for 5 minutes on each of 6 successive days. On the seventh day, they were injected with the radioactive phosphate using the standard double isotope method and allowed to perform in the runway for 5 minutes, after which they were sacrificed. Their brains were homogenized with brains of naive quiet rats.

The results were unexpected, since the prior trained performing rats showed about the same increase in radioactive phosphate as do naive rats when they are first trained. We have not yet been able to determine the minimum amount of reminding necessary to increase the amount of radioactive phosphate in brain NAEP from the prior trained rat, since placing a prior trained rat on the safe platform for 5 minutes could elicit this increase in radioactive phosphate in brain NAEP, as well as merely injecting the prior trained rat and placing it back in its own cage. It seems likely that handling is the cue, but we have no data to support this.

It is of interest that rats that were merely shocked using a fixed schedule of random shocks for 5 minutes per day for 6 days did not show an increase of radioactive phosphate in brain NAEP when shocked or kept quiet on the seventh day. The results are interesting since we are comparing three groups of rats whose treatment is identical, i.e., each is picked up, injected with radioactive phosphate, placed in a quiet cage and sacrificed 30 to 35 minutes later. The differences between the groups lie in their past histories; one group is naive, one group has been trained daily for 5 minutes for 6 days, and one group has been given random shocks daily for 5 minutes for six days. Only the prior trained group shows the increase in radioactive phosphate in brain NAEP.

It is obvious that the discussion of the behavioral trigger by GLASSMAN and WILSON (1970), in which it was concluded that the insight development phase exclusively contained the trigger for increased incorporation of uridine into mouse brain RNA does not apply here. It appears that learning this avoidance conditioning task has a permanent biochemical or physiological effect on the rat so that increased incorporation of phosphate into brain NAEP is triggered by subtle reminders yet to be determined.

It is tempting to speculate that a specific emotional response generated by the training experience is actually the trigger, and that this response can also be evoked when prior trained animals are reminded of their experience at a later time. This would be consistent with the fact that the chemical response appears to take place in the basal forebrain, an area of the brain thought to be involved in emotional responses and arousal. The fact that a rat that is not trained, but has been randomly shocked for 5 minutes per day for 6 days does not show an increase in radioactive phosphate in brain NAEP when reminded of this experience on the seventh day would suggest that the learning aspect of the training experience is also important to the effects of a reminder experience. In the absence of additional data, it would be premature to speculate further on the significance of these phenomena.

LITERATURE

COLEMAN, M.S., J.E. WILSON, and E. GLASSMAN: Brain function
    and macromolecules. VII. Uridine incorporation into
    polysomes of mouse brain during extinction.
    Nature 229, 54-55 (1971).

COLEMAN, M.S., B. PFINGST, J.E. WILSON, and E. GLASSMAN:
    Brain function and macromolecules. VIII. Uridine in-
    corporation into brain polysomes of hypophysectomized
    rats and ovariectomized mice during avoidance con-
    ditioning.
    Brain Research 26, 349-360 (1971).

GLASSMAN, E., and J.E. WILSON: The effect of short experiences
    on macromolecules in the brain.
    In: Biochemistry of Brain and Behavior (R.E. BOWMAN
    and S.P. DATTA, eds.), pp. 279-299, Plenum Press, New
    York (1970).

KAHAN, E., M.R. KRIGMAN, J.E. WILSON, and E. GLASSMAN: Brain
    function and macromolecules. VI. Autoradiographic ana-
    lysis of the effect of a brief training experience on
    the incorporation of uridine into mouse brain.
    Proc. Nat. Acad. Sci. 65, 300-303 (1970).

ZEMP, J.W., J.E. WILSON, and E. GLASSMAN: Brain function and
    macromolecules. II. Site of increased labeling of RNA
    in brains of mice during a short-term training ex-
    perience.
    Proc. Nat. Acad. Sci. 58, 1120-1125 (1967).

BIOCHEMICAL REGULATION OF SYNAPTIC CONNECTIVITY

H. Matthies

Department of Pharmacology and Toxicology
Medical Academy Magdeburg
Magdeburg (GDR)

ABSTRACT

Biochemical and morphological investigations on brightness discrimination in rats led to the conclusion that the
consolidation of a stable memory trace (long-term memory)
is characterized by a nuclear regulation of RNA and protein
synthesis. It is assumed that these synthetic processes
change the structure or the biochemical functions of the
synapses which are activated during acquisition. The preferred pathways thus developed in the neuronal nets are
really the biological basis of information storage in the
CNS. From the results, it can be concluded further, that the
supply of pyrimidine nucleotides seems to be a limiting factor of the velocity and effectivity of this nuclear regulation and the resulting behavioral changes. With respect to
the slow development of the stable memory trace, an additional synaptosomal regulation is proposed as a basis for an
intermediate memory, and is characterized by conformational

changes in enzyme or membrane proteins.

Interest in the biochemical correlates of learning and memory formation has increased markedly during the last decades. Many investigations have been performed and many hypotheses on the biological basis of information storage in the CNS have been elaborated. In general, two principal conceptions are apparent: one proposes that specific macromolecules are the site of memory storage, whereas the other considers that changes in synaptic connections and the neuronal patterns thus formed are of primary importance. For several reasons, which are not discussed here in detail, we suggest that the storage of environmental information in the CNS is principally determined by a change in neuronal connectivity induced by the systemic activity during aquisition and for a short time afterwards. This initial systemic activity is regarded as the basis for the so-called "short-term memory" which is followed by metabolic regulations in some neurons during consolidation. This phase is quite labile and depends on the effectivity of the synthesizing mechanisms of the cells. If the necessary functional or structural changes are accomplished by the synthesizing processes, the memory trace becomes very stable and resistant to disturbances forming the "long-term memory". This general point of view has determined our research strategy and the experimental work of our group to verify or to modify our working hypothesis (MATTHIES, 1972), a schematic representation of which is shown in Fig. 1. The main questions arising from this conception are:

1. Are any changes observable in the RNA and protein metabo-

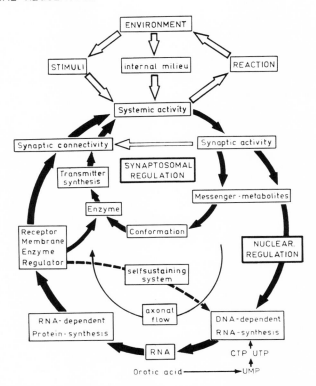

Fig. 1: <u>Schematic representation of the intraneuronal</u>
<u>regulation of synaptic connectivity</u>

    lism which do not arise as a result of stress or neuronal
activity not directly related to learning?

2. Which neuronal cells or structures, if any, are mainly
involved in these metabolic events?

3. Which biochemical changes during acquisition may induce
the synthesizing processes?

4. Which type of protein is formed and into which subcellular
systems does it become integrated?

5. Which mechanisms retain the memory trace before the syn-

thesizing processes have been completed?

These questions can only be answered by simultaneous experimental evaluations using biochemical, physiological and morphological methods, and by a correlation of the results thus obtained. It is obviously impossible to summarize and compare the findings of the numerous authors working in this field, because of the different experimental conditions employed and the resulting contradictions. We therefore decided to perform our experimental work using different methods on a single model of learning: a brightness discrimination in rats (OTT and MATTHIES, 1970). This task is more complex than a simple passive or active avoidance and the formation of new connections can thus be expected to occur in some neocortical areas. Furthermore, the retention can be measured in terms of correct responses as well as in terms of latencies, so that disturbances in performance by experimental interferences will not have such an adverse effect on the measurement of memory.The acceleration or delay of extinction can be ascertained in a relatively short time, and relearning can also be used to determine the memory trace. Although a somewhat longer training period is required, as compared with one-trial experiments, it does not last more than 40 to 45 minutes.

Using this model of a learning process, we evaluated, with numerous microautoradiographic studies, the incorporation of different RNA precursors during acquisition and consolidation. For each experiment, one passive and one active control were used in order to differentiate between bio-

chemical changes underlying learning and processes related
either to stress by punishment or to motor activity. The
active controls received the same number of conditioned
and unconditioned stimuli as the experimental animals. However, these stimuli were given to the active controls in
random order so that they performed the same number of
runs without reaching the criterion.

RNA was labelled by an intraventricular injection of
^{3}H-orotic acid or ^{3}H-uridine-5-monophosphate (UMP) given
prior to training or at different times after acquisition
(Fig. 2). A significantly higher incorporation into the
nuclear RNA of CA cells in the hippocampus and in the pyramidal cells of the visual cortex was observed in trained
animals as compared to active controls. In the thalamic or
hypothalamic regions, at least at this stage of investigations, no significant increases were found (POHLE and
MATTHIES, 1971). Moreover, no characteristic changes were
observed in comparable glial cells. One to two hours after
acquisition the RNA labelled during acquisition seems to
migrate into the cytoplasm, the cytoplasmic-nuclear activity ratio being higher in trained animals at this time.

If the increased RNA synthesis is related to the first
step of memory formation, this metabolic process should be
a key for an experimental interference of learning. There
are two main ways of influencing RNA synthesis: the application (i) of inhibitors, or (ii) of precursors. Since inhibitors of RNA synthesis produce severe side effects which
influence other neuronal functions, we studied the effects
of RNA precursors. The application of a single dose of orotic

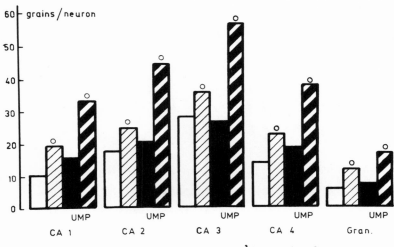

100 μg UMP intraventr. ; Incorporation of ³H-Guanosine 2 h.

Fig. 2: <u>Effect of uridine-5-monophosphate on the incorporation of ³H-guanosine into brain cells of the rat
during the consolidation of a brightness discrimina-
tion</u>

Intraventricular injection of ³H-guanosine and 100
μg UMP or artificial liquor 15 min before training.
Decapitation of the animals 2 hrs after injection.
Histoautoradiographic determination of incorporation. Identification of labelling by RNAse-treat-
ment of slices.
White histograms: active controls without UMP-treatment; lightly hatched histograms: trained animals
without UMP-treatment; black histograms: active controls treated with UMP; heavily hatched histograms:
trained animals treated with UMP; circles: significant increase of incorporation in trained animals
compared to active controls, $p < 0.05$. In all groups,
a significant increase in incorporation is due to
brightness discrimination training. The difference
between treated and untreated active controls is very
slight and not significant. In trained animals, however, treatment with UMP increases the incorporation
markedly in all CA and granular cells of the hippocampus.

acid before training had no effect on acquisition, extinction or relearning. However, treatment with a daily dose of
100 mg/kg for 14 days before training delays extinction
considerably (Fig. 3). This prolonged retention, so far not
achieved by means of any other substance, occurs not only
in our learning experiment but also in both the one-way
shuttle box and the pole-climbing test; furthermore, it is
apparent not only in extinction but also in relearning experiments (OTT and MATTHIES, 1972; MATTHIES and LIETZ,
1967; MATTHIES and KIRSCHNER, 1967). Systematic investigations led to the following results:
1. The effect of orotic acid can also be achieved by using
   the pyrimidine nucleotides UMP and CMP, but only a

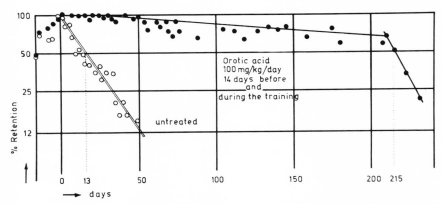

Fig. 3: Effect of orotic acid on the extinction of a conditioned shock avoidance in rats

Open circles: untreated controls; closed circles:
treated animals, 100 mg/kg intraperitoneally
daily, 14 days before training and during the
training period of 14 days.

negligible effect is obtained with the purine nucleotides
AMP and GMP (MATTHIES et al., 1971; OTT et al., 1972).

2. 6-azauracil, which inhibits the transformation of orotic
   acid into UMP, blocks the effect of orotic acid on retention but not that of UMP (OTT and MATTHIES, 1972).

3. The effect of UMP can be counteracted by intraventricular
   injection of cycloheximide, an inhibitor of protein synthesis.

4. The local injection of UMP into the hippocampus has the
   same effect on extinction as oral, intraperitoneal or
   intraventricular application. In contrast, however, injections into the frontal cortex are ineffective (MATTHIES
   and LIETZ, 1967; LÖSSNER and MATTHIES, 1971).

From these results we conclude that RNA synthesis in
specific neurons or cell groups in the rat brain may be involved in the consolidation of long-term memory. The avail-
ability of pyrimidine nucleotides seems to be an important
factor, as already assumed by MANDEL (1969). The limiting
step is probably the synthesis of endogenic orotic acid.
The counteraction of the enhancing effect of applied precursors by an inhibitor of protein synthesis suggests that
the consolidation of memory could be effected by proteins
formed after increased RNA synthesis rather than by the RNA
itself.

If the consolidation of the stable memory trace depends
indirectly on the supply of pyrimidine nucleotides in the
brain, not only the strength and duration of long-term memory
but also the speed of its development should be influenced
by the application of precursors. A common method to deter-

mine the formation of long-term memory is based on the in-
sensitivity of the stable trace to electroconvulsive shock
(ECS). Even though many contradictions exist concerning the
interpretation of the amnesic effect of ECS, no better or
simpler method is available. Following pretreatment with
orotic acid in a dosage which delays the extinction of the
learning task, the amnesic effect of ECS, applied immediately after acquisition and measured 24 hrs later, can be inhibited completely. It was demonstrated that orotic acid
does not influence either the acquisition of the learned
behavior nor the electroconvulsive threshold. Our assumption therefore seems to be confirmed that an increased supply
of pyrimidine nucleotides accelerates the speed of consolidation, so that the memory trace is already formed
during the training period of 40-45 minutes; furthermore,
it can no longer be disturbed by ECS even if given immediately after training. This result confirms not only our
conception concerning the role of RNA and protein synthesis
in learning, but also seems to sustain the assumption that
the gradient of the amnesic effect of ECS may be correlated
with the formation of long-term memory.

In a further series of experiments we studied the
protein metabolism under identical experimental conditions.
^{14}C-leucine was injected intraventricularly or intravenously
and the incorporation into the total protein of different
brain regions during the several phases of the learning experiment was determined. We developed a reliable and repro-
duceable method by means of which the rat brain can be dissected into 15 regions within a few minutes (POPOV et al.,

1968). By this method, a significant increase in labelled proteins was found to occur in the hippocampus of trained animals as compared with active controls. At the same time, a correlation between the increase in incorporation and the number of reinforcements required to reach criterion was observed: the "good learners" showed a higher increase in incorporation than the so-called "bad learners".

These biochemical results were confirmed and supplemented by microautoradiographic studies under the same experimental conditions. In the first hour after acquisition a significant increase was found in the incorporation of ^{3}H-leucine in the CA and granular cells of the hippocampus and also in the pyramidal cells of the visual and cingular cortex (Fig. 4).

It would be of interest to evaluate whether the observed participation of RNA metabolism in the learning process is related to changes in protein synthesis. We therefore determined the incorporation of ^{14}C-leucine into the total protein of the hippocampus after pretreatment with UMP. The results showed a much higher incorporation during acquisition of the brightness discrimination in treated animals than in untreated trained controls. Therefore one can assume that the increased incorporation of RNA precursors during acquisition may be closely related to the subsequent increase in the incorporation of amino acids.

Our present results confirm the conception that the formation of new connections during the elaboration of a new behavior may be realized by synthesizing processes in

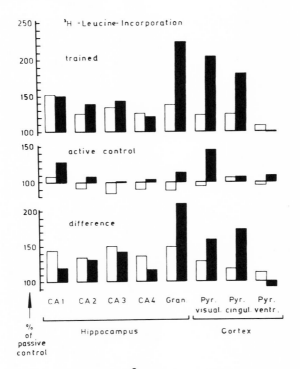

Fig. 4: Incorporation of ^{3}H-leucine into brain cells of the rat during, and one hour after, acquisition of a brightness discrimination

Ordinate: grains/cell in % of the passive controls. White histograms: ^{3}H-leucine injected i.p. 15 min before training; decapitation 1 hour after injection. Black histograms: ^{3}H-leucine injected after training; decapitation 1 hour after injection. The trained animals showed a much higher incorporation rate during and after training. In the hippocampal cells the increase tends to diminish slightly after training, whereas in the cortical cells the increase becomes more pronounced after training. No changes were observed in the pyramidal cells of the ventricular cortex.

neuronal cells, and that this intraneuronal metabolic regulation could be triggered by the synaptic activity of the cells involved during acquisition. The proteins thus synthesized change the functional properties of the previously activated synapses, so that, for a given set of stimuli, preferred pathways to "commanding cells" may be developed. The synthesized proteins are specific in respect to their cellular functions, but not in respect to the information stored by this mechanism. The specificity of information should be reflected in the formation of new preferred pathways in the neuronal network.

Since the induction of protein synthesis by nuclear regulation, and the subsequent formation and incorporation of macromolecules into the entire cell system require some length of time, the well-known latency in the development of long-term memory can be observed in behavioral experiments. From the results of the experiments with ECS we can estimate the duration of the consolidation phase lasting about four to five hours after acquisition. The question then arises, by which mechanisms is memory stored in the meantime? It is commonly supposed that in short-term memory, neurophysiological processes, like posttetanic potentiation or reverberating activity, could play an important role. But it seems doubtful that such processes would be of sufficient duration and in fact no adequate experimental evidence exists supporting this assumption. For this reason we have to postulate an intermediate memory mechanism maintaining the retention between short- and long-term memory. This mechanism must come into operation before the end of short-term memory and must be capable of retaining the effectivity of the preferred

pathways until the stable changes, accomplished by the synthesized macromolecules, have been consolidated in the long-
term memory. Such demands could be realized by conformational
changes in transmitter-synthesizing enzymes or membrane
proteins in the synaptosomal region and may be due to dis-
locations or changes in concentration of transmitters, their
precursors or metabolites, or other substances involved in
synaptical transmission. The onset as well as the duration
of this synaptosomal regulation, as far as is known from
the experimental data, may be sufficient to bridge the gap
between short- and long-term memory.

We have some evidence that in the cholinergic neurons
of the hippocampus at least, considerable changes in the
acetylcholine content occur during this intermediate phase
of memory formation. We were able to measure a six- to eight-
fold increase in the transmitter during acquisition in trained animals as compared with active controls. Further experiments showed that the main contributors to this considerable increase are the free and labile-bound fractions of
acetylcholine, the control values being reached 70 minutes
after the training. At this time, however, the stable-bound
fraction is significantly elevated, but it subsequently decreases as the labile-bound fraction increases again. The
whole picture gives the impression of an oscillating regulatory system and could be taken as an example of a mechanism
assisting intermediate memory (Fig. 5).

Summarizing the results with respect to the working
hypothesis depicted in Fig. 1, it can be stated that some
evidence has been obtained supporting this conception. We

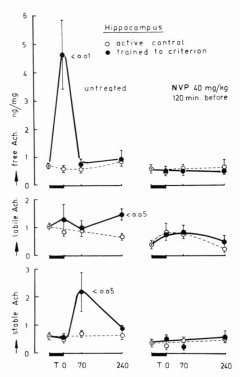

Fig. 5: <u>Changes in the acetylcholine fractions of the rat</u>
        <u>hippocampus during a brightness discrimination</u>

    Abscissa: T = time after training; training period
    of 45 min represented by thick black line. Left:
    untreated animals; right: animals treated 120 min
    before training with 40 mg/kg i.p. of NVP, an in-
    hibitor of cholineacetyltransferase.
    The increase in all fractions during and after train-
    ing can be counteracted by inhibition of acetyl-
    choline synthesis.

can assume the existence of two different intraneuronal
regulatory mechanisms: nuclear regulation characterized by
synthesizing processes which occur during consolidation of
long-term memory, and synaptosomal regulation characterized

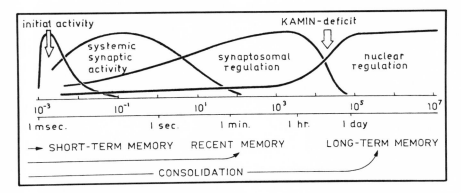

Fig. 6: <u>Schematic representation of the time course of the</u>
<u>intraneuronal processes during acquisition and con-</u>
<u>solidation</u>

Ordinate: intensity of retention. Abscissa: sec in a
logarithmic scale.
Induction of the different processes by the initial
activity during acquisition: systemic activity
responsible for short-term memory, synaptosomal regulation for intermediate or recent memory and nuclear
regulation for permanent or long-term memory. Insufficient overlapping of the single stages may be due
to a transient memory deficit, like the KAMIN-effect.

by conformational changes leading to altered synaptic effectivity subserving the proposed intermediate memory (Fig). But
many questions still remain to be answered. These questions
mainly concern the nature of the so-called "messenger metabol-
ite" which may induce both the synaptosomal and the nuclear
regulation, and the nature of the proteins synthesized during
consolidation which effect the permanent changes in synaptic
connectivity. We believe, however, that our working hypothesis
may be helpful in performing further experiments to close

the gaps in our knowledge.

LITERATURE

LÖßNER, B., and H. MATTHIES: Die Wirksamkeit intraventri-
    kulär applizierten Na-Orotats auf eine bedingte Flucht-
    reaktion der Ratte.
    Acta biol. med. Germ. 27, 221-224 (1971).

MANDEL, P.:
    In: The Future of the Brain Sciences (S. BOGOCH, ed.),
    pp. 197-216, Plenum Press, New York (1969).

MATTHIES, H.: Pharmacological influences on the teaching and
    memorization processes (russ.).
    Farmakol. i. Toxikol. (Moscow) 53, 259-265 (1972).

MATTHIES, H., and W. LIETZ: Der Einfluß von Orotsäure auf
    das Erlöschen einer bedingten Fluchtreaktion der Ratte.
    Acta biol. med. Germ. 19, 785-787 (1967).

MATTHIES, H., and M. KIRSCHNER: Die Wirkung von Orotsäure
    auf den Stabsprungtest der Ratte.
    Acta biol. med. Germ. 19, 789-790 (1967).

MATTHIES, H., and W. LIETZ: Die Bedeutung von Applikations-
    art und Applikationsdauer für die Wirkung von Orot-
    säure auf ein einfaches Modell eines Lernvorgangs.
    Acta biol. med. Germ. 19, 1053-1055 (1967).

MATTHIES, H., C. FÄHSE, and W. LIETZ: Die Wirkung von RNS-
    Präkursoren auf die Erhaltung des Langzeitgedächtnis-
    ses.
    Psychopharmacologia (Berl.) 20, 10-15 (1971).

OTT, T., and H. MATTHIES: Der Einfluß von Orotsäure auf den
Bedeutungswandel des bedingten Reizes.
Acta biol. med. Germ. 25, 181-183 (1970).

OTT, T., and H. MATTHIES: Der Einfluß von 6-Azauridin auf
die Begünstigung des Wiedererlernens durch Ribonuklein-
säureprekursoren.
Psychopharmacologia (Berl.) 23, 272-278 (1972).

OTT, T., B. LÖßNER, and H. MATTHIES: Die Wirkung von Nukleo-
tidmonophosphaten auf die Acquisition und Extinction
bedingter Reaktionen.
Psychopharmacologia (Berl.) 23, 261-271 (1972).

POHLE, W., and H. MATTHIES: The incorporation of ^{3}H-uridine-
monophosphate into the brain during the training period.
A microautoradiographic study.
Brain Research 29, 123-127 (1971).

POPOV, N., W. POHLE, and H. MATTHIES: Der Einfluß von Phenel-
zin und AOAA auf die γ-Aminobuttersäure und α-Ketoglu-
tarsäure-Transaminase in verschiedenen Regionen des Rat-
tenhirns.
Acta biol. med. Germ. 20, 509-516 (1968).

RADIOACTIVE STUDIES OF CHANGES IN PROTEIN METABOLISM BY ADEQUATE AND INADEQUATE STIMULATION IN THE OPTIC TECTUM OF TELEOSTS

Hinrich Rahmann

Department of Neurobiology

Zoological Institute of the University

44 Münster, Hindenburgplatz 55, W. Germany

ABSTRACT

The influence of different external stimuli (motor, electrical and light-pattern stimulation) on the protein metabolism of corresponding nerve structures in the fish brain was investigated by means of autoradiographic and biochemical techniques. 1. Following adequate motor stimulation it was demonstrated that the cerebellum is least affected in its RNA- and protein metabolism, while the valvula cerebelli and the optic tectum are extremely sensitive to an increase or inhibition in mobility.
2. Following electrical stimulation of the brain by implanted electrodes, either directly into the optic tectum or into the optic nerve, large differences could be demonstrated in the incorporation of 3H-histidine into both structural and soluble proteins of the optic tectum.
3. The incorporation of 3H-histidine in the pericaryal layer of the optic tectum was investigated autoradio-

graphically after light-pattern stimulation of one eye.
The exposure of one eye to one narrow vertical slit of
light produced one zone of marked increase of protein
labelling to the contralateral tectum hemisphere; the exposure of two parallel slits caused two zones. These zones
probably represent the extension of corresponding engrams.
4. These results, obtained by a combination of histological and biochemical techniques, make it possible to localize the site of metabolic changes in the brain following
external stimulation. These changes are assumed to be involved in the process of memory formation.

INTRODUCTION

It is of general importance to investigate the manner in which the nervous tissue, in particular the different regions of the CNS, reacts metabolically to different stimuli which may be either adequate or inadequate.
By such means it may be possible to obtain instructive information concerning the formation of engrams.

The correlation between special structures and peripheral functions in the vertebrate brain has been studied
in a rather general manner by means of either behavioral
observations following brain lesions, or electrophysiological methods. However, neither of these methods provides
information about neuronal metabolic reactions in the re-
presentation-fields correlated with functional actions. In
fish, comparatively little is known about the influence of
peripheral function on the brain biochemistry. The inves-

tigations of HYDEN (1943, 1964), and JAKOUBEK and EDSTRÖM
(1965) have shown that in the CNS of fish a significant increase of RNA in the neurons occurs after motor stimulation. In addition, WAWRZYNIAK (1962) and MASAI (1961) have
shown by means of histochemical methods, that under normal
conditions the different brain structures show variations
in enzyme activity and furthermore, that such enzyme patterns may be strongly influenced by adequate stimulations.

In order to investigate the correlation between neuronal metabolism and external function in fish, the optic
system was chosen as a model and a large number of experiments were performed on the metabolic responses to
adequate or inadequate stimuli. Prior to experimentation
the main features of brain metabolism, especially of the
optic tectum, the cerebellum and the valvulae cerebelli,
were examined in order to obtain comparative data. Biochemical and autoradiographical techniques, using radioactive tracers, were chiefly employed in these experiments.

Three series of experiments will be presented in this
paper:
1. The changes in protein and RNA-metabolism in the optic
tectum,the cerebellum and the valvulae cerebelli of the
fish brain, in response to changes in mobility.
2. The changes in protein metabolism in the optic tectum
following electrical stimulation (a functionally inadequate
stimulus).
3. The changes in protein metabolism in the optic tectum
following light-pattern stimulation (a functionally ade-

quate stimulus).

MATERIAL AND METHODS

Animals

In the following experiments three species of fish
were used: Brachydanio rerio (zebra fish), Carassius
carassius (crucian carp) and Scardinius erythrophthalmus
(like Carassius, a Cyprinid fish). Fish were chosen as
the experimental animals because their brain anatomy is
less complicated than that of higher vertebrates. Moreover, the projection pathways of the retina intersect
completely at the optic chiasm and the endings of the
fibres in the optic tectum are well known (RAHMANN, 1967,
1968). Furthermore, the representation of the retinal
image on the optic tectum has already been thoroughly
investigated in fish, using electrophysiological methods,
by BUSER and DUSSARDIER (1953), JACOBSON and GAZE (1964),
SCHWASSMANN and KRÜGER (1964), FIEDLER (1967), MEYER et
al. (1970) and VANEGAS et al. (1971).

As the precursor for proteins we used 3H-L-histidine
(spec. act. 5000 mC/mM), and for RNA, 3H-uridine (spec.
act. 5000 mC/mM; the Radiochemical Centre, Amersham, England). These substances were injected intraperitoneally
in doses of 25 µC/animal.

Procedures

Following injection of the radioactive tracers the animals were subjected to the different stimulative procedures (viz., motor activation or inhibition; electrical stimulation; intermittent light-pattern stimulation). The animals were then sacrificed and the brain samples (both optic hemispheres, cerebellum and the valvula cerebelli) were examined using either autoradiographic or biochemical methods (Fig. 1).

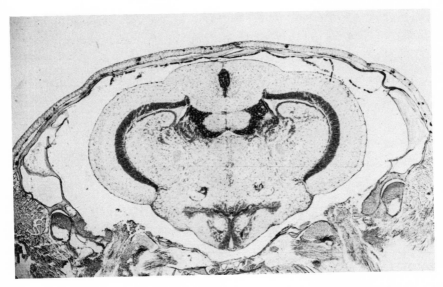

Fig. 1: Histological section through the mesencephalic region of the CNS of Brachydanio rerio, showing the optic tectum, the cerebellum and the valvulae cerebelli

Autoradiographs

These were prepared by means of Agfa dipping emulsion
Nuc 7.15, after the brain slices had been treated with TCA
(5 %, 15 min, 0-2°C). The radioactivity was measured by
counting the silver grains in the photographic emulsion
on top of the brain structures (for details see RENSCH
and RAHMANN, 1966, and RAHMANN and HILBIG, 1972).

Scintillation Counting

In addition to autoradiographic investigations, the
radioactivity of the nerve tissue fractions was determined
in the Packard scintillation spectrometer (model 3320). For
this purpose, the samples were homogenized, treated with
TCA, and finally dissolved by immersion in soluene (Packard).
(For details see RAHMANN et al., 1972).

RESULTS

Autoradiographic Investigations on the Influence
of Changes in Mobility on the Protein and RNA-Label-
ling of Different Brain Structures of Brachydano Rerio

Previous experiments (RAHMANN and HILBIG, 1972) have
shown significant differences in the incorporation of labelled precursors for RNA, proteins, lipids, and polysac-
charides in the optic tectum, cerebellum and valvulae
cerebelli, under normal conditions. On the basis of these

findings, the incorporation of protein and RNA-precursors,
during specific motor stimulation, into these three brain
regions was now investigated.

The fish were tested in 3 groups: one group was
physically immobilized; the second had to fight against
a strong artificial streaming (caused by a 1000 rpm rotor
of 30 mm diameter); the third was kept under normal conditions. The precursors were injected following an adaptation time of 1 hr to each of the mobility situations. After
an incorporation period of 22 min or 3 hrs, during which
the experimental situations were maintained as before, the
animals were sacrificed and the brain slices investigated
autoradiographically (the number of silver grains on top
of the 3 structures was measured in the stimulated animals
and compared with that of untreated animals).The most important results are summarized in Fig. 2.

1. Animals which were forced to be mobile for 1 hr
showed a remarkable increase in whole-brain incorporation
of protein- and RNA-precursors of 26 and 56 % respectively,
following an incorporation period of 22 min. In contrast,
there was a decrease in incorporation in fixed animals of
72 and 42 % respectively.
2. Following stimulation for 4 hrs, with a 3 hr incorporation period, there was a slight decrease in the amount
of whole-brain incorporation of 3H-histidine and 3H-uridine
of 9 and 2 % respectively. However, the total brain activity in immobilized animals was decreased by up to 45 %
for uridine.

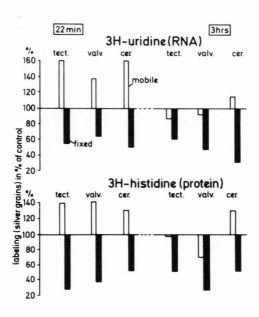

Fig. 2: Incorporation of 3H-uridine and 3H-histidine in
the optic tectum, valvulae cerebelli and cerebellum of Brachydanio following an increased (=
mobile) or inhibited (= fixed) motor stimulation of the animals
Labelling of normal fish = 100 %.

3. By means of autoradiographic investigations we were
able to demonstrate that these changes in tracer incorporation, as a result of either motor stimulation or inhibition, are not the same in all brain structures (Fig. 2).
As can be seen, the RNA- and protein metabolism are least
affected in the cerebellum, especially after a stimulation
period of 4 hrs, whereas the valvula and the tectum react
extremely sensitively to changes in mobility.

The results described above corresponded very well
with those obtained in mammals (WATSON,1965; EDSTRÖM,
1957; CHITRE et al., 1964): usually there is an increase
in brain protein and RNA-metabolism following slight motor stimulation, but a decrease after extreme stimulation
or physical immobilization of the animals. In addition to
these results, our investigations show significant differences in reaction between the different brain regions,
and the optic tectum in particular (which is primarily
concerned with visual stimuli) reacts very sensitively to
motor stimulation.

Quantitative Investigations on the Incorporation
of 3H-Histidine in the Optic Tectum after Electrical
Stimulation

In addition to our radiographic experiments concerning the influence of an adequate motor stimulation on 3H-
histidine and 3H-uridine incorporation, the influence of
direct electrical stimulation on protein metabolism in the
optic tectum was investigated (RAHMANN et al., 1972, in
prep.). We stimulated the fish with microelectrodes implanted either into the mesencephalon (both hemispheres)
or into both optic nerves following removal of the skull
(Fig. 3a and b; cf., VANEGAS et al., 1971). Thereafter a
single hemisphere or nerve was stimulated locally by mono-
phasic, monopolar, cathodic pulses of 0.1 msec duration
with a frequency of 100 pulses per sec over a period of
30 min or 3 hrs. These parameters correspond to the physiological conditions and are comparable with those applied

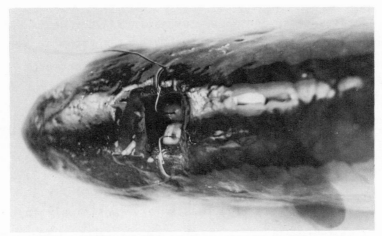

a.

b.

Fig. 3: <u>Electrodes implanted into the optic tectum (a)</u>
<u>and optic nerve (b) of Carassius carassius</u>

in other experiments with fish (FIEDLER, 1967; MEYER et
al., 1970; VANEGAS et al., 1971). Prior to stimulation

the fish were injected intraperitoneally with 3H-histidine
(100 μC), the incorporation period being in all cases 3
hrs.

Following incorporation the fish were decapitated
and both hemispheres rapidly removed and homogenized in
phosphate buffer. After the first centrifugation at 900 g
(to separate the nuclei) the supernatant was centrifuged
again at 105,000 g for 1 hr. The specific radioactivity
of the soluble proteins (i.e. the 105,000 g supernatant)
and the structural and microsomal proteins (i.e., the
105,000 g pellet) - both precipitated with TCA, washed
twice with TCA, extracted with ether and finally dissolved
in 1 N NaOH - was then determined by liquid scintillation
counting and by measuring the protein concentration according to LOWRY et al. (1951). The specific radioactivity
of the proteins, following incorporation of tritiated histidine for 3 hrs, was measured and expressed in counts per
minute per microgram protein (cpm). The average specific
activity for each optic tectum was about 5 cpm. In order
to obtain comparable data, the specific activity of the
stimulated hemisphere was expressed as a percentage of the
specific activity of the control hemisphere. The mean values and deviations for different numbers of animals are
shown in Fig. 4 and Fig. 5.

The results may be summarized as follows:
1. The mere implantation of an electrode into each
hemisphere without any electrical stimulation induced, almost exclusively, a large fluctuation of 3H-histidine incorporation in both structural and soluble proteins. This

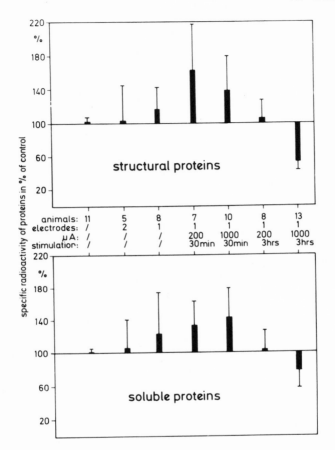

Fig. 4: <u>Incorporation of 3H-histidine into structural and</u>
<u>soluble proteins of the optic tectum following</u>
<u>electrical stimulation with different parameters</u>

Specific radioactivities in per cent of non-stimulated brains.
For details see text.

effect was absent when no electrodes were implanted.

    2. This effect can be seen very clearly by implanting
an electrode into one hemisphere, using the other as the

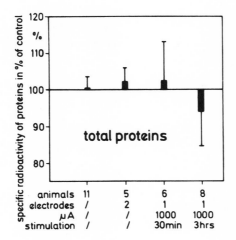

Fig. 5:   Electrical stimulation of the optic nerve
          For details see Fig. 4.

control. Apparently this increase of 3H-histidine incorporation was due to the injury and the subsequent process
of regeneration.

    3. Experiments with electrical stimulation: when the
animals were stimulated for only the first 30 min of a 3
hr incorporation period, an increase in the specific protein labelling could be observed in both the soluble and
the particulate fractions. These changes were much more
pronounced than those obtained following injury. The increase was found following stimulation at 1000 μA and at
only 200 μA.

    4. Weak pulses of 200 μA applied throughout the whole
of the incorporation period (3 hrs) had only a slight inhibitory effect on protein synthesis or had no effect at
all, as compared with controls.

    5. Strong pulses of 1000 μA, however, caused a marked

inhibition of 3H-histidine incorporation, in particular
within the structural proteins. This marked inhibition
after strong and uninterrupted electrical stimulation was
not the result of a decreased uptake of histidine by the
nervous tissue, as proved by additional measurements of
the TCA-soluble radioactivity, but seemed to be a consequence of the impaired protein-synthesizing system itself.

6. In order to show that the above results were comparable with the effects of adequate stimulation, the effect of electrical stimulation of one optic nerve on the
incorporation of 3H-histidine into both protein fractions
of the optic tectum was tested (according to VANEGAS et
al., 1971). Similar results were obtained with both methods.
As can be seen from Fig. 5, the electrical impulses reaching the optic tectum either directly or indirectly by
transmission via the optic nerve cause very effective changes in the protein metabolism.

Autoradiographic Investigations on the Influence
of Light-Pattern Stimulation on the Incorporation of
3H-Histidine into the Optic Tectum

On the basis of the biochemical data from the above
experiments we could only describe the reactions of the affected structures in toto, and we were not able to show if
the tectum as a whole reacts homogeneously or not. This can
be done by means of autoradiography.

In order to obtain more information about the local-

ization of reactions in the CNS caused by defined stimulation, we investigated the influence of a <u>light-pattern</u>
<u>stimulation</u> on the protein metabolism of the optic tectum
of the crucian carp (RENSCH and RAHMANN, 1966; RENSCH et
al., 1968).

Prior to the experiments the animals were kept in total darkness for 10 days. Then, after an intraperitoneal
injection of 30 to 50 µC of 3H-histidine, the fish were
curarized and placed into a special apparatus (Fig. 6),
in which one eye was exposed for a period of 1 hr and 15
min to intermittent light patterns of 15 sec light alternating with 45 sec darkness. Immediately after light stim-

Fig. 6: <u>Apparatus for light-pattern stimulation of a fish (F)</u>
Aquarium (A); Neon lamps (L) for stimulation.

ulation the fish were decapitated and histological sections of the mesencephalon, including the optic tectum,
were subjected to the autoradiographic procedure. The
autoradiograms were exposed for several weeks and then
the density of silver grains (per 440 μ^2) in the peri-
ventricular cell-body layers and the fiber layers of the
stimulated hemisphere was measured and compared with that
of the non-stimulated hemisphere, which served as the control.

The results are presented schematically in Fig. 7.

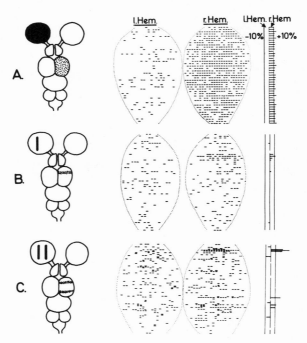

Fig. 7: <u>Schematic diagrams of the incorporation of 3H-
histidine into the optic tectum of Carassius following different light-pattern.stimulation</u>
For details see text.

Those regions in which the density of silver grains per
square unit was at least 10 % above the mean density of
each hemisphere section are indicated by the horizontal
strips. In addition, at the margin of the diagram, all
data in which the mean values of the stimulated hemisphere
(right) were significantly different from those of the non-
stimulated hemisphere (left) are plotted on the vertical
lines.

The most important results (which have since been replicated and confirmed by SKRZIPEK (1969) in fish, and
WEGENER (1970) in frogs) may be summarized as follows:
1. Fish kept in total darkness had the same degree of 3H-
histidine-labelling in both hemispheres of the optic tectum.
2. On the other hand, the incorporation of 3H-histidine was
markedly increased by at least 10 % in the majority of sections from the stimulated hemispheres of those fish, which
were monocularly exposed to a 100 watts lamp for 1 hr and
15 min (Fig. 7A).
3. The exposure of one eye to one small vertical slit of
light (2 mm wide) caused one zone of autoradiographically
marked increase of protein metabolism of about 5 to 10 %
(Fig. 7B).
4. In those fish, which were monocularly exposed to two
parallel, vertical slits of light, two distinct areas with
increased labelling could be found in the contralateral
optic tectum, while the ipsilateral tectum did not show
such an increase (Fig. 7C).
5. Quite remarkably, the zones in the optic tectum showing
a local increase of protein synthesis were found even after
48 hrs providing the tritiated histidine was injected during
stimulation.

6. In all cases, differences in protein labelling were only observed in the pericaryal layers of the contralateral optic tectum. No differences could be found in either the corresponding fiber layers, or the optic layer of the tectum, where the nerve fibers of the optic tract end. This indicates that, under the conditions referred to here, a direct reaction of the nervous tissue can only be demonstrated in a secondary (i.e., postsynaptically) stimulated layer. These data correspond very well with our results obtained by means of biochemical methods following electrical stimulation of the optic nerve.

In addition, our results confirm the findings of RAHMANN (1965), SINGH and TALWAR (1966), TALWAR et al. (1966) and MARAINI et al. (1967), which demonstrated that following light stimulation a significant increase of protein metabolism occurs in the corresponding visual centers of the CNS.

CONCLUSION

The results reported here were obtained using a combination of histological and biochemical techniques. By these means it was possible to localize the site of metabolic changes in the whole brain following external stimulation. It is assumed that these changes may be involved in the processes of memory formation.

LITERATURE

BUSER, P., and M. DUSSARDIER: Organisation des projections
    de la rétine sur le lobe optique, étudiée chez quel-
    ques téléostéens.
    J. Physiol. (Paris) 45, 57-60 (1953).

CHITRE, V.S., S.P. CHOPRA, and G.P. TALWAR: Changes in the
    RNA content of the brain during experimentally in-
    duced convulsions.
    J. Neurochem. 11, 439-448 (1964).

EDSTRÖM, J.E.: Effects of increased motor activity on the
    dimensions and the staining properties of the neuron
    soma.
    J. Comp. Neurol. 107, 295-304 (1957).

FIEDLER, D.: Degenerationen und Verhaltenseffekte nach Elek-
    trokoagulationen im Gehirn von Fischen (Diplodus,
    Crenilabrus - Perciformes).
    Zool. Anz. Suppl. 30, 351-366 (1967).

HYDEN, H.: Protein metabolism in the nerve cell during
    growth and function.
    Acta Physiol. Scand. 17, 1-150 (1943).

HYDEN, H.: RNA - a functional characteristic of the neuron
    and its glia.
    Brain Function II, 29-70 (1964).

JACOBSON, M., and R.M. GAZE: Types of visual response from
    single units in the optic tectum and optic nerve of
    the goldfish.
    Quart. J. exper. Physiol. 49, 199-209 (1964).

JAKOUBEK, B., and J.E. EDSTRÖM: RNA changes in the Mauth-
    ner axon and myelin sheath after increased functional
    activity.
    J. Neurochem. 12, 845-849 (1965).

LOWRY, O.H., N.J. ROSEBROUGH, A.L. FARR, and R.J. RANDALL:
    Proteinmeasurement with the Folin phenol reagent.
    J. Biol. Chem. 193, 265-275 (1951).

MARAINI, G., F. CARTA, R. FRANGUELLI, and M. SANTORI: Ef-
    fect of monocular light-deprivation on leucine uptake
    in the retinae and the optic centres of the newborn
    rat.
    Exp. Eye Res. 6, 299-302 (1967).

MASAI, H.: Comparative neurobiological studies on the gly-
    cogen distribution in the central nervous system of
    submammals.
    Yokohama Med. Bull. 12, 239-260 (1961).

MEYER, D.L., D. SCHOTT, and K.P. SCHAEFER: Reizversuche im
    Tectum opticum freischwimmender Kabeljaue bzw. Dorsche
    (Gadus morrhua L.).
    Pflügers Arch. 314, 240-252 (1970).

RAHMANN, H.: Über den Einfluß adäquater Lichtreizung auf
    die biochemische und morphologische Ausprägung der
    Sehrinde der Maus.
    Z. Zellforsch. 67, 561-574 (1965).

RAHMANN, H.: Darstellung des intraneuronalen Proteintrans-
    ports vom Auge in das Tectum opticum und die Cerebro-
    spinalflüssigkeit von Teleosteern nach intraokularer
    Injektion von 3H-Histidin.
    Naturwissenschaften 54, 174-175 (1967).

RAHMANN, H.: Syntheseort und Ferntransport von Proteinen
im Fischhirn.
Z. Zellforsch. 86, 214-237 (1968).

RAHMANN, H., and R. HILBIG: Autoradiographische Untersuchungen über Stoffwechselunterschiede in verschiedenen Hirnstrukturen von Teleosteern sowie deren Beein-
flußbarkeit nach motorischer Stimulation.
Z. Zellforsch., im Druck.

RAHMANN, H., H. RÖSNER, and R. WILHELM: Einfluß elektri-
scher Reizung auf den Proteinstoffwechsel des ZNS von
Teleostiern.
In Vorber., 1972.

RENSCH, B., and H. RAHMANN: Autoradiographische Untersuchungen über visuelle "Engramm"-Bildung bei Zahnkarp-
fen I.
Pflügers Arch. ges. Physiol. 290, 158-166 (1966).

RENSCH, B., H. RAHMANN, and K.H. SKRZIPEK: Autoradiographische Untersuchungen über visuelle "Engramm"-Bil-
dung bei Fischen II.
Pflügers Arch. ges. Physiol. 304, 242-252 (1968).

SCHWASSMANN, H.O., and L. KRÜGER: Organization of the
visual projection upon the optic tectum of some fresh-
water fish.
J. comp. Neurol. 124, 113-126 (1965).

SINGH, U.B., and G.P. TALWAR: Effect of the flicker frequency of light and other factors on the synthesis of
proteins on the occipital cortex of monkey.
J. Neurochem. 14, 675-680 (1967).

SKRZIPEK, K.H.: Die Proteinsynthese des Tectum opticum in
    Abhängigkeit von der Gestalt intermittierender
    Lichtmuster bei Carassius carassius L. (Pisces).
    J. Hirnforsch. 11, 407-417 (1969).

TALWAR, G.P., S.P. CHOPRA, B.K. GOEL, and B.D. MONTE:
    Correlation of the functional activity of the brain
    with metabolic parameters III.
    J. Neurochem. 13, 109-116 (1966).

VANEGAS, H., E. ESSAYAG-MILLAN, and M. LAUFER: Response of
    the optic tectum to stimulation of the optic nerve in
    the teleost Eugerres plumieri.
    Brain Research 31, 107-118 (1971).

WATSON, W.E.: An autoradiographic study of the incorpora-
    tion of nucleic acid precursors by neurons and glia
    through nerve stimulation.
    J. Physiol. 180, 754-765 (1965).

WAWRZYNIAK, M.: Chemoarchitektonische Studien am Tectum
    opticum von Teleostiern unter normalen und experimen-
    tellen Bedingungen.
    Z. Zellforsch. 58, 234-264 (1962).

WEGENER, G.: Autoradiographische Untersuchungen über ge-
    steigerte Proteinsynthese im Tectum opticum von Frö-
    schen nach optischer Reizung.
    Exper. Brain Res. 10 363-379 (1970).

ENCEPHALOTROPIC DRUGS AND CEREBRAL RNA METABOLISM

Karl Kanig

Abteilung für Neurochemie "J.L.W. Thudichum"
der Universitäts-Nervenklinik

665 Homburg/Saar, Germany (BRD)

ABSTRACT

Substances which are assumed to influence an impaired
memory are also assumed to influence the nucleic acid metabolism of the brain. To investigate this relationship we
used pyritinol, which on the basis of experimental and
clinical observations is considered to be an encephalotrophic drug. The method we used enables the separation of
DNA and four highly purified RNA-fractions. The metabolism
of the "rapidly labelled" RNA-, of the rRNA- and of the
tRNA-fractions was examined in order to investigate the
influence of pyritinol on the rat brain. An oral dosage
of 100 mg/kg, administered over several weeks, caused alterations of ^{32}P-incorporation rate in the single RNA-
fractions. The ^{32}P-incorporation of the "rapidly labelled"
RNA increased, while the turnover of rRNA decreased. The
turnover of tRNA showed no change. To prove the reproducibility of these results and their dependence on dosage

and age of animals, we undertook further experiments with
200 and 400 mg/kg pyritinol. For comparison we examined
the influence of pentobarbital narcosis and of one other
encephalotrophic substance on the RNA metabolism The results of this study and some theoretical implications are
discussed.

Since the publications of HYDEN and coworkers (HYDEN
and LANGE, 1966), it is accepted that nucleic acids are
involved in the function of memory. GLASSMAN (1969) has
summarized the complex of questions in a review article
and concluded that biochemistry alone will not supply the
answers to learning phenomena. Nevertheless, substances
which are assumed to influence an impaired memory, are
also assumed to influence the nucleic acid metabolism of
the brain. To investigate this relationship we used pyritinol (Fig. 1), which on the basis of experimental and
clinical observations is considered to be an encephalotrophic drug. QUADBECK et al. (1962), working with animals, found an increased transport of glucose through
the blood-brain barrier. Using another method, SEILER
and collaborators (1967) were unable to confirm these results. On the other hand, BECKER and HOYER (1966) established that, in patients who had a much decreased glucose
utilization, pyritinol therapy caused a significant increase. KÜNKEL and WESTPHAL (1970) demonstrated that under
oral administration of pyritinol, in an acute experiment
on young healthy subjects, EEG changes at highly significant levels occurred. These alterations were interpreted
as an expression of activation or raised vigilance. PÖL-

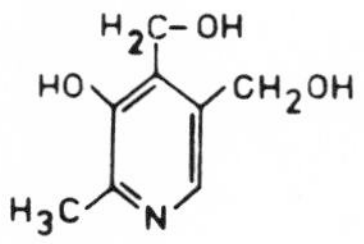

H$_2$C-OH
HO CH$_2$OH
H$_3$C N

Pyridoxol   (Pyridoxin, Vit. B$_6$)

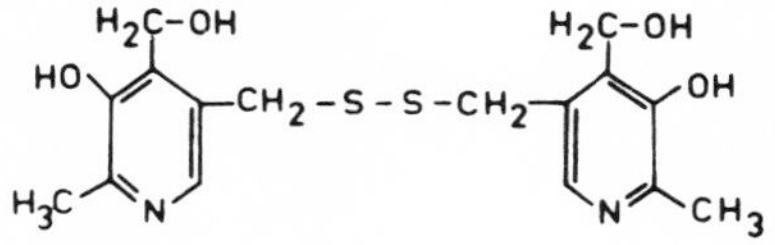

H$_2$C-OH   H$_2$C-OH
HO   OH
CH$_2$-S-S-CH$_2$
H$_3$C N   N CH$_3$

Pyritinol

(Pyrithioxin)   (Encephabol[®])

Fig. 1: <u>Formula of pyritinol (pyrithioxine) in comparison with the formula of pyridoxol</u>

DINGER and coworkers (1970) ascertained that oral administration of pyritinol leads to an earlier improvement in de-
pressive patients who had received chlorprothixen intramuscularly.

Experiments on RNA metabolism are accompanied by great
technical difficulties because the nucleic acid metabolism
differs greatly not only in the various regions of the
brain (MANDEL et al., 1961) but also in a particular tissue,
for example, the glia and the neurons (HYDEN and LANGE,
1966). Moreover, in our opinion, drugs could affect the
various nucleic acid fractions in a number of different
ways, although until now such a diverse mode of action has

not been reported. In this respect, MANDEL and coworkers
(DI CARLO et al., 1970) have shown that amphetamine causes
an increase in the specific activity (S.A.) of all investigated nucleic acid fractions, especially tRNA, 6-10S and
28S RNA. In our own experiments (in collaboration with
SCHÖNECKER, 1971), we found a decrease in the S.A. of re-
sidual-RNA ("rapidly labelled" RNA), rRNA and tRNA following pentobarbital administration (Tab. I).

The method which enables the separation of DNA and
four highly purified RNA-fractions has been described elsewhere (OESTERLE et al., 1970, 1971). The aqueous phase of
the cold extraction contains DNA, tRNA, rRNA and, until
now, an unknown fraction. Because this fraction is not
precipitated by 1.5 M sodium chloride we called it hmssRNA
(high molecular weight salt soluble RNA). We obtained this
RNA only by extraction of the whole tissue and not by extraction of the ribosomal fraction, therefore we assume
that this RNA is not a degraded rRNA. In the subsequent hot
extraction of the residual interphase, we succeeded in obtaining an enriched "rapidly labelled" RNA (res. RNA). It
is highly probable that this fraction contains mRNA (see
also JACOB et al., 1966). Further fractionation and purification were achieved by salt precipitation and filtration
on dextran- and agarose-gels (Fig. 2). We controlled the
rRNA- and the res. RNA-fraction by centrifugation on a
sucrose density gradient (Fig. 3; OESTERLE et al., 1971).
The rRNA consists of 18S and 28S RNA in a ratio of 1:2.6.
The res.RNA is enriched with highly labelled 9-13S RNA. In
this region, MANDEL and his colleagues (JACOB et al., 1966)

Tab. I: <u>Influence of pyritinol and pentobarbital on the</u> <u>^{32}P-incorporation into nucleic acid fractions</u> <u>of the rat brain</u>

Radioactivity was determined 9 h after intracerebral injection of 100 µC ^{32}P per

Age of the animals	Animals per exper.	Number of exper.	Dosage mg/kg (orally)	Number of admin- istration	Specific activity (S.A.) nC/OD$_{260}$			Quotient S.A. res.RNA / S.A. rRNA
					res.RNA	rRNA	tRNA	
60-70 days	2	17	controls	8	13.1 ± 1.1	7.3 ± 0.9	5.8 ± 0.7	1.80
"	2	24	Pyritinol 200	8	19.7 ± 2.3	5.8 ± 0.6	5.9 ± 0.6	3.40
					+ 50 %	- 21 %	+ 2 %	+ 90 %
60-70 days	2	5	Pyritinol 400	8	14.8 ± 1.5	5.2 ± 0.4	5.5 ± 0.5	2.85
					+ 13 %	- 29 %	- 5 %	+ 28 %
60-70 days	2 10	6 3	Pentobarbital *	*	9.2 ± 1.3	5.6 ± 0.7	4.0 ± 0.7	1.66
* 85 mg/kg were injected in 3 portions.					- 31 %	- 23 %	- 31 %	- 8 %
170-180 days	10	5	controls	60	12.9 ± 0.8	7.6 ± 0.3	5.7 ± 0.5	1.68
"	10	5	Pyritinol 100	60	15.4 ± 0.9	6.7 ± 0.3	5.6 ± 0.5	2.30
					+ 19 %	- 11 %	- 2 %	+ 37 %
470 days	3	2	controls	63	7.6	5.5	5.1	1.38
"	3	2	Pyritinol 200	73	9.6	4.0	4.6	2.40
					+ 26 %	- 27 %	- 10 %	+ 74 %

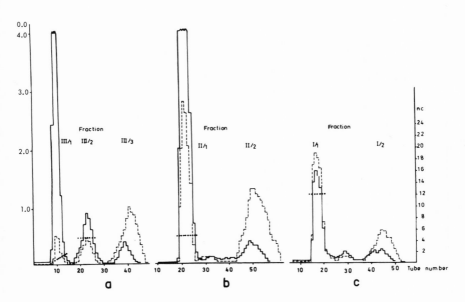

Fig. 2: <u>Filtration of different nucleic acid sediments on</u>
<u>Sephadex gels</u>

(For procedure see OESTERLE et al., 1970, 1971).
Fraction III/1 contains DNA and hmssRNA, fraction
III/2 tRNA, fraction II/1 18S and 28S RNA, and
fraction I/1 the so-called residual RNA (res.RNA)
or "rapidly labelled" RNA.

also found the "rapidly labelled" RNA. The four fractions
showed different ^{32}P-incorporation rates and their turn-
overs were very characteristic (Fig. 4; KANIG and OESTER-
LE, 1972). They also showed differences in base ratios and
in the S.A. of their nucleotide compounds (OESTERLE et
al., 1970, 1971; KANIG et al., 1971).

The metabolism of the res.RNA, rRNA and tRNA was

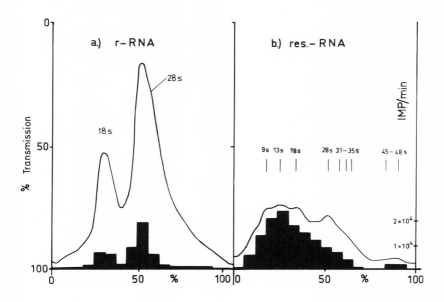

Fig. 3: <u>Sucrose density gradient analysis of rRNA and res.RNA</u>

(For procedure see OESTERLE et al., 1971).

examined in order to investigate the influence of pyritinol on the rat brain. An oral dosage of 100 mg/kg, administered over several weeks, caused a statistically significant change of ^{32}P-incorporation rate in the single
RNA-fractions (Tab. I). Because there was a slight decrease of 7 % in the ^{32}P$_i$-uptake by the brain under pyritinol treatment, we corrected the data obtained from the
labelled RNA accordingly. The ^{32}P-incorporation of res.
RNA increased, while the turnover of the rRNA decreased.
Therefore the ratio of the S.A. of res.RNA to that of
rRNA increases considerably. The turnover of tRNA showed
no change. To prove the reproducibility of these results

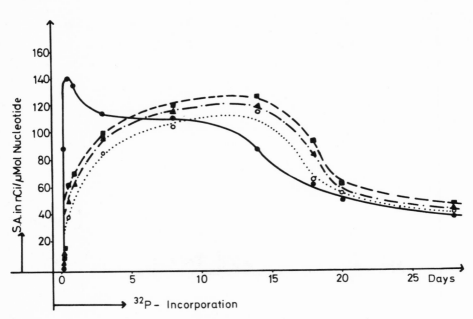

Fig. 4: Incorporation of ^{32}P in various RNA-fractions in
the course of 28 days after a single intracerebral
injection of 100 µC per animal
———— res.RNA; ----- rRNA; ·--·--·-- tRNA;
..... hmssRNA. (From KANIG and OESTERLE, 1972).

and their dependence on dosage, we undertook further experiments with 200 and 400 mg/kg pyritinol. With 200 mg,
the effects described above were still more pronounced
(+ 50 %) and statistically highly significant. The higher
incorporation rate of res.RNA, however, shows greater variance than that of the controls. The ratio of the specific
activity of res.RNA to that of rRNA increases to nearly
double that of the controls. There are also differences
between the two dosage levels of 200 and 400 mg/kg: res.RNA

turnover is highest at the 200 mg dosage, whereas the decrease in rRNA turnover is even lower at 400 mg/kg than at
200 mg/kg. None of these dosages affects the tRNA metabolism.
As pyritinol is principally used for therapy of older patients, a further series of experiments was conducted with
470-day old rats. The results showed an evident decrease
in the incorporation rate of all RNA-fractions even in the
controls. But long lasting treatment with pyritinol had almost the same effect as in younger animals (Tab. I).

A satisfactory explanation of our results is not possible at this time. However, the increase in the res.RNA
(mRNA) ^{32}P-incorporation rate indicates an activation of
protein metabolism or, at least, an increased synthesis of
certain proteins. The role of the decreased rRNA ^{32}P-in-
corporation rate in this context is difficult to explain.
But one can assume that the rRNA is synthesized from high
molecular weight precursors, as indicated by the increase
of the specific radioactivity up to 14 days (Fig. 4). Therefore it may be that the labelling of rRNA, 9 hours after
intracerebral injection of ^{32}P$_i$, is lower if the synthesis
during this time has occurred from the store of not yet
labelled high molecular weight precursors.

It is also difficult to say whether our results apply
only to drugs like pyritinol. This assumption appears to
be valid if one takes into consideration the influence of
amphetamine on the one hand and of pentobarbital on the
other. Preliminary investigations with another drug, Nafti-
drofuryl (Dusodril^R), which is also considered to influence

impaired cerebral metabolism, as well as an impaired memory (EICHHORN, 1969; QUADBECK, 1971; DANIELCZYK, 1971; BOUVIER and CHUPIN, 1971), showed no such effects on the RNA metabolism of the brain.

ACKNOWLEDGEMENTS

I wish to thank Miss D. Scheib, Mr. P. Johann and Mr. N. Rubly for skilful technical assistance. I am indebted to Merck A.G., Darmstadt, W. Germany, and to the Deutsche Forschungsgemeinschaft for supporting this work.

LITERATURE

BECKER, K., and S. HOYER: Hirnstoffwechseluntersuchungen unter der Behandlung mit Pyrithioxin.
Dtsch. Z. Nervenheilk. 188, 200-209 (1966).

BOUVIER, J.P., and B. CHUPIN: Beeinflussung der intellektuellen Leistung chronischer Cerebralsklerotiker durch Dusodril.
Referat: Roland GmbH, Essen 1971

DANIELCZYK, W.: Neue therapeutische Aspekte bei der Rehabilitation zerebraler Gefäßkranker.
Therapiewoche 21, 307-311 (1971).

DI CARLO, R., S. EDEL, H. RANDRIANARISOA, and P. MANDEL: Amphetamine and cerebral RNA metabolism.
VII CINP Congress, Prag 1970, Abstracts, p. 117.

EICHHORN, O.: Zur Behandlung der ischämischen Hirnschädigung.

Med. Welt (N.F.) 20, 2314-2317 (1969).

GLASSMAN, E.: The biochemistry of learning: and evaluation of the role of RNA and protein.

Ann. Rev. Biochem. 38, 605-646 (1969).

HYDEN, H., and P.W. LANGE: A genetic stimulation with production of adenine-uracil rich RNA in neurons and glia learning.

Naturwiss. 53, 64-70 (1966).

JACOB, M., J. STEVENIN, R. JUND, C. JUDES, and P. MANDEL: Rapidly-labelled ribonucleic acids on brain.

J. Neurochem. 13, 619-628 (1966).

KANIG, K., and W. OESTERLE: The influence of encephalotropic substances on the nucleic acid metabolism of brain. Vortrag III. Neurobiologisches Sympos., Magdeburg, 5.-7.5.1971.

In: Biochemical, Physiological and Pharmacological Aspects of Learning Processes. Series "Ergebnisse derDeutschen Gesellschaft für experimentelle Medizin, VEB Verlag Volk und Gesundheit, Berlin 1972, in press.

KANIG, K., W. OESTERLE, and N. RUBLY: Nucleinsäuren im Rattengehirn II. Stoffwechsel einzelner RNA-Fraktionen.

Hoppe-Seyler's Z. Physiol. Chem. 352, 977-983 (1971).

KÜNKEL, H., and M. WESTPHAL: Quantitative EEG analysis of pyrithioxin action.

Pharmakopsychiatrie 3, 41-49 (1970).

MANDEL, P., T. BORKOWSKI, S. HARTH, and R. MARDELL: Incorporation of ^{32}P in ribonucleic acid of subcellular fractions of various regions of the rat central nervous system.
J. Neurochem. **8**, 126-138 (1961).

OESTERLE, W., K. KANIG, W. BÜCHEL, and A.-K. NICKEL: Preparation of DNA and four different RNA species from rat brain. A new RNA fraction and a new characteristic of the various RNAs.
J. Neurochem. **17**, 1403-1419 (1970).

OESTERLE, E., K. KANIG, and P. JOHANN: Nucleinsäuren im Rattengehirn I. In-vivo-Markierungen mit ^{32}P$_i$; Präparation von DNA und vier verschiedenen RNA-Fraktionen in einem Arbeitsgang.
Hoppe-Seyler's Z. Physiol. Chem. **352**, 959-976 (1971).

PÖLDINGER, W., A. GEHRING, and W. SUTTER: Die Beschleunigung des Wirkungseintritts von Psychopharmaka durch Pyrithioxin.
Arzneim. Forsch. (Drug. Res.) **20**, 936-937 (1970).

QUADBECK, G.: Erste experimentelle Untersuchungen zur Frage der Wirkung von Dusodril auf cerebrale Ernährung.
Referat: Roland GmbH, Essen 1971.

QUADBECK, G., H.R. LANDMANN, W. SACHSSE, and I. SCHMIDT: Der Einfluß von Pyrithioxin auf die Blut-Hirn-Schranke.
Med. exper. (Basel) **7**, 144-154 (1962).

SCHÖNECKER, B.: Der Einfluß von Pentobarbital auf den Nucleinsäure-Stoffwechsel des Rattengehirns.
Inaug. Diss. Med. Fak. Homburg/Saar, in preparation.